Springer Series in Operations Research and Financial Engineering

The Springer Series in Operations Research and Financial Engineering publishes monographs and textbooks on important topics in theory and practice of Operations Research, Management Science, and Financial Engineering. The Series is distinguished by high standards in content and exposition, and special attention to timely or emerging practice in industry, business, and government. Subject areas include:

Linear, integer and non-linear programming including applications; dynamic programming and stochastic control; interior point methods; multi-objective optimization; Supply chain management, including inventory control, logistics, planning and scheduling; Game theory Risk management and risk analysis, including actuarial science and insurance mathematics; Queuing models, point processes, extreme value theory, and heavy-tailed phenomena; Networked systems, including telecommunication, transportation, and many others; Quantitative finance: portfolio modeling, options, and derivative securities; Revenue management and quantitative marketing Innovative statistical applications such as detection and inference in very large and/or high dimensional data streams; Computational economics

Pavle Mladenović

Extreme Values In Random Sequences

Springer

Pavle Mladenović
Probability and Statistics
University of Belgrade
Belgrade, Serbia

ISSN 1431-8598 ISSN 2197-1773 (electronic)
Springer Series in Operations Research and Financial Engineering
ISBN 978-3-031-57411-5 ISBN 978-3-031-57412-2 (eBook)
https://doi.org/10.1007/978-3-031-57412-2

Mathematics Subject Classification: 60G70, 60G55, 60G10, 62D10, 60C05, 05B30, 26A12

This Springer imprint is published by the registered company Springer Nature Switzerland AG
The registered company address is: Gewerbestrasse 11, 6330 Cham, Switzerland

Paper in this product is recyclable.

Preface

This book contains four chapters. Chapters 1 and 2 give an overview of basic results related to regularly varying functions and extreme value theory, and provide more examples, exercises, and supplements than previous books related to these topics. Chapters 3 and 4 set the book apart from other books related to extreme value theory.

The notion of regularly varying functions was introduced by Karamata in the 1930s, first in the Romanian journal *Mathematica* (Cluj) in 1930, and then in *Bulletin de la Société Mathématique de France* in 1933. Karamata was a professor at the University of Belgrade and proved basic results related to regularly varying functions including the most-cited theorem on the asymptotic behavior of the integrals of such functions. Several of Karamata's students continued his work. We cite in this book the results of Avakumović, Aljančić, Adamović, Arandjelović, the four As among his students and descendants, and Bojanić, who was Karamata's student as well. Regularly varying functions appear to be a very useful tool in probability theory, especially in the theory of the summation of random variables, probabilistic extreme value theory, modeling of random heavy tail phenomena, and other important topics.

Karamata most likely did not have in mind any applications of the notion of regular variation in probability theory. In 1943, Gnedenko published his seminal paper in which he determined the domains of attraction of the Fréchet and Weibull extreme value distributions in terms of regularly varying functions at infinity and at a fixed point, respectively. Gnedenko did not use the term regular variation, and simply wrote conditions from the definition of regularly varying functions. Gnedenko obtained these conditions without knowing about Karamata's work. Feller was the first to emphasize the importance of regularly varying functions in probability theory in his influential book *An Introduction to Probability Theory and Its Applications*.

Later on, several books or book chapters were devoted to the notion of regular variation. Let us mention here the influential work by de Haan [87], the books written by Seneta [210] and Resnick [192], and the encyclopedic book by Bingham et al. [28]. In this book, we have included more examples and exercises. Some results related to regularly varying functions that are not in the main stream of extreme value theory are provided as supplements at the end of Chap. 1. The notion of multivariate regular variation has also been introduced and extensively studied.

The development of probabilistic extreme value theory started with papers by Fisher and Tippett [66], von Mises [153], and Gnedenko [81]. Fisher and Tippet obtained three possible forms of the limiting distributions of maxima, called extreme value distributions, in the sequences of independent and identically distributed random variables. Von Mises obtained the sufficient conditions for an absolutely continuous distribution function to belong to any of the three possible domains of attraction for maxima. Gnedenko obtained the necessary and sufficient conditions for an arbitrary distribution function to belong to any of the domains of attraction. Extreme value theory was extended to stationary random sequences primarily owing to the paper by Leadbetter [141]; see also Leadbetter et al. [143]. Extreme order statistics, multivariate extremes, extremes in incomplete samples from stationary sequences, extreme values in non-stationary sequences, extremes in random fields, and other topics have also attracted considerable attention. Although the statistical inference of extreme values is not in the focus of this book, let us mention here that a great deal of research has been devoted to the statistics of extreme values. See, for example, the books by Gumbel [86], Beirlant et al. [23], Embrechts et al. [58], Coles [47], Reiss and Thomas [190], de Haan and Ferreira [91], and references therein.

Chapter 3 provides results related to the asymptotic behavior of maxima in incomplete samples from time series. The motivation for this arises naturally given that real data are often incomplete, and therefore the theoretical results used in statistical inference should take this fact into account. The problem was first considered by Mittal [154], who investigated the asymptotic behavior of maxima in complete and incomplete samples from Gaussian sequences. Note here that only in very special cases is a sequence of observed random variables from a stationary sequence stationary as well. Hence, the results provided in this chapter are also related to non-stationary sequences. For weakly dependent stationary sequences, i.e., if conditions $D(u_n, v_n)$ and $D'(u_n)$ are satisfied, it appears that the maxima of observed and non-observed random variables, among the first n terms, are asymptotically independent as n tends to infinity. Moreover, the limiting distribution of a random vector whose components are the maximum in a complete sample and the maximum of observed random variables is uniquely determined by the asymptotic relative frequency of observed random variables. This is not the case unless the conditions of weak dependence are satisfied. The asymptotic relative frequencies of different patterns of observed and non-observed random variables may play a significant role in this regard. Several results in Chap. 3 confirm this fact. A non-trivial combination of combinatorics and point process theory was used to prove some of the results presented in this chapter.

The interplay between combinatorics and probability started in the early history of probability theory, when combinatorics was used to calculate the probabilities of events in discrete models. Probabilistic methods were applied in combinatorics starting with several papers by Erdös and his coauthors; see, for example, [60–62], and the paper by Goncharov [232]. These methods are still used extensively in combinatorics today.

In Chap. 4, we gathered results mainly related to the asymptotic behavior of the extremal characteristics of different combinatorial configurations. These results are related to the following main topics: the extremal characteristics of random permutations, the coupon collector's problem, the polynomial scheme, random trees and random forests, random partitions of finite sets, and the geometric properties of samples of random vectors. The topics and results presented in this chapter provide insight into the natural connections between probability theory and algebra, combinatorics, graph theory and combinatorial geometry. Extreme value distributions, generalized Pareto distributions, and many other distributions as well appear in theorems that describe the asymptotic behavior of the extremal characteristics of the combinatorial configurations considered.

Several results are given here without proof because of space restrictions. For example, proofs of the main results from Sect. 4.2.5 are provided in a 60-page-long paper [16]. For all results stated without proof in the book, references where these proofs can be found are clearly indicated.

The list of references consists of 242 items and can serve as useful reading for master and PhD students as well as researchers. For several years, I have used materials included in this book in my lectures for master and PhD students at the University of Belgrade.

My participation at 10 Conferences on Extreme Value Analysis (EVA), out of 13 organized up to 2023, had a great influence on my decision to write this book. I am grateful to many colleagues from different countries for their very interesting lectures presented at these conferences and for their papers published in scientific journals. My interest in combinatorics and in bridges between probabilistic extreme value theory and other fields of mathematics, including combinatorics, has shaped this book.

Original results and results from joint papers with my former PhD students L. Glavaš, J. Jocković, and J. Vukmirović are included in sections of Chapters 3 and 4. The results of papers written jointly with professors V.I. Piterbarg and G. Samorodnitsky are also included. I am grateful to all of them. Joint work with L. Glavaš on the topic of extreme values in incomplete samples has continued successfully after she finished her PhD studies, resulting in several new papers.

I would like to express my deepest gratitude to the following professors in alphabetical order: A.V. Bulinski, J. Bradić, A.T. Fomenko, T. Mikosch, Y.Y. Nikitin, G.V. Nosovski, V.I. Piterbarg, Z. Pop-Stojanović, and G. Samorodnitsky. They all visited the University of Belgrade, Faculty of Mathematics, and gave lectures at the Seminar of our Department of Probability and Statistics. Some of them visited our Department more than once and presented a series of lectures. We had fruitful discussions about limit theorems related to stochastic processes. With some of these professors, I had inspiring talks about different topics in mathematics in general, and also interesting discussions about music, the arts, literature, poetry, and history.

Finally, I owe my sincere gratitude to Mrs. Alice Copple-Tošić for her careful proofreading of the final version of the book.

Belgrade, Serbia
2024

Pavle Mladenović

Contents

Chapter 1
Regularly Varying Functions

1.1 Definition and Basic Properties

The concept of regular variation was introduced by Karamata in the 1930s, *see* [122–124]. Regularly varying functions appear to be a very useful tool in probability theory and mathematical statistics. The basic results related to the extreme value theory and central limit problem are formulated in terms of regular variation. The importance of regularly varying functions in the theory of the summation of random variables was mentioned by Gnedenko and Kolmogorov [231]. Gnedenko [81] characterized the domains of attraction of the Frechѐt and Weibull extreme value distributions in terms of regularly varying functions. De Haan [87] introduced the classes of Π-varying and Γ-varying functions that play a crucial role in the extreme value theory together with regularly varying functions. Feller promoted the importance of regularly varying functions in his very well-known and influential book [64]. An exhaustive presentation of the theory of regular variation is given in the book by Bingham et al. [28]. Several other books contain chapters or sections that are devoted to regularly varying functions, *see* Ibragimov and Linnik [233], Seneta [210], Geluk and de Haan [73], and Resnick [192].

1.1.1 Regular Variation at Infinity

We shall start this section by some simple examples and then introduce the notion of regular variation.

Example 1.1.1 Let us consider the power function $F(x) = x^\rho$, $x > 0$, where $\rho \in \mathbb{R}$ is a parameter. Then, for every $x > 0$ and $t > 0$,

P. Mladenović, *Extreme Values In Random Sequences*,
Springer Series in Operations Research and Financial Engineering,
https://doi.org/10.1007/978-3-031-57412-2_1

$$\frac{F(tx)}{F(t)} = \frac{(tx)^\rho}{t^\rho} = x^\rho. \ \triangle$$

Example 1.1.2 Consider now the function $F(x) = x^\rho \ln x$, $x > 0$, where $\rho \in \mathbb{R}$. For every $x > 0$, the following equality holds:

$$\lim_{t\to\infty} \frac{F(tx)}{F(t)} = \lim_{t\to\infty} \frac{(tx)^\rho (\ln t + \ln x)}{t^\rho \ln t} = x^\rho. \ \triangle$$

Example 1.1.3 Let $F(x) = \left(\frac{\pi}{2} - \operatorname{arctg} x\right)^{-\rho}$, $x > 1$, $\rho \in \mathbb{R}$. Using the formula $\operatorname{arctg} x = \frac{\pi}{2} - \frac{1}{x} + o\left(\frac{1}{x}\right)$, as $x \to \infty$, we obtain that for every $x > 0$,

$$\lim_{t\to\infty} \frac{F(tx)}{F(t)} = \lim_{t\to\infty} \frac{\left(\frac{1}{tx} + o\left(\frac{1}{t}\right)\right)^{-\rho}}{\left(\frac{1}{t} + o\left(\frac{1}{t}\right)\right)^{-\rho}} = x^\rho. \ \triangle$$

The functions from Examples 1.1.2–1.1.3 behave at infinity like the power function from Example 1.1.1. This leads us to the following definition:

Definition 1.1.4 A measurable function $F : \mathbb{R}_+ \to \mathbb{R}_+$ is *regularly varying* at infinity, with an *exponent of regular variation* $\rho \in \mathbb{R}$, if

$$\lim_{t\to\infty} \frac{F(tx)}{F(t)} = x^\rho \quad \text{for every } x > 0. \tag{1.1.1}$$

We use the notation $F \in RV_\rho$ for a regularly varying function with exponent ρ. If relation (1.1.1) holds with $\rho = 0$, then we say that function F is *slowly varying* at infinity.

Note that the notion introduced by Definition 1.1.4 is a local property of function F at infinity. Therefore, instead of $\mathbb{R}_+$, we can consider any interval $(c, +\infty)$ as the domain of function F.

Example 1.1.5

(a) Function $(x \ln x)^\rho$, $x > 0$, is ρ-varying at infinity.
(b) Functions $\ln x$ and $\operatorname{arctg} x$ are slowly varying at infinity.
(c) If $F(x)$ is ρ-varying at infinity, then function $L(x) = x^{-\rho} F(x)$ is slowly varying at infinity. If $L(x)$ is slowly varying, then $F(x) = x^\rho L(x)$ is ρ-varying.
(d) Let $G_{1,\alpha}(x) = \exp(-x^{-\alpha})$ for $x > 0$, and $G_{1,\alpha}(x) = 0$ for $x \leqslant 0$, be the Fréchet distribution function, where α is a positive parameter. Then we have $1 - G_{1,\alpha}(x) \sim x^{-\alpha}$ as $x \to +\infty$, and hence the tail $1 - G_{1,\alpha}(x)$ is $(-\alpha)$-varying at infinity.
(e) Functions $\sin x$, $\cos x$, $x - [x]$, e^x, and e^{-x} are not regularly varying.
(f) Function $[\ln x]$ is slowly varying, while function $e^{[\ln x]}$ is not regularly varying.
(g) If $\lim\limits_{x\to+\infty} F(x) = c \in \mathbb{R}_+$, then function F is slowly varying at infinity.

(h) Function $\dfrac{(3x^2 + x + 5)\,\mathrm{arctg}\,x}{1 + x^2}$ is slowly varying at infinity. △

An easy consequence of Definition 1.1.4 is the following theorem:

Theorem 1.1.6 *Let functions $F_1 : \mathbb{R}_+ \to \mathbb{R}_+$ and $F_2 : \mathbb{R}_+ \to \mathbb{R}_+$ be given such that $F_1 \in RV_{\rho_1}$ and $F_2 \in RV_{\rho_2}$. Then we have*

$$F_1 \cdot F_2 \in RV_{\rho_1+\rho_2}, \quad F_1/F_2 \in RV_{\rho_1-\rho_2}.$$

1.1.2 Regular Variation at an Arbitrary Point

The notion of regular variation can be extended to an arbitrary point in the domain of function F. From the definition below it follows that it can always be reduced to regular variation at infinity.

Definition 1.1.7 Let $F : \mathbb{R}_+ \to \mathbb{R}_+$ be a measurable function. Function F is regularly varying at 0 if the function $F\left(\frac{1}{x}\right)$ is regularly varying at infinity. Function F is regularly varying at the point $x_0 > 0$ (from the left side or as $x \to x_0 - 0$) if the function $F(x_0 - 1/x)$ is regularly varying at infinity.

1.1.3 Regular Variation and the Cauchy Equation

We shall now discuss the following question: Is it possible to obtain another limit function in (1.1.1) instead of the power function x^α?

Theorem 1.1.8 *A measurable function $F : \mathbb{R}_+ \to \mathbb{R}_+$ is regularly varying if there exists a function $h : \mathbb{R}_+ \to \mathbb{R}_+$, such that, for every $x > 0$,*

$$\lim_{t\to\infty} \frac{F(tx)}{F(t)} = h(x). \tag{1.1.2}$$

Proof

(a) First note that function h is a measurable function as the limit of measurable functions. For every $x > 0$ and $y > 0$, we obtain that

$$\frac{F(txy)}{F(t)} = \frac{F(txy)}{F(ty)} \cdot \frac{F(ty)}{F(t)}. \tag{1.1.3}$$

Using equality (1.1.2), we obtain that (1.1.3) implies that for every $x > 0$ and $y > 0$ the following equality holds:

$$h(xy) = h(x) \cdot h(y). \tag{1.1.4}$$

We shall prove that all measurable functions $h : \mathbb{R}_+ \to \mathbb{R}_+$ that satisfy equation (1.1.4) are given by $h(x) = x^\rho$, where $\rho \in \mathbb{R}$. Using the substitution $g(x) = \ln h(e^x)$, we obtain that equality (1.1.4) holds for all $x, y \in \mathbb{R}_+$ if and only if

$$g(x + y) = g(x) + g(y), \tag{1.1.5}$$

for all $x, y \in \mathbb{R}$, where $g : \mathbb{R} \to \mathbb{R}$ is a measurable function. Equation (1.1.5) is referred to as the *Cauchy equation.*

(b) Let us now prove the statement: *If a measurable function $g : \mathbb{R} \to \mathbb{R}$ satisfies equation* (1.1.5), *then g is a continuous function.*

Let $I = [a, b] \subset \mathbb{R}$ be an arbitrary interval. Since g is measurable, it follows that the restriction of g on the interval I is also measurable. Using the Luzin theorem (*see* Theorem 5.4.9 in Kolmogorov and Fomin [234]), we conclude that for every $\varepsilon > 0$, there exists a closed set $F \subset I$, such that $\lambda(I \setminus F) < \varepsilon$, where λ is the Lebesgue measure, and the restriction of g on the set F is continuous. Hence, for every positive integer n, there exists a closed set F_n, such that

$$\lambda(F_n) > b - a - \frac{1}{n^2} \tag{1.1.6}$$

and the restriction of g on set F_n is continuous. Since F_n is a compact set, it follows that function g is uniformly continuous on F_n. Therefore, there is an $\eta_n > 0$, such that for arbitrarily chosen points $x, x + \delta \in F_n$, where $|\delta| < \eta_n$, the following equality holds:

$$|g(x + \delta) - g(x)| < \frac{1}{n}. \tag{1.1.7}$$

Let δ_n be a fixed real number such that $0 < \delta_n < \min\{\eta_n, 1/n^2\}$. Let us denote

$$A_n = \{x : x \in F_n, x + \delta_n \notin F_n\},$$
$$B_n = \{x : x \in F_n, x + \delta_n \in F_n\}.$$

The sets A_n and B_n are measurable, and since (1.1.6) holds and $A_n \cup B_n = F_n$, it follows that $\lambda(A_n) \leqslant 1/n^2 + \delta_n \leqslant 2/n^2$ and $\lambda(B_n) > b - a - 3/n^2$. Using (1.1.7) we obtain that for $x \in B_n \subset F_n$,

$$|g(x + \delta_n) - g(x)| < \frac{1}{n}.$$

Let $G = \bigcup_{n=1}^{\infty} \bigcap_{k=n}^{\infty} B_k$ be the set of points x that belong to all except maybe the finite number of the sets B_k. Let us denote $D_k = I \setminus B_k$ and $H = I \setminus G$. Then, $H = \bigcap_{n=1}^{\infty} \bigcup_{k=n}^{\infty} D_k$ is the set of points x contained in infinitely many of sets D_k. For each $n \in \mathbb{N}$, the following inequalities hold:

$$\lambda(H) \leqslant \lambda\left(\bigcup_{k=n}^{\infty} D_k\right) \leqslant \sum_{k=n}^{\infty} \frac{3}{k^2}.$$

Hence, $\lambda(H) = 0$, $\lambda(G) = b - a$, and for every $x \in G$,

$$\lim_{n\to\infty} g(x + \delta_n) = g(x). \tag{1.1.8}$$

Let t_0 be an arbitrary real number from interval (a, b), and $t \in G$, such that $a < t < t_0 < b$. For $x = t_0 - t$ and $y = t + \delta_n$, relation (1.1.5) implies that

$$g(t_0 + \delta_n) = g(t_0 - t) + g(t + \delta_n). \tag{1.1.9}$$

Using (1.1.8), (1.1.9), and (1.1.5), we obtain that

$$\lim_{n\to\infty} g(t_0 + \delta_n) = g(t_0 - t) + g(t) = g(t_0).$$

Hence, equality (1.1.5) holds for all $x \in (a, b)$. Note that sequence (δ_n) depends on interval $I = [a, b]$, and recall that $0 < \delta_n < \min\{\eta_n, 1/n^2\}$ holds. Let (ω_n) be a sequence of positive real numbers such that $\omega_n \to 0$ as $n \to \infty$. Then there exists a subsequence (ω_{n_k}), such that $0 < \omega_{n_k} < \min\{\eta_{n_k}, 1/n_k^2\}$ for every k, and

$$\limsup_{n\to\infty} g(t_0 + \omega_n) = \lim_{k\to\infty} g(t_0 + \omega_{n_k}) = g(t_0).$$

Analogously, we prove that $\liminf_{n\to\infty} g(t_0 + \omega_n) = g(t_0)$, and since (ω_n) is an arbitrary sequence of positive real numbers that tends to 0, it follows that function g is right-continuous at point t_0. In order to prove that function g is left-continuous at point t_0, it is sufficient to consider the function $\widetilde{g}(x) = g(-x)$ for which equality $\widetilde{g}(x + y) = \widetilde{g}(x) + \widetilde{g}(y)$ holds and repeat the previous argument proving that $\widetilde{g}$ is right-continuous.

(c) Let us now determine all solutions of Eq. (1.1.5) using the fact that g is continuous. It follows from (1.1.5) that for every positive integer n and all $x_1, x_2, \ldots, x_n \in \mathbb{R}$, the following equality holds:

$$g(x_1 + x_2 + \cdots + x_n) = g(x_1) + g(x_2) + \cdots + g(x_n). \tag{1.1.10}$$

For $x_1 = x_2 = \cdots = x_n = x$, relation (1.1.10) implies that for every positive integer n and every real number x,

$$g(nx) = n \cdot g(x). \tag{1.1.11}$$

For $x = (m/n) \cdot y$, where $y \in \mathbb{R}$ and $m, n \in \mathbb{N}$, we obtain that $g(nx) = g(my)$. Now (1.1.11) implies that $ng(x) = mg(y)$, i.e.,

$$g(x) = \frac{m}{n} g(y). \tag{1.1.12}$$

Let us denote $r = m/n$. For $y = 1$, relation (1.1.12) implies that

$$g(r) = g(1) \cdot r \tag{1.1.13}$$

holds for every positive rational number r. For $x = y = 0$, (1.1.5) implies that $g(0) = 0$, and consequently equality (1.1.13) also holds for $r = 0$. Since g is a continuous function, it follows from (1.1.13) that $g(x) = g(1) \cdot x$ for every $x \geqslant 0$. If we put $y = -x$ in (1.1.5), we obtain that $g(-x) = -g(x) = g(1) \cdot (-x)$. It is easy to check that every linear function $g(x) = c \cdot x$ satisfies equation (1.1.5). Hence, every measurable solution of Eq. (1.1.5) has the form $g(x) = c \cdot x$, where $c \in \mathbb{R}$.

Conclusion Every measurable function $h : \mathbb{R} \to \mathbb{R}$, which is a solution of Eq. (1.1.4), has the form $h(x) = x^\rho$, where $\rho \in \mathbb{R}$. □

Theorem 1.1.9 *Let $F : \mathbb{R}_+ \to \mathbb{R}_+$ be a measurable function, such that*

$$\lim_{t\to\infty} \frac{F(tx)}{F(t)} = h^*(x) \in \mathbb{R}_+, \tag{1.1.14}$$

for every $x \in [a, b]$, where a and b, $0 < a < b < +\infty$, are fixed real numbers. Then there exists a function $h : \mathbb{R}_+ \to \mathbb{R}_+$, such that $h(x) = h^*(x)$ for every $x \in [a, b]$ and that equality (1.1.2) *holds for every $x > 0$.*

Proof Let $x \in [a, b]$ and $s > 0$ be a real number such that $a \leqslant x/s \leqslant b$. Then, $\frac{a}{b} \leqslant \frac{x}{b} \leqslant s \leqslant \frac{x}{a} \leqslant \frac{b}{a}$, and

$$\frac{F(tx)}{F(t)} = \frac{F(s(tx/s))}{F(tx/s)} \cdot \frac{F(tx/s)}{F(t)}. \tag{1.1.15}$$

When $t \to \infty$, we have $u := tx/s \to \infty$ and

$$\frac{F(tx)}{F(t)} \to h^*(x), \quad \frac{F(tx/s)}{F(t)} \to h^*(x/s). \tag{1.1.16}$$

Using (1.1.15) and (1.1.16), we obtain that

$$\lim_{u\to\infty}\frac{F(us)}{F(u)}=\lim_{t\to\infty}\frac{F(s(tx/s))}{F(tx/s)}=\frac{h^*(x)}{h^*(x/s)}>0,$$

for $a/b \leqslant s \leqslant b/a$. By the method of mathematical induction, we prove that the equality

$$\lim_{t\to\infty}\frac{F(tx)}{F(t)}=h(x)>0 \tag{1.1.17}$$

holds for every $x \in [(a/b)^n, (b/a)^n]$ and every positive integer n. Since $a/b < 1$ and $b/a > 1$, it follows that relation (1.1.17) holds for every positive real number x. □

1.1.4 Uniform Convergence Theorem

Theorem 1.1.10 *If* $F \in RV_\rho$, *then for arbitrarily chosen a and b, where* $0 < a < b < +\infty$, *then equality* (1.1.1) *holds uniformly for* $x \in [a, b]$.

Proof Without loss of generality, we can assume that $\rho = 0$. In the opposite case we can consider the function $x^{-\rho}F(x)$.

Let us denote $f(x) = \ln F(e^x)$. For $\rho = 0$, equality (1.1.1) is equivalent to the relation $\ln F(tx) - \ln F(t) \to 0$ as $t \to +\infty$, and the statement of theorem is equivalent to the condition

$$\lim_{t\to+\infty}|f(t+x)-f(t)|=0,$$

uniformly for $x \in J = [\ln a, \ln b]$. Suppose, on the contrary, that there exist $\delta > 0$ and the sequences (t_n) and (x_n), such that $t_n \to +\infty$ as $n \to \infty$, and for every n, $x_n \in J$ and

$$|f(t_n+x_n)-f(t_n)|>\delta.$$

Let us define sets S_{1n} and S_{2n} as follows:

$$S_{1n}=\{s\in J : |f(t_n+s)-f(t_n)|>\delta/2\},$$

$$S_{2n}=\{s\in J : |f(t_n+x_n)-f(t_n+s)|>\delta/2\}.$$

If $|f(t_n+s)-f(t_n)| \leqslant \delta/2$ and $|f(t_n+x_n)-f(t_n+s)| \leqslant \delta/2$ for $s \in J$, then $|f(t_n+x_n)-f(t_n)| \leqslant \delta$, and this contradicts the imposed assumption. If at least one of the inequalities $|f(t_n+s)-f(t_n)| > \delta/2$ and $|f(t_n+x_n)-f(t_n+s)| > \delta/2$ holds for some $s \in J$, then $S_{1n} \cup S_{2n} = J$. It follows that at least one of the inequalities

$$\lambda(S_{1n}) \geqslant \frac{1}{2}\lambda(J), \quad \lambda(S_{2n}) \geqslant \frac{1}{2}\lambda(J)$$

holds true, where λ is the Lebesgue measure. Let us define set U_n as follows:

$$\begin{aligned} U_n := &\{u : x_n - u \in J, \ |f(t_n + x_n) - f(t_n + x_n - u)| > \delta/2\} \\ = &\{u : x_n - u \in S_{2n}\}. \end{aligned}$$

Hence, for every n, we have $\lambda(U_n) = \lambda(S_{2n})$. Now we conclude that at least one of the following two inequalities:

$$\lambda(S_{1n}) \geqslant \frac{1}{2}\lambda(J), \quad \lambda(U_n) \geqslant \frac{1}{2}\lambda(J)$$

holds for infinitely many positive integers n. If $\lambda(S_{1n}) \geqslant \frac{1}{2}\lambda(J)$ for infinitely many n, we obtain that

$$\begin{aligned} \lambda\left(\limsup_{n\to\infty} S_{1n}\right) &= \lambda\left(\bigcap_{k=1}^{\infty}\bigcup_{n=k}^{\infty} S_{1n}\right) \\ &= \lim_{k\to\infty} \lambda\left(\bigcup_{n=k}^{\infty} S_{1n}\right) \geqslant \frac{1}{2}\lambda(J). \end{aligned}$$

In the second case we obtain the analogous inequality for the sequence of sets (U_n). Next, it follows that there exist a real number x_0, such that at least one of the relations

$$x_0 \in \bigcap_{k=1}^{\infty}\bigcup_{n=k}^{\infty} S_{1n}, \quad x_0 \in \bigcap_{k=1}^{\infty}\bigcup_{n=k}^{\infty} U_n$$

holds, and this contradicts the assumption $\lim_{t\to\infty} (f(t + x_0) - f(t)) = 0$. □

1.1.5 Fast Varying Functions

Definition 1.1.11 A measurable function $F : \mathbb{R}_+ \to \mathbb{R}_+$ is *fast varying at infinity* if for every positive real number x,

$$\lim_{t\to+\infty} \frac{F(tx)}{F(t)} = x^{\rho},$$

where $\rho = +\infty$ or $\rho = -\infty$, and

$$x^{+\infty} = \begin{cases} 0, & \text{if } 0 < x < 1, \\ 1, & \text{if } x = 1, \\ +\infty, & \text{if } x > 1, \end{cases}$$

$$x^{-\infty} = \begin{cases} +\infty, & \text{if } 0 < x < 1, \\ 1, & \text{if } x = 1, \\ 0, & \text{if } x > 1. \end{cases}$$

Example 1.1.12 The exponential functions e^x and e^{-x} are fast varying at infinity. $\triangle$

Remark 1.1.13 Fast varying functions are also referred to as regularly varying functions with the exponent of regular variation $\rho = +\infty$ or $\rho = -\infty$.

1.2 Integral Properties of Regularly Varying Functions

1.2.1 Karamata's Theorem

The first part of the following theorem allows us to deduce the asymptotic behavior of the integral of a regularly varying function and conclude that this integral is also regularly varying. The second part of the theorem allows us to conclude that the asymptotic behavior of the integral of some function implies its regular variation. This is one of the basic results proved in Karamata's papers [122, 123], and [124] that represent the beginning of the theory of regularly varying functions. *See* also Karamata [125, 126]. The first part of the theorem is more important for applications in probability theory and mathematical statistics.

Theorem 1.2.1 (Karamata [122]) *Let $F : \mathbb{R}_+ \to \mathbb{R}_+$ be a function that is Lebesgue integrable on finite intervals.*

(a) *If $\rho \geqslant -1$ and $F \in RV_\rho$, then $\int_0^x F(t)\,dt \in RV_{\rho+1}$, and*

$$\lim_{x\to\infty} \frac{xF(x)}{\int_0^x F(t)\,dt} = \rho + 1. \tag{1.2.1}$$

If $\rho < -1$ and $F \in RV_\rho$, then function F satisfies the conditions $\int_x^\infty F(t)\,dt < +\infty$, $\int_x^\infty F(t)\,dt \in RV_{\rho+1}$, and

$$\lim_{x\to\infty} \frac{xF(x)}{\int_x^\infty F(t)\,dt} = -\rho - 1. \tag{1.2.2}$$

If $\rho = -1$ and $\int_x^\infty F(t)\,dt < +\infty$, then equality (1.2.2) *also holds.*

(b) *If function F satisfies condition*

$$\lim_{x\to\infty}\frac{xF(x)}{\int_0^x F(t)\,dt}=\alpha\in(0,+\infty), \tag{1.2.3}$$

then $F\in RV_{\alpha-1}$.

If $\int_x^\infty F(t)\,dt<+\infty$ *and*

$$\lim_{x\to\infty}\frac{xF(x)}{\int_x^\infty F(t)\,dt}=\alpha\in(0,+\infty), \tag{1.2.4}$$

then F *is regularly varying with index* $-\alpha-1$, *i.e.,* $F\in RV_{-\alpha-1}$.

Proof

(a) **Case:** $\rho>-1$, $F\in RV_\rho$. It follows that

$$\lim_{t\to\infty}\frac{F(2t)}{F(t)}=2^\rho>\frac{1}{2}. \tag{1.2.5}$$

Equality (1.2.5) implies that there exists t_0 such that, for all $t>t_0$, the following inequality holds:

$$\frac{F(2t)}{F(t)}>\frac{1}{2}. \tag{1.2.6}$$

If n is a positive integer such that $2^n>t_0$, then (1.2.6) implies that

$$\int_{2^{n+1}}^{2^{n+2}}F(t)\,dt=\int_{2^n}^{2^{n+1}}2F(2t)\,dt>\int_{2^n}^{2^{n+1}}F(t)\,dt. \tag{1.2.7}$$

Let us denote $n_0=\min\{n:2^n>t_0\}$. Then (1.2.7) implies that

$$\int_{t_0}^{\infty}F(t)\,dt>\sum_{n=n_0}^{\infty}\int_{2^{n+1}}^{2^{n+2}}F(t)\,dt>\sum_{n=n_0}^{\infty}\int_{2^{n_0}}^{2^{n_0+1}}F(t)\,dt=+\infty. \tag{1.2.8}$$

For $x>0$ and $0<A<+\infty$ (and $\rho>-1$), we obtain

$$\int_0^y F(tx)\,dt\sim\int_A^y F(tx)\,dt,\quad\text{as } y\to\infty. \tag{1.2.9}$$

Let $\varepsilon>0$. Since $F\in RV_\rho$, it follows that for every $x>0$ there exists $A>0$, such that for all $t>A$,

$$x^\rho(1-\varepsilon)\leqslant\frac{F(tx)}{F(t)}\leqslant x^\rho(1+\varepsilon). \tag{1.2.10}$$

Using (1.2.9) and (1.2.10), we obtain that

$$\begin{aligned}
\limsup_{y\to\infty} \frac{\int_0^{yx} F(t)\,dt}{\int_0^y F(t)\,dt} &= \limsup_{y\to\infty} \frac{x\int_0^y F(tx)\,dt}{\int_0^y F(t)\,dt} \\
&= \limsup_{y\to\infty} \frac{x\int_A^y F(tx)\,dt}{\int_A^y F(t)\,dt} \\
&\leqslant \limsup_{y\to\infty} x^{\rho+1}(1+\varepsilon)\frac{\int_A^y F(t)\,dt}{\int_A^y F(t)\,dt} \\
&= x^{\rho+1}(1+\varepsilon). \qquad (1.2.11)
\end{aligned}$$

Similarly we prove the following inequality:

$$\liminf_{y\to\infty} \frac{\int_0^{yx} F(t)\,dt}{\int_0^y F(t)\,dt} \geqslant x^{\rho+1}(1-\varepsilon). \qquad (1.2.12)$$

Relations (1.2.11) and (1.2.12) imply that $\int_0^x F(t)\,dt \in RV_{\rho+1}$. Let us define function $b : \mathbb{R}_+ \to \mathbb{R}_+$ as follows:

$$b(x) = \frac{xF(x)}{\int_0^x F(t)\,dt}. \qquad (1.2.13)$$

It is easy to see that

$$\int_1^x \frac{b(t)}{t}\,dt = \int_1^x \frac{F(t)\,dt}{\int_0^t F(s)\,ds} = \ln\left\{\int_0^x F(t)\,dt\right\} + c_1,$$

and consequently

$$\int_0^x F(t)\,dt = c\cdot\exp\left\{\int_1^x \frac{b(t)}{t}\,dt\right\}, \qquad (1.2.14)$$

where $c = e^{-c_1}$. It follows from (1.2.13) and (1.2.14) that for every $x > 0$,

$$F(x) = c\cdot\frac{b(x)}{x}\exp\left\{\int_1^x \frac{b(t)}{t}\,dt\right\}. \qquad (1.2.15)$$

Using Theorem 1.1.6 and equality (1.2.13), we obtain that $b(x)$ is slowly varying, i.e., $b \in RV_0$. Let us prove equality (1.2.1), i.e.,

$$\lim_{x\to\infty} b(x) = \rho + 1. \qquad (1.2.16)$$

Now, using Fatou's lemma, we obtain

$$\liminf_{x\to\infty}\frac{1}{b(x)}=\liminf_{x\to\infty}\frac{\int_0^x F(t)\,dt}{xF(x)}=\liminf_{x\to\infty}\frac{\int_0^1 F(xt)\,dt}{F(x)}$$
$$\geqslant\int_0^1\liminf_{x\to\infty}\frac{F(tx)}{F(x)}\,dt=\int_0^1 t^\rho\,dt=\frac{1}{\rho+1}. \tag{1.2.17}$$

Relation (1.2.17) implies that function $b(x)$ is bounded on $[x_0,+\infty)$ for some $x_0>0$. Since $b(x)$ is slowly varying and bounded as $x\to+\infty$, it follows that for every $t>0$,

$$\lim_{x\to\infty}\frac{b(xt)-b(x)}{b(x)}=0,$$

$$\lim_{x\to\infty}(b(xt)-b(x))=0. \tag{1.2.18}$$

Using equality (1.2.18) and Lebegue's theorem of dominant convergence, we obtain that for every $s>0$,

$$\lim_{x\to\infty}\left\{\int_1^s\frac{b(xt)}{t}\,dt-b(x)\ln s\right\}=\lim_{x\to\infty}\int_1^s\frac{b(xt)-b(x)}{t}\,dt=0. \tag{1.2.19}$$

Since $\int_0^x F(t)\,dt\in RV_{\rho+1}$, equality (1.2.14) implies that

$$\exp\left\{\int_1^x\frac{b(t)}{t}\,dt\right\}\in RV_{\rho+1}. \tag{1.2.20}$$

Now, using (1.2.20) we obtain that for every $s>0$,

$$\lim_{x\to\infty}\int_1^s\frac{b(xt)}{t}\,dt=\lim_{x\to\infty}\int_x^{sx}\frac{b(t)}{t}\,dt$$
$$=\lim_{x\to\infty}\ln\left\{\frac{\exp\left(\int_1^{sx}\frac{b(t)}{t}\,dt\right)}{\exp\left(\int_1^{x}\frac{b(t)}{t}\,dt\right)}\right\}$$
$$=(\rho+1)\ln s. \tag{1.2.21}$$

Relations (1.2.19) and (1.2.21) imply equality (1.2.16).
Case: $\rho=-1$, $F\in RV_{-1}$. If $\int_0^\infty F(t)\,dt<+\infty$, then $\int_0^x F(t)\,dt\in RV_{-1+1}=RV_0$. If $\int_0^\infty F(t)\,dt=+\infty$, then the same way as in case $\rho>-1$ we prove that $\int_0^x F(t)\,dt\in RV_{-1+1}=RV_0$. For function $b(x)$ that is given by (1.2.13), we have (*see* (1.2.17))

$$\liminf_{x\to\infty} \frac{1}{b(x)} \geqslant \int_0^1 t^\rho \, dt. \tag{1.2.22}$$

The integral on the right-hand side of (1.2.22) is equal to $+\infty$ for $\rho = -1$. Hence, $b(x) \to 0$ as $x \to \infty$, i.e., equality (1.2.1) holds.
Case: $\rho < -1$, $F \in RV_\rho$, or $\rho = -1$, $\int_0^\infty F(t)\,dt < +\infty$. Let us define function $b_1 : \mathbb{R}_+ \to \mathbb{R}_+$ as follows:

$$b_1(x) = \frac{xF(x)}{\int_x^\infty F(t)\,dt}. \tag{1.2.23}$$

Analogously, as in case $\rho > -1$, we prove that for $x > 0$,

$$F(x) = c \cdot \frac{b_1(x)}{x} \exp\left\{ -\int_1^x \frac{b_1(t)}{t}\,dt \right\}, \tag{1.2.24}$$

and the proof is similar as in case $\rho > -1$.

(b) Suppose that (1.2.3) holds, and let us denote

$$b(x) = \frac{xF(x)}{\int_0^x F(t)\,dt}.$$

Then, $b(x) \to \alpha \in (0, +\infty)$ as $x \to \infty$, and (1.2.15) holds. For every $\varepsilon > 0$ and $s_0 > 0$, there exists $t_0 > 0$, such that for all $t \geqslant t_0$ and $s \geqslant s_0$ the following inequalities hold:

$$\alpha - \varepsilon < b(ts) < \alpha + \varepsilon. \tag{1.2.25}$$

Using (1.2.15) and (1.2.25), we obtain that

$$\frac{F(tx)}{F(t)} = \frac{b(tx)}{xb(t)} \exp\left\{ \int_1^x \frac{b(st)}{s}\,ds \right\} \to x^{\alpha-1},$$

as $t \to +\infty$. Hence, $F \in RV_{\alpha-1}$.

If $\int_x^\infty F(t)\,dt < +\infty$ and (1.2.4) holds, then (1.2.24) also holds, where function $b_1(x)$ is given by (1.2.23). In this case we analogously obtain that $F \in RV_{-\alpha-1}$.

□

Remark 1.2.2 We shall give the following interpretation of the equalities (1.2.1) and (1.2.2). Let $f : (0, +\infty) \to (0, +\infty)$ be a function that is Lebesgue integrable on finite intervals.

If $f(x) = \dfrac{L(x)}{x^\alpha}$, where $\alpha < 1$, and L is a function that is slowly varying at infinity, then we have

$$\int_0^x f(t)\,dt = \int_0^x \frac{L(t)}{t^\alpha}\,dt \sim \frac{1}{1-\alpha}\cdot\frac{L(x)}{x^{\alpha-1}} = \frac{xf(x)}{1-\alpha}, \quad x\to\infty. \tag{1.2.26}$$

If $f(x) = \dfrac{L(x)}{x^\alpha}$, where $\alpha > 1$, and L is a function that is slowly varying at infinity, then we have

$$\int_x^\infty f(t)\,dt = \int_x^\infty \frac{L(t)}{t^\alpha}\,dt \sim \frac{1}{\alpha-1}\cdot\frac{L(x)}{x^{\alpha-1}} = \frac{xf(x)}{\alpha-1}, \quad x\to\infty. \tag{1.2.27}$$

In order to obtain the asymptotic behavior of integrals in (1.2.26) and (1.2.27), we can put a slowly varying function as a factor before the integral and calculate the integral of a power function.

Additionally, if X is a random variable with absolutely continuous probability distribution function $F(x) = P\{X \leqslant x\}$ and density function $f(x) = x^{-\alpha}L(x)$, where $\alpha > 1$, and L is slowly varying at infinity, then relation (1.2.27) implies that

$$1 - F(x) = P\{X > x\} \sim \frac{1}{\alpha-1}\cdot\frac{L(x)}{x^{\alpha-1}}, \quad x\to\infty. \tag{1.2.28}$$

1.2.2 Canonical Representation of Regularly Varying Functions

Theorem 1.2.3 (Karamata [122]) *Let $F : \mathbb{R}_+ \to \mathbb{R}_+$ be a function that is Lebesgue integrable on finite intervals. Then, the following statement holds: $F \in RV_\rho$ if and only if function F can be represented in the form:*

$$F(x) = h(x)\cdot\exp\left\{\int_1^x \frac{a(t)}{t}\,dt\right\}, \tag{1.2.29}$$

where $h : \mathbb{R}_+ \to \mathbb{R}_+$ and $a : \mathbb{R}_+ \to \mathbb{R}$ are functions such that

$$\lim_{x\to\infty} h(x) = h_0 \in \mathbb{R}_+, \tag{1.2.30}$$

$$\lim_{x\to\infty} a(x) = \rho. \tag{1.2.31}$$

Proof

(a) Let F be a function that is Lebesgue integrable on finite intervals and such that the equalities (1.2.29)–(1.2.31) hold. Let $x > 0$ and $\varepsilon > 0$ be fixed real numbers. Then we have

$$\frac{F(tx)}{F(t)} = \frac{h(tx)}{h(t)} \exp\left\{\int_t^{tx} \frac{a(s)}{s}\,ds\right\}. \tag{1.2.32}$$

Condition (1.2.31) implies that there exists $x_0 > 0$ such that for all $s > x_0$,

$$\rho - \varepsilon \leqslant a(s) \leqslant \rho + \varepsilon. \tag{1.2.33}$$

Case: $x \geqslant 1$. It follows from (1.2.33) that for $t \geqslant x_0$ and $tx \geqslant x_0$ the following inequalities hold:

$$(\rho - \varepsilon)\ln x \leqslant \int_t^{tx} \frac{a(s)}{s}\,ds \leqslant (\rho + \varepsilon)\ln x. \tag{1.2.34}$$

Now, using the inequalities (1.2.33) and the equalities (1.2.30) and (1.2.32), relation (1.2.34) implies that

$$x^{\rho-\varepsilon} \leqslant \liminf_{t\to\infty} \frac{F(tx)}{F(t)} \leqslant \limsup_{t\to\infty} \frac{F(tx)}{F(t)} \leqslant x^{\rho+\varepsilon}. \tag{1.2.35}$$

Since $\varepsilon > 0$ can be chosen arbitrarily, inequality (1.2.35) implies equality $\lim_{t\to\infty} \frac{F(tx)}{F(t)} = x^{\rho}$, i.e., $F \in RV_\rho$.

Case: $0 < x < 1$. For $t \geqslant x_0$ and $tx \geqslant x_0$, we obtain

$$(\rho + \varepsilon)\ln x \leqslant \int_t^{tx} \frac{a(s)}{s}\,ds = \int_{tx}^{t} \frac{-a(s)}{s}\,ds \leqslant (\rho - \varepsilon)\ln x. \tag{1.2.36}$$

Similarly, as in the previous case, (1.2.36) implies that $F \in RV_\rho$.

(b) Suppose that $F \in RV_\rho$. For $\rho > -1$, function F can be represented in the form (1.2.15), with $b(x)$ given by (1.2.13). For $x > 0$, we obtain

$$\begin{aligned} F(x) &= c\frac{b(x)}{x} \exp\left\{\int_1^x \frac{b(t)}{t}\,dt\right\} \\ &= cb(x)\exp\left\{\int_1^x \frac{b(t)-1}{t}\,dt\right\}. \end{aligned} \tag{1.2.37}$$

Let us denote $h(x) = cb(x)$, $h_0 = c(\rho+1)$, and $a(x) = b(x) - 1$. Since $b(x) \to \rho + 1$ as $x \to \infty$, *see* (1.2.16), it follows that F can be represented in the form (1.2.29), with conditions (1.2.30) and (1.2.31) satisfied.

If $\rho < -1$, then (1.2.24) and (1.2.23) imply that, for every $x > 0$,

$$F(x) = cb_1(x)\exp\left\{\int_1^x \frac{-b_1(x)-1}{t}\,dt\right\}.$$

Finally, if $\rho = -1$, then Theorem 1.1.6 implies that $xF(x) \in RV_0$, and using the result that is proved for case $\rho > -1$, we obtain

$$xF(x) = h(x)\exp\left\{\int_1^x \frac{a(t)}{t}\,dt\right\}, \qquad x > 0,$$

where $h(x) \to h_0 \in \mathbb{R}_+$ and $a(x) \to 0$ as $x \to \infty$. Hence, if $\rho = -1$, then for every $x > 0$ we have

$$\begin{aligned} F(x) &= \frac{h(x)}{x}\exp\left\{\int_1^x \frac{a(t)}{t}\,dt\right\} \\ &= h(x)\exp\left\{\int_1^x \frac{a(t)-1}{t}\,dt\right\}, \end{aligned}$$

where $h(x) \to h_0 \in \mathbb{R}_+$, and $a(x) - 1 \to -1$ as $x \to \infty$, i.e., $F \in RV_{-1}$.

□

Theorem 1.2.4 *Let us consider function F with the canonical representation* (1.2.29)–(1.2.31). *Function $a(t)$ in equality* (1.2.29) *can be chosen such that it is continuous.*

Proof For every $t > 1$, let us denote

$$\begin{aligned} F_0(t) &= \exp\left\{\int_1^t \frac{a(s)}{s}\,ds\right\}, \\ b_0(t) &= \frac{tF_0(t)}{\int_1^t F_0(s)\,ds}. \end{aligned}$$

Since $F \in RV_\rho$, (1.2.29) and (1.2.30) imply that $F_0 \in RV_\rho$. Using (1.2.37) we obtain that for every $x > 0$,

$$F_0(x) = cb_0(x)\exp\left\{\int_1^x \frac{b_0(t)-1}{t}\,dt\right\},$$

and therefore

$$F(x) = h(x)\cdot cb_0(x)\exp\left\{\int_1^x \frac{b_0(t)-1}{t}\,dt\right\},$$

where $b_0(t) - 1$ is a continuous function. □

We shall give some examples of the canonical representation of regularly varying functions. Since the regular variation is a local property at infinity, any point $x_0 > 0$ may appear as the lower limit of the integral in (1.2.29).

Example 1.2.5 Function $(\ln x)^{\beta}$, $x \geqslant e$, where $\beta \in \mathbb{R}$, has the following canonical representation:

$$(\ln x)^{\beta} = e^{\ln(\ln x)^{\beta}} = e^{\beta \ln \ln x} = \exp\left\{\int_e^x \frac{\beta\, dt}{t\, \ln t}\right\}.$$

Here, $h(x) = 1$ and $a(x) = \beta/\ln t$. △

Example 1.2.6 $\ln \ln x = e^{\ln \ln \ln x} = \exp\left\{\int_{e^e}^x \frac{dt}{t\, \ln t\, \ln \ln t}\right\}, \; x \geqslant e^e.$ △

Example 1.2.7 For $\rho \in \mathbb{R}$, $\beta \in \mathbb{R}$, and $x \geqslant e$, we have

$$x^{\rho}(\ln x)^{\beta} = e^{\rho \ln x + \beta \ln \ln x} = e^{\rho} \cdot \exp\left\{\int_e^x \frac{\rho \ln t + \beta}{t\, \ln t}\, dt\right\}.$$

△

Example 1.2.8 For $x \geqslant \pi/4$, function $\operatorname{arctg} x$ has the following canonical representation:

$$\operatorname{arctg} x = e^{\ln \operatorname{arctg} x} = \exp\left\{\int_{\pi/4}^x \frac{d(\operatorname{arctg} t)}{\operatorname{arctg} t}\right\} = \exp\left\{\int_{\pi/4}^x \frac{dt}{(1+t^2)\operatorname{arctg} t}\right\}.$$

In this case $h(t) = 1$ and $a(t) = \dfrac{t}{(1+t^2)\operatorname{arctg} t}$. △

1.2.3 Modified Form of Karamata's Theorems

The next theorem gives a modified form of the results from the Karamata Theorems 1.2.3–1.2.4.

Theorem 1.2.9 (de Haan [88]) *Let $H : \mathbb{R}_+ \to \mathbb{R}$ be a function such that H and e^H are integrable on finite intervals. Then the following conditions are equivalent:*

(a) *There exists a constant $\rho \in \mathbb{R}$, such that for every $x > 0$,*

$$\lim_{t\to\infty} (H(tx) - H(t)) = \rho \ln x. \tag{1.2.38}$$

(b) *There exist the functions $c : \mathbb{R}_+ \to \mathbb{R}$ and $a : \mathbb{R}_+ \to \mathbb{R}$, such that*

$$\lim_{x\to\infty} c(x) = c_0 \in \mathbb{R}, \tag{1.2.39}$$

$$\lim_{x\to\infty} a(x) = \rho, \tag{1.2.40}$$

$$H(x) = c(x) + \int_1^x \frac{a(t)}{t}\, dt. \tag{1.2.41}$$

(c) *The following equality holds true:*

$$\lim_{x\to\infty} \left\{ H(x) - \frac{1}{x}\int_1^x H(t)\, dt \right\} = \rho. \tag{1.2.42}$$

Proof Let $F(x) = e^{H(x)}$. Equality (1.2.38) holds if and only if $F \in RV_\rho$. Theorems 1.2.3 and 1.2.4 imply that condition (a) is equivalent to condition (b). Let us now prove that conditions (b) and (c) are equivalent as well.

Suppose that (b) holds. Then we have

$$\begin{aligned} H(x) - \frac{1}{x}\int_1^x H(t)\, dt = c(x) &+ \int_1^x \frac{a(t)}{t}\, dt \\ &- \frac{1}{x}\int_1^x c(t)\, dt - \frac{1}{x}\int_1^x \left\{ \int_1^t \frac{a(s)}{s}\, ds \right\} dt. \end{aligned}$$

Fubini's theorem implies that

$$\begin{aligned} \int_1^x \left\{ \int_1^t \frac{a(s)}{s}\, ds \right\} dt &= \int_1^x \left\{ \frac{a(s)}{s} \int_s^x dt \right\} ds = \int_1^x \frac{a(s)}{s}(x-s)\, ds \\ &= x\int_1^x \frac{a(s)}{s}\, ds - \int_1^x a(s)\, ds. \end{aligned}$$

Using (1.2.39) and (1.2.40) we now obtain

$$\begin{aligned} H(x) - \frac{1}{x}\int_1^x H(t)\, dt &= c(x) - \frac{1}{x}\int_1^x c(t)\, dt + \frac{1}{x}\int_1^x a(s)\, ds \\ &\to c_0 - c_0 + \rho = \rho, \quad \text{as } x \to \infty. \end{aligned}$$

Now let us suppose that (c) holds. For each $x > 0$, let us denote

$$g(x) = H(x) - \frac{1}{x}\int_1^x H(t)\, dt.$$

Then we have $\lim_{x\to\infty} g(x) = \rho$, and the following equalities hold:

$$\begin{aligned} \int_1^x \frac{g(t)}{t}\, dt &= \int_1^x \frac{H(t)}{t}\, dt - \int_1^x \left\{ \frac{1}{t^2}\int_1^t H(s)\, ds \right\} dt \\ &= \int_1^x \frac{H(t)}{t}\, dt - \int_1^x \left\{ H(s) \int_s^x \frac{1}{t^2}\, dt \right\} ds \end{aligned}$$

$$
\begin{aligned}
&= \int_1^x \frac{H(t)}{t}\,dt - \int_1^x \frac{H(s)}{s}\,ds + \frac{1}{x}\int_1^x H(s)\,ds \\
&= H(x) - g(x).
\end{aligned}
$$

Therefore

$$
H(x) = g(x) + \int_1^x \frac{g(t)}{t}\,dt, \tag{1.2.43}
$$

and condition (b) easily follows. □

1.2.4 Absolutely Continuous Functions

Theorem 1.2.10 *Let $F : \mathbb{R}_+ \to \mathbb{R}_+$ be an absolutely continuous function with density f, i.e., $F(x) = \int_0^x f(t)\,dt$ for $x > 0$.*

(a) [von Mises] *If*

$$
\lim_{x\to\infty} \frac{xf(x)}{F(x)} = \rho, \tag{1.2.44}
$$

then $F \in RV_\rho$. Additionally, if $\rho > 0$, then $f \in RV_{\rho-1}$.

(b) [Landau, de Haan] *If $F \in RV_\rho$, $\rho \in \mathbb{R}$, and f is a monotone function, then equality* (1.2.44) *holds. If $\rho \neq 0$, then $(\operatorname{sgn}\rho)\cdot f(x) \in RV_{\rho-1}$.*

Proof

(a) Let $b(x) = xf(x)/\int_0^x f(t)\,dt$. Analogously, as in the proof of Theorem 1.2.1, we obtain that $F(x) = F(1)\exp\left\{\int_1^x b(t)\,dt/t\right\}$, i.e., $F \in RV_\rho$.

(b) Let $f(x)$ be an increasing function. (Analogously, we consider the case when $f(x)$ is a decreasing function.) Let us suppose that $0 < a < b$. Then we have

$$
\frac{F(bx) - F(ax)}{F(x)} = \int_{ax}^{bx} \frac{f(t)\,dt}{F(x)}. \tag{1.2.45}
$$

Since $f(t)$ is an increasing function, equality (1.2.45) implies that

$$
\frac{(b-a)x}{F(x)} f(ax) \leqslant \frac{F(bx) - F(ax)}{F(x)} \leqslant \frac{(b-a)x}{F(x)} f(bx). \tag{1.2.46}
$$

Using the first of the inequalities (1.2.46) and the condition $F \in RV_\rho$, we obtain that

$$
\limsup_{x\to\infty} \frac{(b-a)x}{F(x)} f(ax) \leqslant \lim_{x\to\infty} \frac{F(bx) - F(ax)}{F(x)} = b^\rho - a^\rho,
$$

and therefore

$$\limsup_{x\to\infty} \frac{xf(ax)}{F(x)} \leqslant \frac{b^\rho - a^\rho}{b-a}. \tag{1.2.47}$$

When $b \downarrow a$, inequality (1.2.47) implies that

$$\limsup_{x\to\infty} \frac{xf(ax)}{F(x)} \leqslant \rho a^{\rho-1}. \tag{1.2.48}$$

Analogously, using the second of the inequalities (1.2.46), we obtain that

$$\liminf_{x\to\infty} \frac{xf(bx)}{F(x)} \geqslant \rho b^{\rho-1}. \tag{1.2.49}$$

If we put $a = 1$ in (1.2.48) and $b = 1$ in (1.2.49), we obtain that

$$\rho \leqslant \liminf_{x\to\infty} \frac{xf(x)}{F(x)} \leqslant \limsup_{x\to\infty} \frac{xf(x)}{F(x)} \leqslant \rho,$$

and equality (1.2.44) follows.

□

1.2.5 Additional Properties of Regularly Varying Functions

Lemma 1.2.11 *If $F \in RV_\rho$, where $\rho \in \mathbb{R}$, then we have*

$$\lim_{x\to\infty} \frac{\ln F(x)}{\ln x} = \rho, \tag{1.2.50}$$

$$\lim_{x\to\infty} F(x) = \begin{cases} 0, & \text{if } \rho < 0, \\ +\infty, & \text{if } \rho > 0. \end{cases} \tag{1.2.51}$$

Proof Using the canonical representation (1.2.29) of regularly varying function F and the equalities (1.2.30) and (1.2.31), we obtain that

$$\frac{\ln F(x)}{\ln x} = \frac{\ln h(x)}{\ln x} + \frac{\int_1^x \frac{a(t)}{t}\,dt}{\ln x}.$$

Let $\varepsilon > 0$. There exists $A > 0$, such that for $t \geqslant A$ and $x \geqslant A$, we have $\rho - \varepsilon \leqslant a(t) \leqslant \rho + \varepsilon$, and

$$(\rho - \varepsilon)(\ln x - \ln A) \leqslant \int_A^x \frac{a(t)}{t}\,dt \leqslant (\rho + \varepsilon)\ln x,$$

and equality (1.2.50) follows easily. Equality (1.2.51) follows from (1.2.50). □

Lemma 1.2.12 *If $L : \mathbb{R}_+ \to \mathbb{R}$ is a slowly varying function at infinity, then the limit $\lim_{x\to\infty} L(x)$ does not have to exist.*

Proof Let $x_0 = 1, x_1, x_2, x_3, \ldots$ be an increasing sequence of real numbers, such that

$$y_k := \int_{x_{k-1}}^{x_k} \frac{dt}{t} = k, \quad \text{for } k = 1, 2, 3, \ldots$$

It is easy to see that $x_0 = e^0$, $x_1 = e$, $x_2 = e^3$ $x_3 = e^6$, $x_4 = e^{10}$, $x_5 = e^{15}, \ldots,$ and generally $x_n = e^{0+1+2+\cdots+n}$. Let us define the function $a : [1, +\infty) \to \mathbb{R}$ as follows:

$$a(t) = \frac{(-1)^{k-1}}{k}, \quad \text{for } x_{k-1} \leqslant t < x_k, \quad k = 1, 2, \ldots$$

Then, for every positive integer k,

$$z_k := \int_{x_{k-1}}^{x_k} \frac{a(t)}{t}\, dt = (-1)^{k-1}.$$

Let $L(x) = \exp\left\{\int\limits_1^x \frac{a(t)}{t}\, dt\right\}$. Since $a(t) \to 0$ as $t \to \infty$, it follows that L is a slowly varying function, and

$$L(x_{2k}) = e^0 = 1, \quad \text{for } k = 0, 1, 2, \ldots;$$
$$L(x_{2k+1}) = e^1 = e, \quad \text{for } k = 0, 1, 2, \ldots.$$

Since (x_n) is an increasing sequence such that $x_n \to \infty$ as $n \to \infty$, it follows that $\lim_{x\to\infty} L(x)$ does not exist. □

Lemma 1.2.13 *Let $F \in RV_\rho$, $\rho \in \mathbb{R}$, and let (a_n) and (b_n) be two sequences of positive real numbers such that*

$$\lim_{n\to\infty} a_n = \lim_{n\to\infty} b_n = \infty,$$
$$\lim_{n\to\infty} \frac{a_n}{b_n} = c \in \mathbb{R}_+.$$

Then, $\lim_{n\to\infty} \frac{F(a_n)}{F(b_n)} = c^\rho.$

Proof Using the Uniform Convergence Theorem 1.1.10 applied on interval $[c - \varepsilon, c + \varepsilon]$, we obtain that

$$\lim_{n\to\infty}\frac{F(a_n)}{F(b_n)}=\lim_{n\to\infty}\frac{F\left(b_n\cdot(a_n/b_n)\right)}{F(b_n)}=\lim_{n\to\infty}\frac{F(b_nc)}{F(b_n)}=c^{\rho}.$$

□

Lemma 1.2.14 *Let $F\in RV_\rho$, where $\rho\in\mathbb{R}$. Then for every $\varepsilon>0$ there exists t_0, such that for all $t\geqslant t_0$ and $x\geqslant 1$ the following inequalities hold:*

$$(1-\varepsilon)x^{\rho-\varepsilon}<\frac{F(tx)}{F(t)}<(1+\varepsilon)x^{\rho+\varepsilon}.$$

Proof Karamata's Theorem 1.2.3 implies that

$$\begin{aligned}\frac{F(tx)}{F(t)}&=\frac{h(tx)}{h(t)}\cdot\exp\left\{\int_t^{tx}\frac{a(u)}{u}\,du\right\}\\&=\frac{h(tx)}{h(t)}\cdot\exp\left\{\int_1^{x}\frac{a(ts)}{s}\,ds\right\}.\end{aligned}$$

Relations (1.2.30) and (1.2.31) imply that for every $\varepsilon>0$ there exists t_0, such that for all $t\geqslant t_0$ and $s\geqslant 1$ the following inequalities hold:

$$\begin{aligned}&\rho-\varepsilon<a(ts)<\rho+\varepsilon,\\&1-\varepsilon<\frac{h(tx)}{h(t)}<1+\varepsilon.\end{aligned}$$

The statement of the lemma now follows easily. □

Lemma 1.2.15 *If $F_1\in RV_{\rho_1}$, $F_2\in RV_{\rho_2}$, and $F_2(x)\to\infty$ as $x\to+\infty$, then we have $F_1\circ F_2\in RV_{\rho_1\cdot\rho_2}$.*

Proof The following equalities hold:

$$\lim_{t\to\infty}\frac{F_1\circ F_2(tx)}{F_1\circ F_2(t)}=\lim_{t\to\infty}\frac{F_1(F_2(tx))}{F_1(F_2(t))}=\lim_{t\to\infty}\frac{F_1\left(F_2(t)\cdot\dfrac{F_2(tx)}{F_2(t)}\right)}{F_1(F_2(t))}.$$

Using Theorem 1.1.10 and notation $u=F_2(t)$, we obtain that

$$\lim_{t\to\infty}\frac{F_1\circ F_2(tx)}{F_1\circ F_2(t)}=\lim_{t\to\infty}\frac{F_1(ux^{\rho_2})}{F_1(u)}=x^{\rho_1\rho_2}.$$

□

Lemma 1.2.16 *If $F_1\in RV_{\rho_1}$ and $F_2\in RV_{\rho_2}$, then*

$$F_1+F_2\in RV_{\max\{\rho_1,\rho_2\}}.$$

Proof The statement of Lemma 1.2.16 is a consequence of the Uniform Convergence Theorem. □

The properties of regularly varying functions were investigated in papers by Aljančić and Karamata [8], Adamović [1], Adamović [2], de Haan [89], and many others. Avakumović [15] generalized the notion of regularly varying functions, and Karamata [125] obtained the properties of the functions that were introduced in Avakumović [15]. In the paper by Aljančić and Arandjelović [7], these functions are called O-regularly varying, and some of their new properties were obtained. Bari and Stechkin [228] introduced monotone O-regularly varying functions that tend to zero and developed their theory.

Regularly varying functions appeared to be very useful in probability theory, especially in the theory of the summation of random variables and the extreme value theory. Buldygin, Klesov, and Steinebach [35] used O-regularly varying functions while investigating the strong law of large numbers for generalized renewal processes.

1.3 Monotone Regularly Varying Functions

Definition 1.3.1 Let A be a nonempty subset of the set of real numbers $\mathbb{R}$ and $F : A \to \mathbb{R}$ be a function defined on A. Function F is *increasing* if

$$x_1 < x_2 \quad \text{implies that} \quad F(x_1) \leqslant F(x_2),$$

for all $x_1, x_2 \in A$. Function F is *decreasing* if $x_1 < x_2$ implies that $F(x_1) \geqslant F(x_2)$ for all $x_1, x_2 \in A$. Function F is *strictly increasing* if $x_1 < x_2$ implies the strict inequality $F(x_1) < F(x_2)$ for all $x_1, x_2 \in A$. Function F is *strictly decreasing* if $x_1 < x_2$ implies that $F(x_1) > F(x_2)$ for all $x_1, x_2 \in A$. The increasing and decreasing functions are called *monotone functions*, while strictly increasing and decreasing functions are *strictly monotone functions*.

1.3.1 Criteria for the Regular Variation of Monotone Functions

The properties of regularly varying functions can be more easily investigated under the assumption that the underlying functions are monotone. Moreover, condition (1.1.1) can be replaced by a much simpler condition. For example, it is sufficient to assume that equality (1.1.1) holds only for a single positive real number t_0. We shall give this result in the next theorem and then provide some more properties of regularly varying monotone functions. *See* also de Haan [87].

Theorem 1.3.2 *Let $L : \mathbb{R}_+ \to \mathbb{R}_+$ be a monotone function. If there exists a positive real number $x_0 \neq 1$, such that*

$$\lim_{t\to\infty} \frac{L(tx_0)}{L(t)} = 1,$$

then L is slowly varying at infinity.

Proof First note that regularly varying functions are measurable. Using Theorems 1.1.8 and 1.1.9, we conclude that it is sufficient to prove that for every $x \in [a, b]$, where $0 < a < b < \infty$, the following equality holds:

$$\lim_{t\to\infty} \frac{L(tx)}{L(t)} = 1.$$

Suppose that L is an increasing function. If $x_0 > 1$, then for all sufficiently large values of t and all $x \in [1, x_0]$ we have

$$1 \leqslant \frac{L(tx)}{L(t)} \leqslant \frac{L(tx_0)}{L(t)},$$

and therefore $\lim\limits_{t\to\infty} \dfrac{L(tx)}{L(t)} = 1$, for all $x \in [1, x_0]$. If $0 < x_0 < 1$, then for all sufficiently large values of t and all $x \in [x_0, 1]$ we have

$$\frac{L(tx_0)}{L(t)} \leqslant \frac{L(tx)}{L(t)} \leqslant 1,$$

and consequently $\lim\limits_{t\to\infty} \dfrac{L(tx)}{L(t)} = 1$, for all $x \in [x_0, 1]$. Analogously we consider the case when L is a decreasing function. □

Theorem 1.3.3 *A monotone function $F : \mathbb{R}_+ \to \mathbb{R}_+$ is regularly varying at infinity if there exist two positive integers m_1 and m_2 and a real number ρ, such that $\dfrac{\ln m_1}{\ln m_2}$ is an irrational real number and*

$$\lim_{n\to\infty} \frac{F(nm_i)}{F(n)} = m_i^{\rho}, \quad \textit{for } i \in \{1, 2\}, \tag{1.3.1}$$

where n tends to infinity over the set of positive integers.

Proof Let $\mathbb{Z}$ be the set of integers, and $\mathbb{N}_0 = \{0, 1, 2, \dots\}$. Let us suppose that F is an increasing function, and let the conditions of the theorem be satisfied. Then, $\rho \geqslant 0$. From condition (1.3.1), it follows that

$$\lim_{n\to\infty} \frac{F(nm)}{F(n)} = m^{\rho}, \quad \text{for all } m \in S_0, \tag{1.3.2}$$

where $S_0 = \{m_1^k m_2^l : k, l \in \mathbb{N}_0\}$. From the Kronecker theorem (*see* Chapter XXIII in Hardy and Wright [102]), it follows that if $\frac{\ln m_1}{\ln m_2}$ is an irrational number, then set $S_1 = \{m_1^k m_2^l : k, l \in \mathbb{Z}\}$ is dense in $\mathbb{R}_+$. Let us first prove the following equality:

$$\lim_{n\to\infty} \frac{F(n+1)}{F(n)} = 1. \tag{1.3.3}$$

Let us suppose that equality (1.3.3) does not hold. Then there exists a sequence of positive integers (n_i) such that $n_i \to \infty$ as $i \to \infty$, and

$$\lim_{i\to\infty} \frac{F(n_i+1)}{F(n_i)} = c \in (1, +\infty]. \tag{1.3.4}$$

Since set S_1 is dense in $\mathbb{R}_+$, it follows that there exist $r, s \in S_0$, such that

$$1 < \frac{r}{s} < c^{1/\rho}. \tag{1.3.5}$$

Let us denote $k_i = [n_i/s]$, $i \in \{1, 2, \dots\}$. For sufficiently large n_i, we obtain

$$k_i s \leqslant n_i < (k_i+1)s < k_i(s+1) \leqslant k_i r. \tag{1.3.6}$$

Using (1.3.2), (1.3.4), (1.3.5), and (1.3.6), we obtain

$$c = \lim_{i\to\infty} \frac{F(n_i+1)}{F(n_i)} \leqslant \lim_{i\to\infty} \frac{F(k_i r)}{F(k_i s)} = \left(\frac{r}{s}\right)^{\rho} < c,$$

and this is an obvious contradiction. Hence, equality (1.3.3) holds.

Let us now prove that $F \in RV_\rho$. Let x and ε be two real numbers such that $0 < \varepsilon < x$. From the condition that set S_1 is dense in $\mathbb{R}_+$, it follows that there exist real numbers $r_0, r_1, r_2 \in S_1$, such that

$$x - \varepsilon < \frac{r_1}{r_0} < x < \frac{r_2}{r_0} < x + \varepsilon. \tag{1.3.7}$$

For every $t > 0$, let us denote $p_t = [t/r_0]$. Then we have

$$\frac{t}{r_0} - 1 < p_t \leqslant \frac{t}{r_0}. \tag{1.3.8}$$

Using the fact that F is a monotone function and inequalities (1.3.7) and (1.3.8), we obtain that

$$\frac{F(p_t r_1)}{F((p_t+1)r_0)} \leqslant \frac{F(tx)}{F(t)} \leqslant \frac{F((p_t+1)r_2)}{F(p_t r_0)}. \tag{1.3.9}$$

Note that the first and the last fraction in (1.3.9) can be represented in the form

$$\frac{F(p_t r_1)}{F((p_t+1)r_0)} = \frac{F(p_t r_1)}{F(p_t)} \cdot \frac{F(p_t)}{F(p_t+1)} \cdot \frac{F(p_t+1)}{F((p_t+1)r_0)}, \tag{1.3.10}$$

$$\frac{F((p_t+1)r_2)}{F(p_t r_0)} = \frac{F((p_t+1)r_2)}{F(p_t+1)} \cdot \frac{F(p_t+1)}{F(p_t)} \cdot \frac{F(p_t)}{F(p_t r_0)}. \tag{1.3.11}$$

Using limit values (1.3.2) and (1.3.3) and equalities (1.3.10) and (1.3.11), we obtain from (1.3.9) that

$$(x-\varepsilon)^\rho \leqslant \liminf_{t\to\infty} \frac{F(tx)}{F(t)} \leqslant \limsup_{t\to\infty} \frac{F(tx)}{F(t)} \leqslant (x+\varepsilon)^\rho. \tag{1.3.12}$$

When $\varepsilon \downarrow 0$, inequality (1.3.12) implies that $\lim\limits_{t\to\infty} \dfrac{F(tx)}{F(t)} = x^\rho$ for $x > 0$. □

Theorem 1.3.4 *A monotone function $F : \mathbb{R}_+ \to \mathbb{R}_+$ is regularly varying at infinity if and only if there exist two sequences of positive real numbers (a_n) and (λ_n), such that the following three conditions are satisfied:*

(a) $\lim\limits_{n\to\infty} \dfrac{\lambda_{n+1}}{\lambda_n} = 1$.
(b) $\lim\limits_{n\to\infty} a_n = \infty$.
(c) *For every $x > 0$, there exists $\lim\limits_{n\to\infty} \lambda_n F(a_n x)$ as a finite positive real number.*

Moreover, if $\kappa(x) = \lim\limits_{n\to\infty} \lambda_n F(a_n x)$ for $x > 0$ and if ρ is the index of regular variation of function F, then for every $x > 0$ we have

$$\frac{\kappa(x)}{\kappa(1)} = x^\rho.$$

Proof Suppose that F is a decreasing function. Analogously we can consider the case when F is an increasing function.
CONDITIONS ARE SUFFICIENT If $\lim\limits_{x\to\infty} F(x) = c_0 > 0$, the theorem holds. Indeed, for $\rho = 0$, $\lambda_n = 1$, and $a_n = n$, all three conditions are satisfied and $\lim\limits_{t\to\infty} \dfrac{F(tx)}{F(t)} = 1$, i.e., F is a slowly varying function.

Let us now suppose that $\lim\limits_{x\to\infty} F(x) = 0$. For every $t > 0$, let $n = n(t)$ be the smallest positive integer such that $a_{n+1} > t$. Then we have $a_n \leqslant t < a_{n+1}$. Using the fact that F is a decreasing function, we obtain that for every $x > 0$ the following inequalities hold:

$$\frac{F(a_{n+1}x)}{F(a_n \cdot 1)} \leqslant \frac{F(tx)}{F(t \cdot 1)} \leqslant \frac{F(a_n x)}{F(a_{n+1} \cdot 1)}. \tag{1.3.13}$$

Conditions (a)–(c) and inequalities (1.3.13) imply that for every $x > 0$,

$$\lim_{t\to\infty} \frac{F(tx)}{F(t)} = \frac{\kappa(x)}{\kappa(1)}.$$

Now, using Theorem 1.1.9, we conclude that F is a regularly varying function, and $\kappa(x)/\kappa(1) = x^\rho$, where ρ is the exponent of regular variation.

CONDITIONS ARE NECESSARY Suppose that $F \in RV_\rho$. Since F decreases, it follows that $\rho \leqslant 0$. Let us define a function $G : \mathbb{R}_+ \to \mathbb{R}_+$ as follows:

$$G(y) = \inf\{x : F(x) \leqslant y\}.$$

It is easy to see that

$$\lim_{y\downarrow 0} G(y) = +\infty, \tag{1.3.14}$$

$$\frac{F(G(y)+0)}{F(G(y))} \leqslant \frac{y}{F(G(y))} \leqslant \frac{F(G(y)-0)}{F(G(y))}. \tag{1.3.15}$$

Let $x \in (0, 1)$ be a fixed real number. Then, there exists $t_0 = t_0(x)$, such that for all $t \geqslant t_0$,

$$1 \leqslant \frac{F(t-0)}{F(t)} \leqslant \frac{F(tx)}{F(t)}. \tag{1.3.16}$$

From (1.3.16) it follows that

$$1 \leqslant \liminf_{t\to\infty} \frac{F(t-0)}{F(t)} \leqslant \limsup_{t\to\infty} \frac{F(t-0)}{F(t)} \leqslant x^\rho. \tag{1.3.17}$$

As $x \uparrow 1$, (1.3.17) implies that $\lim\limits_{t\to\infty} \dfrac{F(t-0)}{F(t)} = 1$. Analogously we prove that $\lim\limits_{t\to\infty} \dfrac{F(t+0)}{F(t)} = 1$. Using the last equality and equalities (1.3.14) and (1.3.15), we obtain

$$\lim_{y\downarrow 0} \frac{F(G(y))}{y} = 1. \tag{1.3.18}$$

Let us denote $\lambda_n = n$ and $a_n = G(1/n)$. Then condition (a) obviously holds, and (b) follows from (1.3.14). Let us prove that condition (c) holds as well. For $x > 0$, we have

$$\lim_{n\to\infty} \lambda_n F(a_n x) = \lim_{n\to\infty} \lambda_n F(a_n) \cdot \frac{F(a_n x)}{F(a_n)}$$
$$= \lim_{n\to\infty} \frac{F(G(1/n))}{1/n} \cdot \frac{F(a_n x)}{F(a_n)}.$$

Now, using (1.3.18) and the fact that $F \in RV_\rho$, equality $\lim_{n\to\infty} \lambda_n F(a_n x) = x^\rho$ follows easily. □

1.3.2 *Criteria for the Fast Variation of Monotone Functions*

Theorem 1.3.5 *A monotone function $F : \mathbb{R}_+ \to \mathbb{R}_+$ is fast varying at infinity with the exponent $\rho = +\infty$ or $\rho = -\infty$, if and only if there exist two sequences of positive real numbers (a_n) and (λ_n) and a positive real number c, such that the following three conditions are satisfied:*

(a) $\lim_{n\to\infty} \frac{\lambda_{n+1}}{\lambda_n} = 1.$
(b) $\lim_{n\to\infty} a_n = \infty.$
(c) $\lim_{n\to\infty} \lambda_n F(a_n x) = \left(\frac{x}{c}\right)^\rho.$

Proof As in the proof of Theorem 1.3.4 we shall suppose that F is a decreasing function. The case when F is an increasing function can be considered analogously. CONDITIONS ARE SUFFICIENT Let us suppose that $x > 1$, and let us choose real numbers c_1 and c_2, such that $0 < c_1 < c < c_2 < +\infty$ and $x = c_2/c_1$. Similarly as in the proof of Theorem 1.3.4 for every $t > 0$, let us denote $n = n(t) = \min\{m : a_{m+1} > t\}$. Then we have $a_n \leqslant t < a_{n+1}$ and

$$\frac{F(tc_2)}{F(tc_1)} \leqslant \frac{F(a_n c_2)}{F(a_{n+1} c_1)}. \tag{1.3.19}$$

Using (1.3.19) and conditions (a)–(c), we obtain that

$$0 \leqslant \lim_{t\to\infty} \frac{F(tx)}{F(t)} = \lim_{t\to\infty} \frac{F(tc_2/c_1)}{F(t)} = \lim_{t\to\infty} \frac{F(tc_2)}{F(tc_1)}$$
$$\leqslant \lim_{n\to\infty} \frac{F(a_n c_2)}{F(a_{n+1} c_1)} = \lim_{n\to\infty} \frac{\lambda_{n+1}}{\lambda_n} \cdot \frac{\lambda_n F(a_n c_2)}{\lambda_{n+1} F(a_{n+1} c_1)}$$
$$= \frac{(c_2/c)^{-\infty}}{(c_1/c)^{-\infty}} = 0 = \left(\frac{c_2}{c_1}\right)^{-\infty} = x^{-\infty}.$$

If $0 < x < 1$, we choose real numbers c_1 and c_2, such that $0 < c_1 < c < c_2 < +\infty$ and $x = c_1/c_2$, and then we proceed as in the previous case.

CONDITIONS ARE NECESSARY Let F be a fast varying function. Since F decreases, we have $\rho = -\infty$. Let us denote $G(y) = \inf\{x : G(x) \leqslant y\}$. Then, for every $y > 0$, we have

$$F(G(y)+0) \leqslant y \leqslant F(G(y)-0). \tag{1.3.20}$$

Using (1.3.20) and $\lim_{y\downarrow 0} G(y) = \infty$, we obtain for $0 < x < 1$,

$$\begin{aligned}\limsup_{y\downarrow 0} \frac{y}{F(xG(y))} &\leqslant \limsup_{y\downarrow 0} \frac{F(G(y)-0)}{F(xG(y))}\\ &\leqslant \lim_{y\downarrow 0} \frac{F(\sqrt{x}G(y))}{F(xG(y))} = (\sqrt{x})^{+\infty} = 0.\end{aligned}$$

Similarly, for $x > 1$, we have

$$\begin{aligned}\liminf_{y\downarrow 0} \frac{y}{F(xG(y))} &\geqslant \liminf_{y\downarrow 0} \frac{F(G(y)+0)}{F(xG(y))}\\ &\geqslant \lim_{y\downarrow 0} \frac{F(\sqrt{x}G(y))}{F(xG(y))} = (\sqrt{x})^{+\infty} = +\infty.\end{aligned}$$

Hence, $\lim_{y\downarrow 0} \dfrac{y}{F(xG(y))} = x^{+\infty}$ for every positive real number $x \neq 1$, i.e.,

$$\lim_{y\downarrow 0} \frac{F(xG(y))}{y} = x^{-\infty}. \tag{1.3.21}$$

Let us denote $\lambda_n = n$ and $a_n = G(1/n)$. It is easy to see that conditions (a) and (b) hold. Suppose that $x \in \mathbb{R}_+ \setminus \{1\}$. Using (1.3.21) we finally obtain

$$\begin{aligned}\lim_{n\to\infty} \lambda_n F(a_n x) &= \lim_{n\to\infty} nF(xG(1/n))\\ &= \lim_{n\to\infty} \frac{F(xG(1/n))}{1/n} = x^{-\infty}.\end{aligned}$$

□

1.4 Generalized Inverses

1.4.1 Definition and Basic Properties

Now we shall define the generalized inverse of a monotone function that is not necessarily bijective and then provide some properties of the generalized inverses.

Definition 1.4.1 Let $F : \mathbb{R} \to \mathbb{R}$ be an arbitrary increasing function, and $\overline{\mathbb{R}} = \mathbb{R} \cup \{+\infty\} \cup \{-\infty\}$. The *generalized inverse* of function F is the function $F^{-1} : \mathbb{R} \to \overline{\mathbb{R}}$, given by

$$F^{-1}(y) = \inf\{x : F(x) \geqslant y\}, \tag{1.4.1}$$

where $\inf\{\} = +\infty$.

Theorem 1.4.2 *Function F^{-1} given by* (1.4.1) *is left-continuous.*

Proof Let y_0 be a fixed real number, and let us consider a sequence of real numbers (y_n) such that $y_n \uparrow y_0$ as $n \to \infty$. Then we have

$$F^{-1}(y_n) \uparrow F^{-1}(y_0 - 0) := \lim_{y \uparrow y_0} F^{-1}(y) \leqslant F^{-1}(y_0).$$

Suppose that $F^{-1}(y_0 - 0) < F^{-1}(y_0)$. Then there exist real numbers $\varepsilon > 0$ and x, such that for every positive integer n,

$$F^{-1}(y_n) < x < F^{-1}(y_0) - \varepsilon < F^{-1}(y_0). \tag{1.4.2}$$

From the definition of function F^{-1} and inequality (1.4.2), it follows that $F(x) \geqslant y_n$ for every positive integer n, and therefore $F(x) \geqslant y_0$. Consequently we obtain that $x \geqslant F^{-1}(y_0)$, and this contradicts the fact that $x < F^{-1}(y_0) - \varepsilon$. □

The next simple theorem gives some important properties of an increasing function that is right-continuous. Note that the distribution function $F(x) = P\{X \leqslant x\}$ of any random variable X is right-continuous.

Theorem 1.4.3 *If a function F is increasing and right-continuous, then the following statements hold:*

$$F(F^{-1}(y)) \geqslant y; \tag{1.4.3}$$

$$F^{-1}(F(x)) \leqslant x; \tag{1.4.4}$$

$$F^{-1}(y) \leqslant x \quad \textit{if and only if} \quad y \leqslant F(x); \tag{1.4.5}$$

$$x < F^{-1}(y) \quad \textit{if and only if} \quad F(x) < y. \tag{1.4.6}$$

Proof All statements are immediate consequences of the assumptions that function F is increasing and right-continuous. □

1.4.2 Convergence of a Sequence of Monotone Functions

In order to introduce the notion of the vague convergence of a sequence of monotone functions (particularly, a sequence of distribution functions or the sequence of their generalized inverses), we need the following notation: For a given function $F : \mathbb{R} \to \overline{\mathbb{R}}$, let $C(F)$ be the set of all continuity points of function F, i.e.,

$$C(F) = \{x \in \mathbb{R} : F(x) \in \mathbb{R} \text{ and } F \text{ is continuous at point } x\}.$$

Definition 1.4.4 A sequence of distribution functions $(F_n)_{n \geqslant 1}$ *converges vaguely* to a function F_0 (notation: $F_n(x) \xrightarrow{v} F_0(x)$ or $F_n \xrightarrow{v} F_0$), if for every $x \in C(F_0)$ we have $\lim_{n\to\infty} F_n(x) = F_0(x)$.

Remark 1.4.5 Function F_0 from definition 1.4.4 need not be a distribution function. For example, if F_n is the distribution function of the uniform distribution over the interval $[-n, n]$, then it is easy to see that $\lim_{n\to\infty} F_n(x) = 1/2$ for every $x \in \mathbb{R}$.

Theorem 1.4.6 *Let F_0, F_1, F_2, ... be a sequence of distribution functions. If $F_n \xrightarrow{v} F_0$, then $F_n^{-1} \xrightarrow{v} F_0^{-1}$.*

Proof Let $y_0 \in C(F_0^{-1})$. We shall prove that

$$\lim_{n\to\infty} F_n^{-1}(y_0) = F_0^{-1}(y_0). \tag{1.4.7}$$

It is sufficient to prove that

$$\limsup_{n\to\infty} F_n^{-1}(y_0) \leqslant F_0^{-1}(y_0) \leqslant \liminf_{n\to\infty} F_n^{-1}(y_0). \tag{1.4.8}$$

Let $\varepsilon > 0$ be a fixed positive real number. Since the set of all discontinuity points of a monotone function is countable, it follows that there exists $x \in \mathbb{R}$ such that

$$F_0^{-1}(y_0) - \varepsilon < x < F_0^{-1}(y_0), \quad x \in C(F_0). \tag{1.4.9}$$

Property (1.4.6) implies that $F_0(x) < y_0$. Since $x \in C(F_0)$, we conclude that $\lim_{n\to\infty} F_n(x) = F_0(x)$, and therefore there exists $n_0 \in \mathbb{N}$, such that $F_n(x) < y_0$ for all $n \geqslant n_0$. Now we obtain that

$$F_0^{-1}(y_0) - \varepsilon < x \leqslant F_n^{-1}(y_0) \quad \text{for } n \geqslant n_0,$$

and therefore $F_0^{-1}(y_0) \leqslant \liminf_{n\to\infty} F_n^{-1}(y_0)$, i.e., the second of the inequalities (1.4.8) holds.

Let us choose $y > y_0$. Then there exists $t \in C(F_0)$, such that

$$F_0^{-1}(y) < t < F_0^{-1}(y) + \varepsilon. \tag{1.4.10}$$

It follows from property (1.4.5) that $y_0 < y \leqslant F_0(t)$. Since $t \in C(F_0)$, we have $\lim_{n\to\infty} F_n(t) = F_0(t)$, and hence there exists $n_1 \in \mathbb{N}$, such that $y_0 \leqslant F_n(t)$ for all

$n \geqslant n_1$, and $F_n^{-1}(y_0) \leqslant t$. Using (1.4.10) we obtain that $F_n^{-1}(y_0) \leqslant t < F_0^{-1}(y)+\varepsilon$ for all $n \geqslant n_1$, and therefore

$$\limsup_{n\to\infty} F_n^{-1}(y_0) \leqslant F_0^{-1}(y). \tag{1.4.11}$$

Since $y > y_0$ is an arbitrary real number and $y_0 \in C(F_0)$, we conclude that relation (1.4.11) implies that $\limsup_{n\to\infty} F_n^{-1}(y_0) \leqslant F_0^{-1}(y_0)$. □

1.4.3 Properties of the Generalized Inverses of Monotone Regularly Varying Functions

Theorem 1.4.7 *Let $F : \mathbb{R}_+ \to \mathbb{R}_+$ be an increasing function, $F(\infty) = \infty$, and $F \in RV_\rho$, where $0 \leqslant \rho \leqslant \infty$. Then, $F^{-1} \in RV_{1/\rho}$.*

Proof Suppose that the theorem does not hold. Then there exist a positive real number $x \neq 1$ and a sequence of positive real numbers (t_n) such that

$$\lim_{n\to\infty} t_n = +\infty, \tag{1.4.12}$$

$$\lim_{n\to\infty} \frac{F^{-1}(t_n x)}{F^{-1}(t_n)} = c \neq x^{1/\rho}. \tag{1.4.13}$$

Since F is an increasing function, the definition of F^{-1} implies that

$$\frac{F(F^{-1}(t_n x) - 1)}{F(F^{-1}(t_n) + 1)} \leqslant \frac{t_n x}{t_n} \leqslant \frac{F(F^{-1}(t_n x) + 1)}{F(F^{-1}(t_n) - 1)}. \tag{1.4.14}$$

Condition $F(\infty) = \infty$ and equality (1.4.12) imply that

$$\lim_{n\to\infty} F^{-1}(t_n x) = \lim_{n\to\infty} F^{-1}(t_n) = \infty. \tag{1.4.15}$$

Now from (1.4.13) and (1.4.15), it follows that

$$\lim_{n\to\infty} \frac{F^{-1}(t_n x) - 1}{F^{-1}(t_n) + 1} = \lim_{n\to\infty} \frac{F^{-1}(t_n x) + 1}{F^{-1}(t_n) - 1} = c. \tag{1.4.16}$$

Using (1.4.13), (1.4.16), Lemma 1.2.13 (in the case $c \in \mathbb{R}_+$), and Exercise 1.11 (in the cases $c = 0$ and $c = \infty$), we obtain that

$$\lim_{n\to\infty} \frac{F(F^{-1}(t_n x) - 1)}{F(F^{-1}(t_n) + 1)} = \lim_{n\to\infty} \frac{F(F^{-1}(t_n x) + 1)}{F(F^{-1}(t_n) - 1)}$$

$$= \lim_{n\to\infty} \frac{F(F^{-1}(t_n x))}{F(F^{-1}(t_n))} = c^{\rho}. \tag{1.4.17}$$

From (1.4.14) and (1.4.17), it follows that $x = c^{\rho}$, i.e., $c = x^{-1/\rho}$, and this contradicts (1.4.13). □

Theorem 1.4.8 *Let F_1 and F_2 be two increasing functions, and $F_1, F_2 \in RV_{\rho}$, where $0 < \rho < \infty$. Then, for $0 \leqslant c \leqslant \infty$, the equality*

$$\lim_{x\to\infty} \frac{F_1(x)}{F_2(x)} = c \tag{1.4.18}$$

holds if and only if

$$\lim_{x\to\infty} \frac{F_1^{-1}(x)}{F_2^{-1}(x)} = c^{-1/\rho}. \tag{1.4.19}$$

Proof Let us suppose that condition (1.4.18) is satisfied, and assume that (1.4.19) does not hold. Then, there exists a sequence of positive real numbers (x_n) such that $\lim_{n\to\infty} x_n = \infty$, and

$$\lim_{n\to\infty} \frac{F_1^{-1}(x_n)}{F_2^{-1}(x_n)} = l \neq c^{-1/\rho}. \tag{1.4.20}$$

Using (1.4.18) we obtain that

$$\lim_{n\to\infty} \frac{F_1(F_1^{-1}(x_n))}{F_1(F_2^{-1}(x_n))} = \lim_{n\to\infty} \frac{F_1(F_1^{-1}(x_n))}{F_2(F_2^{-1}(x_n))} \cdot \frac{F_2(F_2^{-1}(x_n))}{F_1(F_2^{-1}(x_n))}$$

$$= \frac{1}{c} \cdot \lim_{n\to\infty} \frac{F_1(F_1^{-1}(x_n))}{F_2(F_2^{-1}(x_n))},$$

and since $F_i(F_i^{-1}(x_n)) \sim x_n$ as $n \to \infty$, $i \in \{1, 2\}$, it follows that

$$\lim_{n\to\infty} \frac{F_1(F_1^{-1}(x_n))}{F_1(F_2^{-1}(x_n))} = \frac{1}{c}. \tag{1.4.21}$$

On the other hand, from Lemma 1.2.13 and relation (1.4.20), it follows that

$$\lim_{n\to\infty} \frac{F_1(F_1^{-1}(x_n))}{F_1(F_2^{-1}(x_n))} = l^{\rho}. \tag{1.4.22}$$

It follows from (1.4.21) and (1.4.22) that $l^\rho = 1/c$, i.e., $l = c^{-1/\rho}$, and this fact contradicts (1.4.20). Hence, (1.4.18) implies (1.4.19). The statement that (1.4.19) implies (1.4.18) can also be proved analogously using the fact that $F(x) \sim \inf\{y : F^{-1}(y) \geqslant x\}$ as $x \to \infty$. □

Theorem 1.4.9 *Every regularly varying function with exponent $\rho \neq 0$ is asymptotically equivalent to some absolutely continuous and strictly monotone function (which is regularly varying with the same exponent ρ).*

Proof Let us suppose that $F \in RV_\rho$, where $\rho > 0$, and denote

$$G(x) = \int_1^x \frac{F(t)}{t}\,dt, \quad x \geqslant 1.$$

Function G is absolutely continuous, and $F(t)/t \in RV_{\rho-1}$. Using Karamata's theorem we obtain that

$$\lim_{x\to\infty} \frac{F(x)}{G(x)} = \rho.$$

For $\rho > 0$, we have $F(x) \to \infty$ as $x \to \infty$, and therefore function G is strictly increasing. Similarly we consider the case when $\rho < 0$. □

1.4.4 *Khinchin's Theorem*

Theorem 1.4.10 (Khinchin) *Let $(F_n)_{n\geqslant 1}$ be a sequence of distribution functions such that there exists a nondegenerate distribution function G and two sequences of constants $a_n > 0$ and $b_n \in \mathbb{R}$, $n \in \mathbb{N}$, for which the following relation holds as $n \to \infty$:*

$$F_n(a_n x + b_n) \longrightarrow G(x), \quad \textit{for every } x \in C(G). \tag{1.4.23}$$

Let G_* be a nondegenerate distribution function, and $\alpha_n > 0$, $\beta_n \in \mathbb{R}$, $n \in \mathbb{N}$, be two sequences of constants.

(a) *If the following relation holds*

$$F_n(\alpha_n x + \beta_n) \longrightarrow G_*(x), \quad \textit{for every } x \in C(G_*), \tag{1.4.24}$$

then there exist $a > 0$ and $b \in \mathbb{R}$, such that

$$\lim_{n\to\infty} \frac{\alpha_n}{a_n} = a, \quad \lim_{n\to\infty} \frac{\beta_n - b_n}{a_n} = b; \tag{1.4.25}$$

$$G_*(x) = G(ax + b) \quad \textit{for all } x \in \mathbb{R}. \tag{1.4.26}$$

(b) *If equalities* (1.4.25) *hold, then relations* (1.4.24) *and* (1.4.26) *hold.*

Proof If F is a distribution function and $H(x) = F(ax + b)$, where $a > 0$ and $b \in \mathbb{R}$, then it is easy to see that $H^{-1}(y) = (F^{-1}(y) - b)/a$.

(a) Theorem 1.4.6 and relations (1.4.23) and (1.4.24) imply that

$$\lim_{n\to\infty} \frac{F_n^{-1}(y) - b_n}{a_n} = G^{-1}(y), \quad \text{for all } y \in C(G^{-1}); \tag{1.4.27}$$

$$\lim_{n\to\infty} \frac{F_n^{-1}(y) - \beta_n}{\alpha_n} = G_*^{-1}(y), \quad \text{for all } y \in C(G_*^{-1}). \tag{1.4.28}$$

Since G and G_* are nondegenerate distribution functions, it follows that there exist real numbers $y_0, y_1 \in C(G^{-1}) \cap C(G_*^{-1})$, such that $y_0 < y_1$, and

$$G^{-1}(y_0) < G^{-1}(y_1), \quad G_*^{-1}(y_0) < G_*^{-1}(y_1). \tag{1.4.29}$$

From (1.4.27) and (1.4.28), it follows that for $i \in \{0, 1\}$,

$$\lim_{n\to\infty} \frac{F_n^{-1}(y_i) - b_n}{a_n} = G^{-1}(y_i), \tag{1.4.30}$$

$$\lim_{n\to\infty} \frac{F_n^{-1}(y_i) - \beta_n}{\alpha_n} = G_*^{-1}(y_i). \tag{1.4.31}$$

Relations (1.4.30) and (1.4.31) imply that

$$\lim_{n\to\infty} \frac{F_n^{-1}(y_1) - F_n^{-1}(y_0)}{a_n} = G^{-1}(y_1) - G^{-1}(y_0),$$

$$\lim_{n\to\infty} \frac{F_n^{-1}(y_1) - F_n^{-1}(y_0)}{\alpha_n} = G_*^{-1}(y_1) - G_*^{-1}(y_0),$$

$$\lim_{n\to\infty} \frac{\alpha_n}{a_n} = \frac{G^{-1}(y_1) - G^{-1}(y_0)}{G_*^{-1}(y_1) - G_*^{-1}(y_0)} = a > 0. \tag{1.4.32}$$

Using (1.4.32) and (1.4.31) we obtain that

$$\begin{aligned} \lim_{n\to\infty} \frac{F_n^{-1}(y_0) - \beta_n}{a_n} &= \lim_{n\to\infty} \frac{\alpha_n}{a_n} \cdot \frac{F_n^{-1}(y_0) - \beta_n}{\alpha_n} \\ &= a \cdot G_*^{-1}(y_0). \end{aligned} \tag{1.4.33}$$

From (1.4.30) and (1.4.33), it follows that

$$\lim_{n\to\infty} \frac{\beta_n - b_n}{a_n} = G^{-1}(y_0) - a \cdot G_*^{-1}(y_0),$$

and hence (1.4.25) holds with $b = G^{-1}(y_0) - a \cdot G_*^{-1}(y_0)$.

It remains to prove (1.4.26). Let X and X_n, $n \in \mathbb{N}$, be random variables with the distribution functions G and F_n, $n \in \mathbb{N}$, respectively. Let us denote $X_n^* = (X_n - b_n)/a_n$. Relation (1.4.23) implies that X_n^* converges in distribution to X as $n \to \infty$. Using Skorohod's theorem [*see* Billingsley [26], Theorem 25.6], we conclude that there exist random variables Y and Y_n^* that are defined on the probability space $((0, 1), \mathcal{B}(0, 1), \lambda)$, where λ is the Lebesgue measure on $\mathcal{B}(0, 1)$, and such that:

- $Y \overset{d}{=} X$, and $Y_n^* \overset{d}{=} X_n^*$ for each $n \in \mathbb{N}$.
- $Y_n^* \to Y$ a.s. as $n \to \infty$.

Let us denote $Y_n = a_n Y_n^* + b_n$. Then, $Y_n \overset{d}{=} X_n$, and

$$\frac{Y_n - \beta_n}{\alpha_n} = \frac{a_n}{\alpha_n} Y_n^* - \frac{a_n}{\alpha_n} \cdot \frac{\beta_n - b_n}{a_n}$$
$$\to \frac{Y - b}{a}, \quad \text{a.s. as } n \to \infty.$$

Therefore, the random variable $(X_n - \beta_n)/\alpha_n$ converges in distribution to $(X - b)/a$ as $n \to \infty$. Since (1.4.24) implies that the sequence $(X_n - \beta_n)/\alpha_n$ converges in distribution to a random variable with the distribution function G_*, it follows that (1.4.26) holds.

(b) Similarly, as in the previous item, we prove that (1.4.23) and (1.4.25) imply that $(X_n - \beta_n)/\alpha_n$ converges in distribution to $(X - b)/a$ as $n \to \infty$, and consequently (1.4.24) and (1.4.26) hold.

□

1.5 Π-Variation and Γ-Variation

We have defined function $F : \mathbb{R}_+ \to \mathbb{R}_+$ as regularly varying or fast varying at infinity if for every $x > 0$ there exists

$$\lim_{t\to\infty} \frac{F(tx)}{F(t)} = x^\rho,$$

where $\rho \in \mathbb{R}$ for regularly varying functions, and $\rho = +\infty$ or $\rho = -\infty$ for fast varying functions. In both cases the asymptotic behavior of function F is considered in the neighborhood of infinity, and the condition from the definition can be reformulated as follows: For some function $a(t) > 0$ and every $x > 0$, there exists $\lim\limits_{t\to\infty} F(tx)/a(t)$.

In this section we shall introduce two new types of asymptotic behavior of a function in the neighborhood of infinity that will be called Π-variation and Γ-variation. The sets of functions that have these types of asymptotic behavior will be denoted by Π and Γ. The notions of Π-variation and Γ-variation were introduced by de Haan [87, 88]. We shall restrict our consideration here to monotone functions.

1.5.1 Π-Varying Functions

Definition 1.5.1 An increasing function $H : (c, +\infty) \to \mathbb{R}$ is Π-*varying* (at infinity) if there exist a nonnegative function $a : (c, +\infty) \to \mathbb{R}_+$ and a function $b : (c, +\infty) \to \mathbb{R}$, such that for every $x > 0$,

$$\lim_{t\to\infty} \frac{H(tx) - b(t)}{a(t)} = \ln x. \tag{1.5.1}$$

In this case we use notation $H \in \Pi$ or $H \in \Pi(a)$ and say that a is an *auxiliary function* for Π-varying function H.

Theorem 1.5.2

(a) Functions $a(t)$ and $b(t)$ from (1.5.1) can be chosen such that $b(t) = H(t)$ and $a(t) = H(te) - H(t)$.
(b) Function $a(t)$ from (1.5.1) *is slowly varying at infinity.*

Proof If we put $x = 1$ in (1.5.1), then we get $(H(t) - b(t))/a(t) \to \ln 1$ as $t \to \infty$, and therefore

$$\begin{aligned}\frac{H(tx) - H(t)}{a(t)} &= \frac{H(tx) - b(t)}{a(t)} - \frac{H(t) - b(t)}{a(t)} \\ &\to \ln x - \ln 1 = \ln x.\end{aligned} \tag{1.5.2}$$

Hence, we can take $b(t) = H(t)$. Moreover, if we put $x = e$ in (1.5.2), we obtain that auxiliary function a may be defined by $a(t) = H(te) - H(t)$.
(b) The statement follows from the following relations:

$$\begin{aligned}\frac{a(tx)}{a(t)} &= \frac{a(tx)}{a(t)} \cdot \frac{H(txe) - H(tx)}{H(txe) - H(tx)} \\ &= \left\{\frac{H(txe) - H(t)}{a(t)} - \frac{H(tx) - H(t)}{a(t)}\right\} \frac{a(tx)}{H(txe) - H(tx)} \\ &\to \frac{\ln(xe) - \ln x}{\ln e} = 1.\end{aligned}$$

□

In the next two theorems we shall formulate some equivalent conditions of the notion of Π-variation. Theorem 1.5.3 provides such conditions in the case when H

is a strictly increasing function, while Theorem 1.5.4 gives the equivalent conditions when H is an arbitrary increasing function. For the results related to the measurable functions, *see* Geluk and de Haan [73].

Theorem 1.5.3 (de Haan [88]) *Let $H : \mathbb{R}_+ \to \mathbb{R}$ be a strictly increasing function. Then the following five conditions are equivalent:*

(a) $H \in \Pi$.
(b) *For arbitrarily chosen positive real numbers x and $y \neq 1$, the following equality holds:*

$$\lim_{t\to\infty} \frac{H(tx) - H(t)}{H(ty) - H(t)} = \frac{\ln x}{\ln y}.$$

(c) *The function $H(x) - \dfrac{1}{x}\displaystyle\int_1^x H(t)dt$ is slowly varying as $x \to \infty$.*
(d) *There exists a slowly varying function $g(x)$ such that*

$$H(x) = g(x) + \int_1^x \frac{g(t)}{t}\, dt.$$

(e) *For every $x > 0$, it holds*

$$\lim_{t\to\infty} \frac{H(tx) - H(t)}{H(t) - \frac{1}{t}\int_1^t H(s)\, ds} = \ln x.$$

Proof It is obvious that condition (a) is equivalent to condition (b).

Proof that (b) implies (c). Let us denote $h(t) = H(t) - H(tx)$, where $0 < x < 1$. Then, for every $y > 0$,

$$\begin{aligned}
\lim_{t\to\infty} \frac{h(ty)}{h(t)} &= \lim_{t\to\infty} \frac{H(ty) - H(txy)}{H(t) - H(tx)} \\
&= \lim_{t\to\infty} \left\{ \frac{H(t) - H(txy)}{H(t) - H(tx)} - \frac{H(t) - H(ty)}{H(t) - H(tx)} \right\} \\
&= \frac{\ln(xy)}{\ln x} - \frac{\ln y}{\ln x} = 1,
\end{aligned}$$

i.e., function $h(t)$ is slowly varying at infinity. Using Karamata's Theorem 1.2.1, we obtain that

$$\lim_{t\to\infty} \frac{\int_1^t h(s)\, ds}{th(t)} = \lim_{t\to\infty} \frac{\frac{1}{t}\int_1^t H(s)\, ds - \frac{1}{tx}\int_1^{tx} H(s)\, ds}{H(t) - H(tx)} = 1,$$

and therefore

$$\lim_{t\to\infty}\left\{\frac{H(t)-\frac{1}{t}\int_1^t H(s)\,ds}{H(t)-H(tx)}-\frac{H(tx)-\frac{1}{tx}\int_1^{tx} H(s)\,ds}{H(t)-H(tx)}\right\}=0. \tag{1.5.3}$$

Note that the following relations hold:

$$\begin{aligned}\frac{H(t)-\frac{1}{t}\int_1^t H(s)\,ds}{H(t)-H(tx)}&=\frac{H(t)-\int_{1/t}^1 H(tu)\,du}{H(t)-H(tx)}\\ &=\frac{\frac{1}{2}H(t)-\int_{1/t}^{1/2} H(tu)\,du}{H(t)-H(tx)}+\int_{1/2}^1\frac{H(t)-H(tu)}{H(t)-H(tx)}\,du\\ &\geqslant\frac{\frac{1}{2}H(t)-\frac{1}{2}H\left(\frac{t}{2}\right)}{H(t)-H(tx)}+\int_{1/2}^1\frac{H(t)-H(tu)}{H(t)-H(tx)}\,du.\end{aligned} \tag{1.5.4}$$

Using Fatou's lemma and condition (b), we obtain that (1.5.4) implies

$$\liminf_{t\to\infty}\frac{H(t)-\frac{1}{t}\int_1^t H(s)\,ds}{H(t)-H(tx)}\geqslant\frac{\frac{1}{2}\ln\frac{1}{2}}{\ln x}+\int_{1/2}^1\frac{\ln u}{\ln x}\,du>0. \tag{1.5.5}$$

From (1.5.3) and (1.5.5), it follows that

$$\lim_{t\to\infty}\frac{\left\{H(t)-\frac{1}{t}\int_1^t H(s)\,ds\right\}-\left\{H(tx)-\frac{1}{tx}\int_1^{tx} H(s)\,ds\right\}}{H(t)-\frac{1}{t}\int_1^t H(s)\,ds}=0,$$

and now it is easy to conclude that function $H(t)-\frac{1}{t}\int_1^t H(s)\,ds$ is slowly varying at infinity.

Proof that (c) implies (d). If we denote

$$g(x)=H(x)-\frac{1}{x}\int_1^x H(t)\,dt,$$

then function $g(x)$ is slowly varying at infinity, and similarly as in the proof of Theorem 1.2.9, we obtain that

$$H(x)=g(x)+\int_1^x\frac{g(t)}{t}\,dt.$$

Proof that (d) implies (e). Since $g(x)$ is a slowly varying function, it follows $\lim_{t\to\infty}\frac{g(tx)}{g(t)}=1$ for every $x>0$, and we easily obtain

$$\frac{H(tx) - H(t)}{H(t) - \frac{1}{t}\int_1^t H(s)\,ds} = \frac{g(tx)}{g(t)} - 1 + \int_1^x \frac{g(ts)}{sg(t)}\,ds \to \ln x, \quad \text{as } t \to \infty.$$

Finally, it is obvious that (e) implies (a). □

Theorem 1.5.4 (Geluk and de Haan [73]) *Let $H : \mathbb{R}_+ \to \mathbb{R}$ be an increasing function, $c \geqslant 0$, and $g : [c, +\infty) \to \mathbb{R}$ be a function given by*

$$g(x) = H(x) - \frac{1}{x}\int_c^x H(t)\,dt. \tag{1.5.6}$$

Then, the following four conditions are equivalent:

(a) $H \in \Pi$.
(b) *Function $g : [c, +\infty) \to \mathbb{R}$ is well defined for some $c \geqslant 0$, and for every $x > 0$ the following equality holds:*

$$\lim_{t\to\infty} \frac{H(tx) - H(t)}{g(t)} = \ln x. \tag{1.5.7}$$

(c) *Function $g : [c, +\infty) \to \mathbb{R}$ is well defined for some $c \geqslant 0$ and is slowly varying at infinity.*
(d) *There exists a slowly varying function $a(t)$, such that*

$$H(x) = a(x) + \int_c^x \frac{a(t)}{t}\,dt. \tag{1.5.8}$$

Proof Similar to the proof of Theorem 1.5.3. □

Example 1.5.5 Let us denote $H(x) = -\ln\left(-\ln\left(1 - \frac{1}{x}\right)\right)$, $x > 1$. For $x > 1$ and $t > 1$, we obtain that

$$\begin{aligned} H(tx) - H(t) &= -\ln\left(-\ln\left(1 - \frac{1}{tx}\right)\right) + \ln\left(-\ln\left(1 - \frac{1}{t}\right)\right) \\ &= \ln\frac{-\ln\left(1 - \frac{1}{t}\right)}{-\ln\left(1 - \frac{1}{tx}\right)} = \ln\frac{\frac{1}{t} + o\left(\frac{1}{t}\right)}{\frac{1}{tx} + o\left(\frac{1}{t}\right)} \to \ln x, \quad t \to \infty; \end{aligned}$$

$$\frac{H(tx) - H(t)}{H(te) - H(t)} \to \frac{\ln x}{\ln e} = \ln x, \quad t \to \infty.$$

Therefore, $H \in \Pi$. △

1.5.2 Relation Between Π-Variation and Slow Variation

Theorem 1.5.6 *Let $H : \mathbb{R}_+ \to \mathbb{R}$ be an increasing function such that $H \in \Pi$, and let us suppose that there exists $\lim_{x\to\infty} H(x) = H(\infty) \leqslant +\infty$.*

(a) *If $H(\infty) = +\infty$, then $H \in RV_0$.*
(b) *If $H(\infty) < +\infty$, then $H(\infty) - H(x) \in RV_0$.*

Proof

(a) Without loss of generality we can suppose that $c = 1$ in Theorem 1.5.4. Then we have

$$H(x) = a(x) + \int_1^x \frac{a(t)}{t}\,dt. \tag{1.5.9}$$

Equality (1.5.9) can be written in the following form:

$$\frac{H(x)}{\int_1^x \frac{a(t)}{t}\,dt} = 1 + \frac{a(x)}{\int_1^x \frac{a(t)}{t}\,dt}. \tag{1.5.10}$$

Since function $a(t)$ is slowly varying at infinity, Lemma 1.1.6 implies that $a(x)/x \in RV_{-1}$. Using Karamata's Theorem 1.2.1 we obtain that the second addend on the right-hand side of equality (1.5.10) tends to $-1 + 1 = 0$ as $x \to \infty$, and function $\int_1^x \frac{a(t)}{t}\,dt$ is slowly varying at infinity. Hence, $H(x) \sim \int_1^x \frac{a(t)}{t}\,dt \in RV_0$.

(b) Let us suppose that $H(+\infty) < +\infty$. Then from equality (1.5.10) it follows that $H(\infty) = \int_1^\infty \frac{a(t)}{t}\,dt$, and therefore

$$\begin{aligned} H(\infty) - H(x) &= -H(x) + \int_1^\infty \frac{a(t)}{t}\,dt \\ &= -a(x) - \int_1^x \frac{a(t)}{t}\,dt + \int_1^\infty \frac{a(t)}{t}\,dt = -a(x) + \int_x^\infty \frac{a(t)}{t}\,dt, \end{aligned}$$

i.e.,

$$\frac{H(\infty) - H(x)}{\int_x^\infty \frac{a(t)}{t}\,dt} = 1 - \frac{a(x)}{\int_x^\infty \frac{a(t)}{t}\,dt}. \tag{1.5.11}$$

Using Karamata's Theorem 1.2.1, we obtain from (1.5.11) that

$$H(\infty) - H(x) \sim \int_x^\infty \frac{a(t)}{t}\,dt \in RV_0.$$

□

1.5.3 Γ-Varying Functions

Definition 1.5.7 Let $H : (-\infty, x_0) \to \mathbb{R}_+$ be a function that is increasing in some neighborhood of x_0. We say that H is Γ-*varying* as $x \uparrow x_0$ if the following two conditions are satisfied:

(a) $\lim\limits_{x \uparrow x_0} H(x) = +\infty$.
(b) There exists a function $g : (-\infty, x_0) \to \mathbb{R}_+$, such that for every x,

$$\lim_{t \uparrow x_0} \frac{H(t + xg(t))}{H(t)} = e^x. \tag{1.5.12}$$

In this case we use notation $H \in \Gamma$ or $H \in \Gamma(g)$, and function g is called an *auxiliary function* for Γ-varying function H.

Example 1.5.8 For function $H : \mathbb{R} \to \mathbb{R}$ that is given by $H(x) = |x|e^{x^2/2}$ and an auxiliary function defined as follows:

$$g(t) = \begin{cases} 1, & \text{if } t \leqslant 1, \\ 1/t, & \text{if } t > 1, \end{cases}$$

the following equality holds:

$$\lim_{t \to \infty} \frac{H(t + xg(t))}{H(t)} = e^x, \tag{1.5.13}$$

and therefore $H \in \Gamma$. △

Example 1.5.9 Let $H(x) = \dfrac{1}{1 - \exp(-e^{-x})}$, $x \in \mathbb{R}$. For $g(t) = 1$, equality (1.5.13) holds, and hence $H \in \Gamma$. △

Theorem 1.5.10 *The auxiliary function from Definition* 1.5.7 *is determined uniquely up to the asymptotic equivalence, i.e., if* g_1 *and* g_2 *are two auxiliary functions that satisfy (1.5.12), then* $\lim\limits_{t \uparrow x_0} \dfrac{g_1(t)}{g_2(t)} = 1$.

Proof Let us consider the case $x_0 = \infty$, and suppose that H is increasing on $[0, \infty)$. For $t \geqslant 0$ and $x \geqslant 0$, let us define

$$F_t(x) = 1 - \frac{H(t)}{H(t + x)}.$$

For every t, function $F_t(x)$ is a distribution function. If g_1 and g_2 are two auxiliary functions, then for $i \in \{1, 2\}$ we have

$$F_t(xg_i(t)) = 1 - \frac{H(t)}{H(t + xg_i(t))} \to 1 - e^{-x}, \quad t \to \infty.$$

Using Khinchin's theorem we obtain that $g_1(t) \sim g_2(t)$ as $t \to \infty$.

Conversely, if g_1 is an auxiliary function satisfying (1.5.12) and g_2 is a function such that $g_1(t) \sim g_2(t)$ as $t \uparrow x_0$, then it is easy to prove that function g_2 satisfies (1.5.12) as well. □

1.5.4 Relations Between Γ-Variation and Π-Variation

Class Γ consists of functions whose generalized inverses belong to class Π. More precisely, the following theorem holds.

Theorem 1.5.11 *If $H \in \Gamma$ and $g(t)$ is an auxiliary function for H, then $H^{-1} \in \Pi$ with auxiliary function $a(t) = g \circ H^{-1}(t)$.*

Proof Condition $H \in \Gamma$ implies that for every $x \in \mathbb{R}$,

$$\lim_{t \uparrow x_0} \frac{H(t + xg(t))}{H(t)} = e^x. \tag{1.5.14}$$

Let us denote $y = e^x$ ($x = \ln y$). Equality (1.5.14) now has the form

$$\lim_{t \uparrow x_0} \frac{H(t + g(t) \ln y)}{H(t)} = y. \tag{1.5.15}$$

Using Theorem 1.4.6 we obtain that (1.5.15) implies that

$$\lim_{t \uparrow x_0} \frac{H^{-1}(yH(t)) - t}{g(t)} = \ln y. \tag{1.5.16}$$

Let us denote $t = H^{-1}(x)$. Then, $x \to +\infty$ as $t \uparrow x_0$, and equality (1.5.16) can be written in the following form:

$$\lim_{x \to +\infty} \frac{H^{-1}(yH(H^{-1}(x))) - H^{-1}(x)}{g(H^{-1}(x))} = \ln y. \tag{1.5.17}$$

Let us prove that $H(H^{-1}(x)) \sim x$ as $x \to \infty$. First, note that the following statements hold:

$$y < H^{-1}(x) \text{ implies that } H(y) < x;$$

$$y > H^{-1}(x) \text{ implies that } H(y) \geqslant x.$$

Let us suppose that $\varepsilon > 0$ is a given number, and let us denote

$$y_1 := H^{-1}(x) - \varepsilon g(H^{-1}(x)), \quad y_2 := H^{-1}(x) + \varepsilon g(H^{-1}(x)).$$

Then the following relations hold:

$$\begin{aligned}\frac{H(y_1)}{H(H^{-1}(x))} &= \frac{H(H^{-1}(x) - \varepsilon g(H^{-1}(x)))}{H(H^{-1}(x))} \leqslant \frac{x}{H(H^{-1}(x))} \\ &\leqslant \frac{H(H^{-1}(x) + \varepsilon g(H^{-1}(x)))}{H(H^{-1}(x))} = \frac{H(y_2)}{H(H^{-1}(x))}.\end{aligned}$$

Using the assumption that $H \in \Gamma$, we obtain that

$$e^{-\varepsilon} \leqslant \liminf_{x\to+\infty} \frac{x}{H(H^{-1}(x))} \leqslant \limsup_{x\to+\infty} \frac{x}{H(H^{-1}(x))} \leqslant e^{\varepsilon}. \tag{1.5.18}$$

Since $\varepsilon > 0$ can be arbitrarily chosen, (1.5.18) implies that

$$\lim_{x\to+\infty} \frac{x}{H(H^{-1}(x))} = 1. \tag{1.5.19}$$

From (1.5.17) and (1.5.19), it follows that for every $y > 0$,

$$\lim_{x\to\infty} \frac{H^{-1}(xy) - H^{-1}(x)}{g(H^{-1}(x))} = \ln y,$$

i.e., $H^{-1} \in \Pi$, and $a(x) = g(H^{-1}(x))$ is an auxiliary function for H^{-1}. □

Theorem 1.5.12 *If $H \in \Pi$ and $a(t)$ is an auxiliary function for H, then $H^{-1} \in \Gamma$ and $g(t) = a \circ H^{-1}(t)$ is an auxiliary function for H^{-1}.*

Proof Assumption $H \in \Pi$ implies that

$$\lim_{t\to+\infty} \frac{H(tx) - H(t)}{a(t)} = \ln x, \quad x > 0. \tag{1.5.20}$$

Let us denote $t = H^{-1}(s)$. Then $s \to H(\infty)$ as $t \to +\infty$, and equality (1.5.20) can be written in the following form:

$$\lim_{s\to H(\infty)} \frac{H(xH^{-1}(s)) - H(H^{-1}(s))}{a(H^{-1}(s))} = \ln x, \quad x > 0,$$

i.e., for every $x > 0$,

$$\lim_{s \to H(\infty)} \left(\frac{H(xH^{-1}(s)) - s}{a(H^{-1}(s))} + \frac{s - H(H^{-1}(s))}{a(H^{-1}(s))} \right) = \ln x. \tag{1.5.21}$$

Let us prove that $\dfrac{s - H(H^{-1}(s))}{a(H^{-1}(s))} \to 0$ as $s \to H(+\infty)$. Since $t = H^{-1}(s)$, it follows that for every $\varepsilon > 0$ the inequalities $H(t(1-\varepsilon)) \leqslant s \leqslant H(t(1+\varepsilon))$ hold. Hence

$$\begin{aligned} \frac{H(t(1-\varepsilon)) - H(t)}{a(t)} &= \frac{H(t(1-\varepsilon)) - H(H^{-1}(s))}{a(t)} \leqslant \frac{s - H(H^{-1}(s))}{a(H^{-1}(s))} \\ &\leqslant \frac{H(t(1+\varepsilon)) - H(H^{-1}(s))}{a(H^{-1}(s))} = \frac{H(t(1+\varepsilon)) - H(t)}{a(t)}. \end{aligned} \tag{1.5.22}$$

Using (1.5.20) and (1.5.22) we obtain as $t \to +\infty$ the following inequalities:

$$\begin{aligned} \ln(1-\varepsilon) \leqslant \liminf_{s \to H(\infty)} & \frac{s - H(H^{-1}(s))}{a(H^{-1}(s))} \\ &\leqslant \limsup_{s \to H(\infty)} \frac{s - H(H^{-1}(s))}{a(H^{-1}(s))} \leqslant \ln(1+\varepsilon). \end{aligned} \tag{1.5.23}$$

Since ε is an arbitrary positive number, (1.5.23) implies that

$$\lim_{s \to H(\infty)} \frac{s - H(H^{-1}(s))}{a(H^{-1}(s))} = 0. \tag{1.5.24}$$

It follows from (1.5.21) and (1.5.24) that for every $x > 0$,

$$\lim_{s \to H(\infty)} \frac{H(xH^{-1}(s)) - s}{a(H^{-1}(s))} = \ln x.$$

Let us denote $y = \ln x$ for $x > 0$. Then $x = e^y$, and using Theorem 1.4.6 we obtain that

$$\lim_{s \to H(\infty)} \frac{H^{-1}(s + y \cdot a(H^{-1}(s)))}{H^{-1}(s)} = e^y,$$

i.e., $H^{-1} \in \Gamma$, and $g(s) = a(H^{-1}(s))$ is an auxiliary function for H^{-1}. □

1.5.5 Relation Between Γ-Variation and Fast Variation

Class Γ is a subset of the class of fast varying functions. More precisely, the following theorem holds.

Theorem 1.5.13 (Geluk and de Haan [73]) *If $H : (-\infty, x_0) \to \mathbb{R}_+$ is a function such that $H \in \Gamma$, then H is fast varying as $x \uparrow x_0$. If $x_0 = \infty$, then the following statements hold:*

(a) $\lim_{t\to\infty} \dfrac{\ln H(t)}{\ln t} = \infty.$

(b) $\lim_{t\to\infty} \dfrac{H(tx)}{H(t)} = \begin{cases} 0, & \text{if } 0 < x < 1, \\ 1, & \text{if } x = 1, \\ \infty, & \text{if } x > 1. \end{cases}$

Proof The theorem follows from Theorems 1.4.7, 1.5.11, and 1.5.12. □

1.5.6 Mejzler–de Haan Theorem

In Chap. 2 we shall investigate the limiting distributions of maxima in sequences of independent and identically distributed random variables. The class of regularly varying functions and classes Π and Γ will play the central role. The following theorem gives additional relations between the classes of Π-varying and Γ-varying functions and will be very useful.

Theorem 1.5.14 (Mejzler [150], de Haan [87])
Let F be a distribution function and H be the function that is given by

$$H(x) = \frac{1}{1 - F(x)}, \quad \textit{for } x < x_F = \sup\{t : F(t) < 1\}. \tag{1.5.25}$$

Then the following three conditions are equivalent:

(a) *There exist two sequences of real numbers $a_n > 0$ and $b_n \in \mathbb{R}$, $n \in \mathbb{N}$, such that for every $x \in \mathbb{R}$ the following equality holds:*

$$\lim_{n\to\infty} F^n(a_n x + b_n) = \exp(-e^{-x}). \tag{1.5.26}$$

(b) $H \in \Gamma$.

(c) $H^{-1} \in \Pi$.

Proof *See* Mejzler [150], de Haan [87], and Resnick [192]. □

1.6 Regularly Varying Random Variables

1.6.1 Definition and Basic Properties

Definition 1.6.1 Let X be a nonnegative random variable with the distribution function F. Random variable X and its distribution are regularly varying with an exponent of regular variation $\alpha \geqslant 0$ if the right tail $1 - F(x) = P\{X > x\}$ is a regularly varying function with exponent $-\alpha$, i.e., for every $x > 0$,

$$\lim_{t\to\infty} \frac{1 - F(tx)}{1 - F(t)} = x^{-\alpha}. \tag{1.6.1}$$

Remark 1.6.2 Usually the tail $1 - F$ of a regularly varying random variable X with exponent $\alpha \geqslant 0$ is represented in the form $1 - F(x) = x^{-\alpha} L(x)$, where L is a slowly varying function at infinity. Definition 1.6.1 can be adapted to the random variables that can take negative values as well.

Theorem 1.6.3 *Let F be a distribution function that satisfies the condition: $x_F = \sup\{t : F(t) < 1\} = \infty$. Then the following statements hold:*

(a) *Let us suppose that F is absolutely continuous with density f, such that for some $\alpha > 0$,*

$$\lim_{x\to\infty} \frac{xf(x)}{1 - F(x)} = \alpha. \tag{1.6.2}$$

Then, $f \in RV_{-1-\alpha}$, and therefore $1 - F \in RV_{-\alpha}$.

(b) *Let F be absolutely continuous with density f, and $f \in RV_{-1-\alpha}$ for some $\alpha > 0$. Then, (1.6.2) holds.*

(c) *Let F be absolutely continuous with density f, and $1 - F \in RV_{-\alpha}$ for some $\alpha > 0$, and let us suppose that f is monotone on some interval $(c, +\infty)$. Then, relation (1.6.2) holds.*

(d) *Let us suppose that $1 - F \in RV_{-\alpha}$ for some $\alpha > 0$, and $\beta \geqslant \alpha$. Then the following equality holds:*

$$\lim_{x\to\infty} \frac{x^{\beta}(1 - F(x))}{\int_0^x t^{\beta}\, dF(t)} = \frac{\beta - \alpha}{\alpha}. \tag{1.6.3}$$

If $\beta > \alpha$, then relation (1.6.3) implies that $1 - F \in RV_{-\alpha}$.

If $\beta = \alpha$, then we can only conclude that $1 - F(x) = o(x^{-\alpha} L(x))$ as $x \to \infty$, for some slowly varying function L.

(e) *Function $\widetilde{F}(x) = \int_0^x t^2\, dF(t)$ is slowly varying if and only if*

$$1 - F(x) = o\left(\frac{1}{x^2} \int_0^x t^2\, dF(t)\right) \quad \text{as } x \to \infty. \tag{1.6.4}$$

Proof The statements follow from Karamata's Theorem 1.2.1. □

Theorem 1.6.4 *Let X be a nonnegative random variable that is regularly varying with index $\alpha > 0$. Then,*

$$E(X^\beta) \begin{cases} < \infty, \text{ if } \beta < \alpha, \\ = \infty, \text{ if } \beta > \alpha. \end{cases} \tag{1.6.5}$$

Proof Relation (1.6.5) is an immediate consequence of Theorem 1.2.3. *See* also Exercise 1.9. □

1.6.2 Sum of Regularly Varying Random Variables

Theorem 1.6.5 *Let X and Y be independent, nonnegative, and regularly varying random variables with the same index $\alpha \geqslant 0$. Then, $X + Y$ is a regularly varying random variable with index α, and the following asymptotic relation holds as $x \to \infty$:*

$$P\{X + Y > x\} \sim P\{X > x\} + P\{Y > x\}. \tag{1.6.6}$$

Proof For every $x > 0$, we have $\{X > x\} \cup \{Y > x\} \subset \{X + Y > x\}$, and consequently $P\{X + Y > x\} \geqslant P\{X > x\} + P\{Y > x\} + o(1)$ as $x \to \infty$. On the other hand, for $0 < \varepsilon < 1/2$, it is obvious that

$$\{X \leqslant (1-\varepsilon)x\} \cap \{Y \leqslant (1-\varepsilon)x\} \cap (\{X \leqslant \varepsilon x\} \cup \{Y \leqslant \varepsilon x\}) \subset \{X + Y \leqslant x\}.$$

Consequently, the following relation holds:

$$\{X + Y > x\} \subset \{X > (1-\varepsilon)x\} \cup \{Y > (1-\varepsilon)x\} \cup \{X > \varepsilon x, Y > \varepsilon x\},$$

and, using the property of regular variation, we obtain that

$$P\{X + Y > x\} \leqslant P\{X > (1-\varepsilon)x\} + P\{Y > (1-\varepsilon)x\} + o(1) \quad \text{as} \quad x \to \infty.$$

Now, using the fact that $P\{X > (1-\varepsilon)x\}/P\{X > x\} \sim (1-\varepsilon)^{-\alpha}$ as $x \to \infty$, we conclude that

$$\begin{aligned} 1 &\leqslant \liminf_{x\to\infty} \frac{P\{X + Y > x\}}{P\{X > x\} + P\{Y > x\}} \\ &\leqslant \limsup_{x\to\infty} \frac{P\{X + Y > x\}}{P\{X > x\} + P\{Y > x\}} \leqslant (1-\varepsilon)^{-\alpha}, \end{aligned}$$

and letting $\varepsilon \to 0$ relation (1.6.6) follows. □

Remark 1.6.6 Note that random variables X and Y from Theorem 1.6.5 are not necessarily identically distributed.

Corollary 1.6.7 *Let $(X_n)_{n\geqslant 1}$ be a sequence of independent nonnegative and regularly varying random variables with the same distribution function F, and let us denote*

$$S_n = X_1 + \cdots + X_n, \qquad M_n = \max\{X_1, \ldots, X_n\}, \qquad n \geqslant 1. \tag{1.6.7}$$

Then for every $n \geqslant 2$, the following asymptotic relations hold as $x \to \infty$:

$$P\{S_n > x\} \sim nP\{X_1 > x\} \sim P\{M_n > x\}. \tag{1.6.8}$$

The first asymptotic relation in (1.6.8) *follows immediately from Theorem* 1.6.5, *whereas the second relation can easily be obtained using the equality $P\{M_n \leqslant x\} = (F(x))^n$.*

Definition 1.6.8 Let $(X_n)_{n\geqslant 1}$ be a sequence of independent nonnegative random variables with the same distribution function F, and let S_n and M_n be given by (1.6.7). Distribution function F belongs to the class $\mathcal{S}$ of *subexponential distributions* if for every $n \geqslant 2$ the asymptotic relations (1.6.8) hold as $x \to \infty$.

Remark 1.6.9 Corollary 1.6.7 implies that distribution function F with regularly varying tail belongs to class $\mathcal{S}$. An example of subexponential distribution function F with tail $1 - F$ that is not regularly varying is given by $1 - F(x) \sim e^{-x^{1/2}}$ as $x \to \infty$, *see* Embrechts et al. [58] and Embrechts and Omey [59] for details and further discussion of long-tailed distributions. *See* also Example 1.6.16.

1.6.3 Product of Random Variables

In many statistical models we have to deal with the product of random variables, say X and Y. In such cases the following question arises naturally: If random variables X and Y are both regularly varying (or perhaps only one of them is regularly varying), what can we say about the product XY? We shall formulate here some basic results related to this question.

Theorem 1.6.10 *Let X and Y be independent and nonnegative random variables.*

(a) *Let X be a regularly varying random variable with index $\alpha > 0$ and $E(Y^{\alpha+\varepsilon}) < +\infty$ for some $\varepsilon > 0$. Then product XY is regularly varying with the same index α, and the following asymptotic relation holds:*

$$P\{XY > x\} \sim E(Y^{\alpha}) \cdot P\{X > x\}, \qquad x \to \infty. \tag{1.6.9}$$

(b) *If X and Y are both regularly varying with index $\alpha > 0$, then XY is also a regularly varying random variable with the same index α.*

Remark 1.6.11 Part (a) of Theorem 1.6.10 was proved firstly by Breiman [34]. Part (b) is a consequence of the results from Cline [46].

Further extensions of Breiman's theorem are given in Denisov and Zwart [54]. They investigated the question whether the set of conditions

$$P\{X > x\} = L(x)x^{-\alpha}, \quad L \text{ is slowly varying}, \quad \alpha \geqslant 0, \tag{1.6.10}$$

$$E(Y^{\alpha}) < \infty, \tag{1.6.11}$$

$$P\{Y > x\} = o(P\{X > x\}) \text{ as } x \to \infty \tag{1.6.12}$$

implies relation (1.6.9). In general, conditions (1.6.10)–(1.6.12) are not enough for (1.6.9) to hold, and some additional assumptions should be imposed. In order to formulate the results, we need to introduce certain notions and an auxiliary result.

Definition 1.6.12 A random variable X is long-tailed if, for any y, the asymptotic relation $P\{X > x\} \sim P\{X > x + y\}$ holds as $x \to \infty$.

Definition 1.6.13 A nonnegative function f belongs to class $\mathcal{S}_d$ if it satisfies the property

$$\lim_{x\to\infty} \int_0^x \frac{f(x-y)}{f(x)} f(y)\,dy = 2\int_0^{\infty} f(u)\,du < \infty.$$

If $f(x) = P\{U > x\}$ for some random variable U, we say that $f \in \mathcal{S}^*$.

Classes $\mathcal{S}_d$ and $\mathcal{S}^*$ were introduced by Klüppelberg [132, 133]. For an absolutely continuous random variable X with distribution function F and probability density function f, the *hazard rate function* is defined by $h(x) = \frac{f(x)}{1-F(x)}$.

Lemma 1.6.14 (Denisov and Zwart [54]) *Let L be a slowly varying function. Then L admits precisely one of the following four representations:*

(a) $L(x) = c(x)$.

(b) $L(x) = \dfrac{c(x)}{P\{V > \ln x\}}$.

(c) $L(x) = c(x)P\{U > \ln x\}$.

(d) $L(x) = c(x)\dfrac{P\{U > \ln x\}}{P\{V > \ln x\}}$.

In all representations $c(x)$ is a function such that $c(x) \to c \in (0, \infty)$, while U and V are two independent long-tailed random variables with hazard rates converging to 0.

Theorem 1.6.15 (Denisov and Zwart [54])

(a) *Assume that, in addition to* (1.6.10)–(1.6.12),

$$\limsup_{x\to\infty} \sup_{1\leqslant y\leqslant x} L(y)/L(x) < \infty.$$

Then (1.6.9) *holds.*

(b) *Assume, in addition to* (1.6.10)–(1.6.12), *that* $\alpha > 0$ *and that* L *admits representation* (c) *or* (d) *of Lemma* 1.6.14. *If either* $L(e^x) \in \mathcal{S}_d$ *or the auxiliary random variable* $U \in \mathcal{S}^*$ *and* $P\{Y > y\} = o(y^{-\alpha} P\{U > \ln y\})$, *as* $y \to \infty$, *then* (1.6.9) *holds.*

(c) *Let us define* $m(x) = \int_0^x t^\alpha \, dP\{X \leqslant t\}$, *and suppose that* $E(U) = \infty$. *(Then* $E(X^\alpha) = \infty$, *and* $m(x) \to \infty$ *as* $x \to \infty$.*) Assume, in addition to* (1.6.10)–(1.6.12), *that* $\alpha > 0$ *and that*

$$\frac{P\{Y > x\}}{P\{X > x\}} m(x) \to 0, \quad \limsup_{x\to\infty} \sup_{\sqrt{x}\leqslant y\leqslant x} \frac{L(y)}{L(x)} < \infty.$$

Then (1.6.9) *holds.*

Example 1.6.16 Let X and Y be two independent random variables with the same exponential distribution with parameter 1, and let H be the distribution function of the product XY. It was proved in Tang [218] that $H \in \mathcal{S}$. It is easy to see that $1 - H$ is not regularly varying. First note that

$$1 - H(t) = \int_0^\infty e^{-u-t/u} \, du.$$

We shall use the following two formulas:

$$\int_0^\infty x^{\alpha-1} e^{-px-q/x} \, dx = 2\left(\frac{q}{p}\right)^{\alpha/2} K_\alpha(2\sqrt{pq}), \quad p > 0, \ q > 0,$$

where K_α is the modified Bessel function, *see* p. 344 in Prudnikov et al. [240], and

$$K_\alpha(z) \sim \sqrt{\frac{\pi}{2z}} e^{-z} \left\{ 1 + \frac{\mu - 1}{8z} + \frac{(\mu-1)(\mu-9)}{2!(8z)^2} - \frac{(\mu-1)(\mu-9)(\mu-25)}{3!(8z)^3} + \dots \right\},$$

where $\mu = 4\alpha^2$, $|\arg z| < 3\pi/2$, and $|z| \to \infty$, *see* [101]. Using the previous two formulas, we obtain that $1 - H(t) = 2t^{1/2} K_1(2t^{1/2}) \sim \pi^{1/2} t^{1/4} e^{-2t^{1/2}}$ as $t \to \infty$. △

1.6.4 Regular Variation of Random Vectors

Now we shall state one of several existing equivalent definitions of regular variation for the tail probabilities of random vectors.

Definition 1.6.17 The d-dimensional random vector $\mathbf{X} = (X_1, \ldots, X_d)$ and its distribution are *regularly varying* with index $\alpha \geqslant 0$ if there exists a random vector Θ with values in $\mathbb{S}^{d-1}$, where $\mathbb{S}^{d-1}$ is the unit sphere in $\mathbb{R}^d$ with respect to the norm $|\cdot|$, such that for every $t > 0$,

$$\frac{P\left\{|\mathbf{X}| > tu, \dfrac{\mathbf{X}}{|\mathbf{X}|} \in \cdot\right\}}{P\{|\mathbf{X}| > u\}} \xrightarrow{v} t^{-\alpha} P\{\Theta \in \cdot\}, \quad \text{as } u \to \infty. \tag{1.6.13}$$

Remark 1.6.18 The symbol $\xrightarrow{v}$ in (1.6.13) is the notation for the vague convergence of measures on $\mathbb{S}^{d-1}$. More generally, let $\mu, \mu_1, \mu_2, \ldots$ be a sequence of finite measures on metric space M, and let $\mathfrak{M}$ be σ-algebra generated by open sets in M. By definition, $\mu_n \xrightarrow{v} \mu$ as $n \to \infty$ if $\lim_{n\to\infty} \mu_n(A) = \mu(A)$ for all $A \in \mathfrak{M}$ such that $\mu(\partial A) = 0$. *See* Billingsley [27] and Kallenberg [120].

Condition A There exist $\alpha > 0$ and a slowly varying function L such that for every $\mathbf{x} = (x_1, \ldots, x_n) \in \mathbb{R}^n$ there exists the limit value

$$\lim_{t\to\infty} \frac{P\{(\mathbf{x}, \mathbf{X}) > t\}}{t^{-\alpha} L(t)} = w(\mathbf{x}),$$

and there exists $\mathbf{x}_0 \in \mathbb{R}^n$ such that $\mathbf{x}_0 \neq (0, \ldots, 0)$, and $w(\mathbf{x}_0) > 0$.

The following theorem provides a characterization of multivariate regular variation and was proved in Basrak et al. [20].

Theorem 1.6.19 *Let* $\mathbf{X} = (X_1, \ldots, X_b)$ *be a random vector. The following statements hold:*

(a) *If random vector* $\mathbf{X}$ *is regularly varying with index* $\alpha > 0$*, then Condition A holds with the same* α.
(b) *If* $\mathbf{X}$ *satisfies Condition A, where* $\alpha \in (0, +\infty) \setminus \mathbb{N}$*, then* $\mathbf{X}$ *is regularly varying with index* $\alpha > 0$*, and the distribution of* Θ *is uniquely determined.*
(c) *If* $\mathbf{X}$ *takes values in* $[0, \infty)^d$ *with probability* 1 *and satisfies Condition A for* $\mathbf{x} \in [0, \infty)^d \setminus \{\mathbf{0}\}$*, where* $\alpha \in (0, +\infty) \setminus \mathbb{N}$*, then* $\mathbf{X}$ *is regularly varying with index* α*, and the distribution of* Θ *is uniquely determined.*
(d) *If* $\mathbf{X}$ *takes values in* $[0, \infty)^d$ *with probability* 1 *and satisfies Condition A, where* α *is an odd positive integer, then* $\mathbf{X}$ *is regularly varying with index* α*, and the distribution of* Θ *is uniquely determined.*

Exercises and Supplements

Properties of Regularly Varying Functions

1.1 Prove the statement of Example 1.1.5 (f): Function $[\ln x]$ is slowly varying, while function $e^{[\ln x]}$ is not regularly varying.

1.2 Prove that function $-\ln x$ is slowly varying as $x \downarrow 0$.

1.3 Prove that function $\ln x$ is regularly varying as $x \downarrow 1$ with the exponent of regular variation -1.

1.4 Prove that function $L : \mathbb{R}_+ \to \mathbb{R}_+$ is slowly varying at infinity if and only if $\lim_{t\to\infty} L(tx)/L(t) = 1$ for every $x \geqslant 1$.

1.5 *Generalization of Theorem* 1.1.9 Let $F : \mathbb{R}_+ \to \mathbb{R}_+$ be a measurable function, such that $\lim_{t\to\infty} F(tx)/F(t) = h^*(x) \in \mathbb{R}_+$, for all $x \in A$, where A is a subset of $\mathbb{R}$ with a positive Lebesgue measure. Then there exists a function $h : \mathbb{R}_+ \to \mathbb{R}_+$, such that $h(x) = h^*(x)$ for all $x \in A$, and equality (1.1.2) holds for every $x > 0$.

1.6 *Generalization of Theorem* 1.1.10.

(a) Let us suppose that $F \in RV_\rho$, where $\rho > 0$. Then, for every $b > 0$, equality (1.1.1) holds uniformly for $x \in (0, b]$.
(b) Let us suppose that $F \in RV_\rho$, where $\rho < 0$. Then, for every $a > 0$, equality (1.1.1) holds uniformly for $x \in [a, +\infty)$.

1.7 Prove the following statements:

(a) The functions e^x and $1 + e^x$ are fast varying at infinity.
(b) Function e^{-x} is fast varying at infinity, but function $1 + e^{-x}$ is slowly varying at infinity.

1.8 Represent in the canonical form the following functions that are regularly varying at infinity:

(a) $F_1(x) = \left(1 + \dfrac{1}{x^2}\right) \ln x$.
(b) $F_2(x) = x^2 \operatorname{arctg} x$.

1.9 Let $L : \mathbb{R}_+ \to \mathbb{R}_+$ be a slowly varying function and $\alpha > 0$. Then $x^\alpha L(x) \to +\infty$ and $x^{-\alpha} L(x) \to 0$ as $x \to +\infty$.

1.10 Let $L_1(x) = \ln x$ and $L_n(x) = L_{n-1}(\ln x)$ for $n \geqslant 2$. Function $L_n(x)$ is defined for $x > x_n$, where $x_1 = 0$ and $x_n = \exp(x_{n-1})$ for $n \geqslant 2$. Prove that the following statement does not hold: *For every function $L(x)$ that is slowly varying at infinity, there exists $n \in \mathbb{N}$ such that*

$$\lim_{x\to\infty} \frac{L(x)}{L_n(x)} = +\infty.$$

Give a counterexample using the functions $L_n(x), n = 1, 2, \ldots$

1.11 Let $F \in RV_\rho$, $\rho \in \mathbb{R}\setminus\{0\}$, and let (a_n) and (b_n) be two sequences of positive real numbers, such that $a_n \to \infty$, $b_n \to \infty$, and $a_n/b_n \to c \in \{0, +\infty\}$, as $n \to \infty$. Prove that $F(a_n)/F(b_n) \to c^\rho$, as $n \to \infty$, where $\infty^\rho = \infty$ for $\rho > 0$, and $\infty^\rho = 0$ for $\rho < 0$.

1.12 Aljančić and Karamata [8]: Let F be a function that is integrable on interval $[0, A]$ for every $A > 0$, and let

$$\lim_{x\to\infty} \frac{1}{x} \int_0^x F(t)\, dt = \rho \in \mathbb{R}.$$

Prove that $\exp\left\{\int_0^x \frac{F(t)}{t}\, dt\right\} \in RV_\rho$.

1.13 Karamata [126]: Let $L : \mathbb{R}_+ \to \mathbb{R}_+$ be a slowly varying function at infinity, and $\alpha > 0$. For $x > 0$, let us define the functions:

$$L_1(x) = x^{-\alpha} \sup_{0<t\leqslant x} t^\alpha L(t), \qquad L_2(x) = x^\alpha \sup_{x\leqslant t<+\infty} t^{-\alpha} L(t),$$

$$L_3(x) = x^\alpha \inf_{0<t\leqslant x} t^{-\alpha} L(t), \qquad L_4(x) = x^{-\alpha} \inf_{x\leqslant t<+\infty} t^\alpha L(t).$$

Prove that functions $L_i(x)$, $i \in \{1, 2, 3, 4\}$, are slowly varying at infinity, and $L_i(x) \sim L(x)$ as $x \to \infty$, for $i \in \{1, 2, 3, 4\}$.

1.14 Adamović [1]: Let $L_1(x), \ldots, L_n(x)$ be slowly varying functions, and let $R(x_1, \ldots, x_n)$ be a rational function with positive coefficients. Prove that $R(L_1(x), \ldots, L_n(x))$ is a slowly varying function.

1.15 Adamović [1]: Let $L_1(x)$ and $L_2(x)$ be slowly varying functions, such that $L_1(x)$ is continuous, and $L_2(x) \to \infty$ as $x \to \infty$. Prove that $L(x) := L_1(L_2(x))$ is a slowly varying function.

1.16 Parameswaran [178]: Let $L : \mathbb{R}_+ \to \mathbb{R}_+$ be a slowly varying function and $F(x)$ be the function defined by

$$F(x) = \int_1^x \frac{L(t)}{t}\, dt, \quad x \geqslant 1.$$

Prove that $F(x)$ is a slowly varying function such that

$$\lim_{x\to\infty} \frac{F(x)}{L(x)} = +\infty.$$

1.17 Parameswaran [178]: Let $L : \mathbb{R}_+ \to \mathbb{R}_+$ be a slowly varying function. Prove that for every $\varepsilon > 0$ there exists $A > 0$, such that for all $x > y \geqslant A$ the following inequality holds: $\dfrac{xL(x)}{yL(y)} > 1 - \varepsilon$.

1.18 Karamata [122]; Galambos and Seneta [72]; Bojanić and Seneta [29]: Let $c_0, c_1, c_2, \ldots$ be a sequence of positive real numbers. The sequence (c_n) is *regularly varying* if there exists $\rho \in \mathbb{R}$, such that for every $x > 0$,

$$\lim_{n\to\infty} \frac{c_{[nx]}}{c_n} = x^{\rho}.$$

Prove that a sequence (c_n) is regularly varying if and only if there exists a sequence a_n, $n = 0, 1, 2, \ldots$, of positive real numbers such that

$$\lim_{n\to\infty} \frac{c_n}{a_n} = 1, \qquad \lim_{n\to\infty} n\left(1 - \frac{a_{n-1}}{a_n}\right) = \rho.$$

Monotone Regularly Varying Functions

1.19 Let F be a monotone and fast varying function at infinity, and let (a_n) and (b_n) be sequences of positive real numbers, such that $a_n \to \infty$, $b_n \to \infty$, and $a_n/b_n \to c \in [0, +\infty] \setminus \{1\}$, as $n \to \infty$. Prove that

$$\lim_{n\to\infty} \frac{F(a_n)}{F(b_n)} = c^{\rho},$$

where $\rho = +\infty$ or $\rho = -\infty$, and $c^{+\infty}$ and $c^{-\infty}$ are given by Definition 1.1.11.

1.20 Prove that every monotone function $F : \mathbb{R} \to \mathbb{R}$ has countable many discontinuity points.

1.21 Cheong and Teugels [44]: Let $F : \mathbb{R}_+ \to \mathbb{R}_+$ be an increasing function, such that $\lim\limits_{x\to\infty} F(x) = \infty$ and $F \in RV_{\rho}$, where $0 < \rho < +\infty$. Prove that for every $A > 0$, the following equality

$$\lim_{t\to\infty} \frac{F(x+t) - F(t)}{F(x)F(t)} = 0$$

holds uniformly over $x \geqslant A$.

O-Regularly Varying Functions

Definition A positive, finite, and measurable function $F : [a, \infty) \to \mathbb{R}$, $a > 0$, is said to be *O-regularly varying* at infinity if

$$\limsup_{x\to\infty} \frac{F(tx)}{F(x)} = r(t)$$

is finite for each $t > 0$.

Notation $f(x) \asymp g(x)$, $x \to \infty$, means that there exist two numbers $0 < m < M < \infty$ such that $m \leqslant f(x)/g(x) \leqslant M$ holds for large enough x. If these inequalities hold for $x \geqslant a$, we write $f(x) \asymp g(x)$ on $I_a = [a, \infty)$.

1.22 Karamata [125]: The condition that function F is O-regularly varying at infinity is equivalent to each of the following conditions:

(a) There exist measurable and bounded real functions α and β defined on $[b, +\infty)$, for some $b \geqslant a$ such that for $x \geqslant b$,

$$F(x) = \exp\left\{\alpha(x) + \int_b^x \beta(t)\frac{dt}{t}\right\}.$$

(b) There exist four numbers $0 < m < M < \infty$ and $\sigma < \tau$ such that for all $y \geqslant x \geqslant b$ the following inequalities hold:

$$m\left(\frac{y}{x}\right)^{\sigma} \leqslant \frac{F(y)}{F(x)} \leqslant M\left(\frac{y}{x}\right)^{\tau}.$$

(c) There exists a real number σ such that

$$\int_b^x t^{-\sigma} F(t)\frac{dt}{t} \asymp x^{-\sigma} F(x) \quad \text{on} \quad I_c \quad \text{for each} \quad c > b.$$

1.23 Seneta [210]: The condition that function F is O-regularly varying at infinity holds if and only if there exists a real number τ such that

$$\int_x^{\infty} t^{-\tau} F(t)\frac{dt}{t} \asymp x^{-\tau} F(x) \quad \text{on} \quad I_b \quad \text{for some} \quad b \geqslant a.$$

1.24 Aljančić and Arandjelović [7]: Let F be an O-regularly varying function, and let r be a positive and finite function on $(0, \infty)$ defined by the equality $\limsup_{x\to\infty} F(tx)/F(x) = r(t)$. Prove that the limits

$$p := \lim_{t\to 0+} \frac{\ln r(t)}{\ln t}, \quad \text{and} \quad q := \lim_{t\to\infty} \frac{\ln r(t)}{\ln t}$$

exist, and $-\infty < p \leqslant q < \infty$. Moreover, $r(t) \geqslant \max\{t^p, t^q\}$ for each $t > 0$. Prove that for any pair of real numbers p' and q', where $p' < p$ and $q' > q$, there exists a real number $M \geqslant 1$ such that $r(t) \leqslant M \max\{t^{p'}, t^{q'}\}$ for each $t > 0$. [Numbers p and q are the *lower* and the *upper index* of O-regularly varying function F. If $p = q$, we say that F is of index p. If $p = q = 0$, F is called a slow O-regularly varying function.]

Generalized Inverses

1.25 Let $F : \mathbb{R}_+ \to \mathbb{R}_+$ be a decreasing function, such that $F(\infty) = 0$ and $F \in RV_\rho$, where $-\infty \leqslant \rho \leqslant 0$. Prove that $F^{-1} \in RV_{-1/\rho}$.

Π-Varying and Γ-Varying Functions

1.26 de Haan [87]: Let $H : \mathbb{R}_+ \to \mathbb{R}$ be a strictly increasing function, such that for all $x > 1$ and $y > 1$ there exists

$$\lim_{t\to\infty} \frac{H(tx) - H(t)}{H(ty) - H(t)} = H_0(x, y),$$

where $H_0(x, y)$ is a function that is strictly increasing with respect to x and strictly decreasing with respect to y. Prove that function $H_0(x, y)$ has one of the following forms:

(a) $H_0(x, y) = \dfrac{x^\rho - 1}{y^\rho - 1}$, $\quad x > 1, y > 1, \rho \in \mathbb{R} \setminus \{0\}$.

(b) $H_0(x, y) = \dfrac{\ln x}{\ln y}$. In this case the limit $H_0(x, y)$ exists for all positive x and y, where $y \neq 1$.

1.27 Geluk and de Haan [73]: Let $H : \mathbb{R}_+ \to \mathbb{R}$ be a measurable function for which there exists a function $a : \mathbb{R}_+ \to \mathbb{R}_+$, such that for every $x > 0$ there exists

$$\lim_{t\to\infty} \frac{H(tx) - H(t)}{a(t)} = F_0(x) \in \mathbb{R}.$$

Let us suppose that function F_0 is not a constant. Then there exists a real number $c \neq 0$ such that for every $x > 0$ the value of $F_0(x)$ is given by $F_0(x) = c(x^\rho - 1)/\rho$ for some $\rho \neq 0$, or $F_0(x) = c \ln x = \lim\limits_{\rho\to 0} c(x^\rho - 1)/\rho$.

1.28 Geluk and de Haan [73]: Let $H : \mathbb{R}_+ \to \mathbb{R}$ be a measurable function such that the conditions of the previous exercise hold, $F_0(x) = c(x^\rho - 1)/\rho$, for some $\rho \neq 0$, and $c > 0$. Prove that the following statements hold:

(a) If $\rho > 0$, then $H \in RV_\rho$.
(b) If $\rho < 0$, then $H(+\infty) - H(x) \in RV_\rho$.

1.29 Let $H : \mathbb{R}_+ \to \mathbb{R}$ be an increasing function that is Π-varying, and $0 < x_1 < x_2 < +\infty$. Then

$$\lim_{t\to\infty} \frac{H(tx) - H(t)}{a(t)} = \ln x$$

holds uniformly on $x \in [x_1, x_2]$.

1.30 Let $H_1 : \mathbb{R}_+ \to \mathbb{R}$ and $H_2 : \mathbb{R}_+ \to \mathbb{R}$ be two increasing functions. Prove that if $H_1 \in \Pi$ and $H_2 \in \Pi$, then $H_1 + H_2 \in \Pi$.

1.31 Let functions $H : \mathbb{R}_+ \to \mathbb{R}$ and $F : \mathbb{R}_+ \to \mathbb{R}$ be given. If H increases and $H \in \Pi$ and $F \in RV_\rho$, where $\rho > 0$, then $H \circ F(x) = H(F(x)) \in \Pi$. Prove that.

1.32 Let functions $H : \mathbb{R}_+ \to \mathbb{R}$ and $F : \mathbb{R}_+ \to \mathbb{R}$ be given. If H increases, $H \in \Pi$, and $\lim_{t\to\infty} H(t) = \infty$, while F is differentiable and $F' \in RV_\rho$, where $\rho > -1$, then $F \circ H \in \Pi$. Prove that.

1.33 Let $H : \mathbb{R}_+ \to \mathbb{R}$ be an increasing function such that $H \in \Pi(a)$, and H is integrable on finite intervals. Prove that if

$$H_1(x) = \frac{1}{x} \int_0^x H(t)\, dt, \qquad x > 0,$$

then $H_1 \in \Pi(a)$. Prove also that if $H_1 \in \Pi(a)$ and H is an increasing function, then $H \in \Pi(a)$.

1.34 Give an example of a function that is slowly varying at infinity but does not belong to class Π.

1.35 Geluk and de Haan [73]: Let $H : \mathbb{R} \to \mathbb{R}_+$ be an increasing function. Prove that the following three conditions are equivalent:

(a) $H \in \Gamma$.
(b) There exists a differentiable function $h : \mathbb{R}_+ \to \mathbb{R}_+$, such that $h'(x) \to 0$ as $x \to \infty$, and

$$H(x) \sim \exp\left\{ \int_1^x \frac{dt}{h(t)} \right\}.$$

(c) The following equality holds:

$$\lim_{t\to\infty} \frac{H(t) \int_0^t \int_0^x H(s)\, ds\, dx}{\left\{ \int_0^t H(s)\, ds \right\}^2} = 1.$$

1.36 Let $H : \mathbb{R} \to \mathbb{R}$ be a function such that $H \in \Gamma(g)$. Prove that equality

$$\lim_{t\to\infty} \frac{H(t + xg(t))}{H(t)} = e^x$$

holds uniformly on every finite interval $[x_1, x_2] \subset \mathbb{R}$.

1.37 If $H \in \Gamma(g)$, prove that the following asymptotic relation holds:

$$g(t) \sim \frac{1}{H(t)} \int_0^t H(s)\, ds, \qquad \text{as } t \to \infty.$$

1.38 If $H \in \Gamma(g)$, prove that $\int_0^x H(t)\,dt \in \Gamma(g)$. Conversely, if $\int_0^x H(t)\,dt \in \Gamma(g)$ and H is an increasing function, then $H \in \Gamma(g)$.

1.39 If $H \in \Gamma(g)$, prove that $\lim\limits_{t\to\infty} \dfrac{g(t)}{t} = 0$.

1.40 If $H_1 \in \Gamma(g_1)$ and $H_2 \in \Gamma(g_2)$, prove that $H_1H_2 \in \Gamma(g)$, where

$$g(t) = \frac{g_1(t)g_2(t)}{g_1(t) + g_2(t)}.$$

1.41 If $H \in \Gamma$, prove that $H(x) \sim (H^{-1})^{-1}(x)$ as $x \uparrow x_0$.

1.42 Let functions $H : \mathbb{R}_+ \to \mathbb{R}$ and $F : \mathbb{R}_+ \to \mathbb{R}$ be given. If $H \in \Gamma$ and $F \in RV_\rho$, where $\rho > 0$, prove that $F \circ H \in \Gamma$.

1.43 Let the functions $H : \mathbb{R}_+ \to \mathbb{R}$ and $F : \mathbb{R}_+ \to \mathbb{R}$ be given. If $H \in \Gamma$ and F is a differentiable function such that $F' \in RV_\rho$, where $\rho > -1$, prove that $H \circ F \in \Gamma$.

Regularly Varying Random Variables

1.44 Let X be a nonnegative regularly varying random variable with index $\alpha \geqslant 0$. Prove that X^β, where $\beta > 0$, is regularly varying with index α/β.

1.45 *The log-gamma distribution* is determined by the density function

$$f(x) = \frac{\alpha^\beta}{\Gamma(\beta)}(\ln x)^{\beta-1}x^{-\alpha-1}, \quad x > 0 \ (\alpha > 0,\ \beta > 0),$$

and $f(x) = 0$ for $x \leqslant 0$. Prove that the tail $1 - F(x) = \int_x^\infty f(t)\,dt$, $x \geqslant 0$, is a regularly varying function.

1.46 Let $X_1, \ldots, X_n$ be independent identically distributed nonnegative regularly varying random variables with index $\alpha \geqslant 0$ and $c_1, \ldots, c_n$ be nonnegative constants. Prove that the following asymptotic relation holds as $x \to \infty$:

$$P\{c_1X_1 + \cdots + c_nX_n > x\} \sim (c_1^\alpha + \cdots + c_n^\alpha) \cdot P\{X_1 > x\}.$$

1.47 Let $F(x)$ be the distribution function related to the Peter and Paul paradox that is given by

$$F(x) = \sum_{k\in\mathbb{N}:\, 2^k \leqslant x} 2^{-k}, \quad x \geqslant 0.$$

Prove that the tail function $1 - F$ is not regularly varying. If X is a random variable with distribution function F, prove that $E(X) = +\infty$. Moreover, $E(X^\delta) = +\infty$ for $\delta \geqslant 1$, and $E(X^\delta) < +\infty$ for $\delta < 1$.

Chapter 2
Basic Results of Extreme Value Theory

2.1 Extreme Value Distributions

2.1.1 Introduction

Investigations of quantities such as temperature, river water levels, wind speed, concentration of ozone in the atmosphere, currency exchange rate, etc. are based on observed time series of the form $x_1, x_2, \ldots, x_n$, where the data represent recorded values of the quantity considered. The data are usually recorded with some frequency, for example, data may be collected as daily, monthly or annual observations. Special attention is given to extreme values in time series, not only the single extreme value but also all values that exceed some high level (when we are interested in maxima). Note also that observed values $x_1, x_2, \ldots, x_n$ are considered to be the realized values of random variables $X_1, X_2, \ldots, X_n$.

The probabilistic extreme value theory deals with the asymptotic properties of random variables

$$M_n = \max\{X_1, X_2, \ldots, X_n\}, \tag{2.1.1}$$

$$m_n = \min\{X_1, X_2, \ldots, X_n\}, \tag{2.1.2}$$

as $n \to \infty$, where $X_1, X_2, \ldots, X_n$ are in general dependent random variables with some joint probability distribution.

We shall first consider the case when $(X_n)_{n\geqslant 1}$ is a sequence of independent random variables, and some extensions to the case of dependent random variables will be considered in the last section of this chapter and also in the next chapters. Since

$$m_n = \min\{X_1, \ldots, X_n\} = -\max\{-X_1, \ldots. -X_n\}, \tag{2.1.3}$$

P. Mladenović, *Extreme Values In Random Sequences*,
Springer Series in Operations Research and Financial Engineering,
https://doi.org/10.1007/978-3-031-57412-2_2

an investigation of minima can always be reduced to an investigation of maxima, and we shall concentrate our attention on the maxima.

Let us suppose that $X_1, X_2, \ldots$ are independent identically distributed (i.i.d.) random variables with common probability distribution function $F(x) = P\{X_1 \leqslant x\}$. The distribution function of the maximum M_n is then given by

$$P\{M_n \leqslant x\} = (F(x))^n. \tag{2.1.4}$$

The question related to the limiting distribution of linearly normed maximum M_n can be formulated as follows: Is it possible to find constants $a_n > 0$ and b_n, $n \in \mathbb{N}$, such that equality

$$\lim_{n\to\infty} P\left\{\frac{M_n - b_n}{a_n} \leqslant x\right\} = G(x) \tag{2.1.5}$$

holds for every continuity point x of some nondegenerate distribution function $G(x)$? If the answer to this question is positive, then such constants a_n and b_n are called *norming constants*, and $G(x)$ determines the limiting distribution of the linearly normed maximum M_n as $n \to \infty$. If we denote $u_n = u_n(x) := a_n x + b_n$, then probability $P\{(M_n - b_n)/a_n \leqslant x\}$ can be rewritten as

$$P\{X_1 \leqslant u_n, X_2 \leqslant u_n, \ldots, X_n \leqslant u_n\} \tag{2.1.6}$$

and obviously represents the value of the distribution function of random vector $(X_1, X_2, \ldots, X_n)$ at point $(u_n, u_n, \ldots, u_n)$, and

$$P\left\{\frac{M_n - b_n}{a_n} \leqslant x\right\} = \{F(a_n x + b_n)\}^n. \tag{2.1.7}$$

Definition 2.1.1 If the equality

$$\lim_{n\to\infty} \{F(a_n x + b_n)\}^n = G(x) \tag{2.1.8}$$

holds for some nondegenerate distribution function G and constants $a_n > 0$ and $b_n \in \mathbb{R}$, $n \in \mathbb{N}$, then we say that distribution function F belongs to the *domain of attraction* of distribution function G, and we use the notation $F \in DA(G)$.

In what follows the answer to the following questions will be given:

(a) Which distribution functions $G(x)$ can appear in (2.1.5) as the limiting distribution functions of the linearly normed maximum of n i.i.d. random variables as $n \to \infty$?
(b) If F belongs to some domain of attraction for maxima, how can we determine the corresponding distribution function G such that $F \in DA(G)$?
(c) How can norming constants $a_n > 0$ and b_n be determined?

(d) If G is one of the possible limiting distribution functions in (2.1.5), what are the conditions on F under which $F \in DA(G)$?

We shall discuss some other important topics such as:

(e) The rate of convergence in (2.1.5)
(f) The asymptotic distribution of the k-th maximum
(g) The possibility of obtaining the limiting distribution for maximum M_n under nonlinear norming, etc.

2.1.2 The α-Parametrization and the γ-Parametrization

In this subsection we introduce three parametric families of distributions that will be used to model the actual distribution of a maximum of random variables. These families are the *Gumbel distributions*, the *Fréchet distributions*, and the *Weibull distributions*. We shall refer to these families as *extreme value distributions*. There are two different parametrizations for these classes: *α-parametrization* and *γ-parametrization*.

THE α-PARAMETRIZATION The standard representatives of the families mentioned above are defined as follows:

- The Gumbel distribution:

$$G_0(x) = \exp(-e^{-x}), \qquad -\infty < x < +\infty \ . \tag{2.1.9}$$

- The Fréchet distribution with parameter $\alpha > 0$:

$$G_{1,\alpha}(x) = \begin{cases} 0, & \text{if } x \leqslant 0, \\ \exp(-x^{-\alpha}), & \text{if } x > 0, \end{cases} \quad \text{for some } \alpha > 0. \tag{2.1.10}$$

- The Weibull distribution with parameter $\alpha > 0$:

$$G_{2,\alpha}(x) = \begin{cases} \exp\{-(-x)^{\alpha}\}, & \text{if } x < 0, \\ 1, & \text{if } x \geqslant 0, \end{cases} \quad \text{for some } \alpha > 0. \tag{2.1.11}$$

We shall refer to the parameter α in (2.1.10) and (2.1.11) as the *shape parameter*. Extreme value distributions were introduced firstly in Fisher and Tippett [66], and facts about their properties can be found in many books and papers related to extreme value theory. *See*, for example, Gnedenko [81], Gumbel [86], de Haan and Ferreira [91], Kendall and Stuart [127], Galambos [71], Leadbetter et al. [143], Kotz and Nadarajah [135], Reiss and Thomas [190], Resnick [192], Falk et al. [63], and Novak [174].

If a random variable X has the Gumbel distribution function $G_0(x)$, then the distribution function of the random variable $\sigma X + \mu$, where $\sigma > 0$ and $\mu \in \mathbb{R}$, is given by

$$G_{0,\mu,\sigma}(x) = G_0\left(\frac{x-\mu}{\sigma}\right) = \exp\left\{-\exp\left(-\frac{x-\mu}{\sigma}\right)\right\}. \tag{2.1.12}$$

The distribution determined by $G_{0,\mu,\sigma}(x)$ is called the Gumbel distribution with *location parameter* μ and *scale parameter* $\sigma > 0$. In a similar way we define Fréchet and Weibull distribution functions with location and scale parameters μ and $\sigma > 0$:

$$G_{i,\alpha,\mu,\sigma}(x) = G_{i,\alpha}\left(\frac{x-\mu}{\sigma}\right), \quad i \in \{1, 2\}. \tag{2.1.13}$$

The densities of standard extreme value distributions are given as follows:

- The Gumbel distribution:

$$g_0(x) = e^{-x} \cdot \exp(-e^{-x}), \quad -\infty < x < +\infty. \tag{2.1.14}$$

- The Fréchet distribution with parameter $\alpha > 0$:

$$g_{1,\alpha}(x) = \alpha x^{-(1+\alpha)} G_{1,\alpha}(x), \quad x \geqslant 0. \tag{2.1.15}$$

- The Weibull distribution with parameter $\alpha > 0$:

$$g_{2,\alpha}(x) = \alpha(-x)^{\alpha-1} G_{2,\alpha}(x), \quad x \leqslant 0. \tag{2.1.16}$$

THE γ-PARAMETRIZATION Let us introduce new parameter $\gamma = 1/\alpha$ for the Fréchet distributions and $\gamma = -1/\alpha$ for the Weibull distributions. By an appropriate choice of the location and scale parameters, the extreme value distributions can be represented as a parametric family that depends on the shape parameter γ as follows:

$$G_0(x) = \exp(-e^{-x}), \quad -\infty < x < +\infty; \tag{2.1.17}$$

$$G_\gamma(x) = \exp\left\{-(1+\gamma x)^{-1/\gamma}\right\}, \quad 1 + \gamma x > 0, \quad \gamma \neq 0. \tag{2.1.18}$$

This way we obtain the γ-parametrization of the Gumbel, Féchet, and Weibull distribution functions. Let us note the following:

(a) Equality (2.1.17) gives the standard Gumbel distribution, and since $(1 + \gamma x)^{-1/\gamma} \to e^{-x}$ as $\gamma \to 0$, it follows that for every real x the following equality holds:

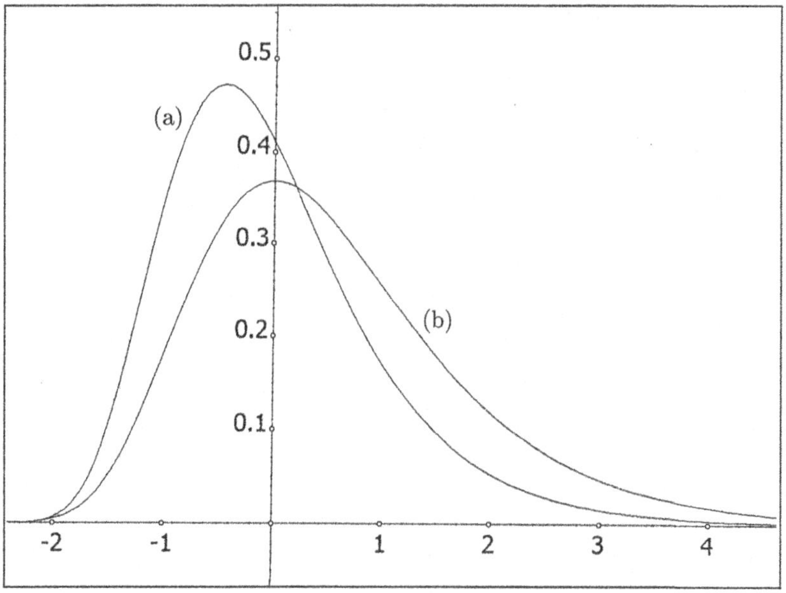

Fig. 2.1 Probability density functions of Gumbel distributions: (**a**) $\mu = -0.45006$, $\sigma = 0.77970$ and (**b**) $\mu = 0$, $\sigma = 1$

$$\lim_{\gamma \to 0} G_\gamma(x) = G_0(x).$$

(b) If $\gamma > 0$, then function $G_\gamma(x)$ given by (2.1.18) represents the Fréchet distribution function with the support of distribution $[-1/\gamma, +\infty)$, where $-1/\gamma < 0$. For $\gamma = 1/\alpha > 0$, we have $G_\gamma(x) = G_{1,\alpha,-\alpha,\alpha}(x)$.

(c) If $\gamma < 0$, then function $G_\gamma(x)$ from (2.1.18) represents the Weibull distribution function with support $(-\infty, -1/\gamma]$, where $-1/\gamma > 0$. For $\gamma = -1/\alpha < 0$, we have $G_\gamma(x) = G_{2,\alpha,\alpha,\alpha}(x)$.

(d) The density that corresponds to $G_\gamma(x)$ in (2.1.18) is given by

$$g_\gamma(x) = (1 + \gamma x)^{-(1+1/\gamma)} G_\gamma(x), \quad 1 + \gamma x > 0, \quad \gamma \neq 0. \tag{2.1.19}$$

It is easy to check that $\lim_{\gamma \to 0} g_\gamma(x) = g_0(x)$, where $g_0(x)$ is given by (2.1.14) (Fig. 2.1).

MOMENTS OF THE GUMBEL DISTRIBUTION Let X be a random variable with Gumbel distribution function $G_{0,\mu,\sigma}$ and corresponding density

$$g_{0,\mu,\sigma}(x) = \frac{1}{\sigma} e^{-(x-\mu)/\sigma} \exp\left\{-e^{-(x-\mu)/\sigma}\right\}. \tag{2.1.20}$$

Then the density of random variable $X_0 = (X - \mu)/\sigma$ is given by

$$g_0(x) = g_{0,0,1}(x) = e^{-x} \cdot \exp\{-e^{-x}\}. \tag{2.1.21}$$

Random variable $Z = e^{-(X-\mu)/\sigma} = e^{-X_0}$ has standard exponential distribution with the density given by

$$f_Z(x) = \begin{cases} e^{-x}, & \text{if } x \geqslant 0, \\ 0, & \text{if } x < 0. \end{cases} \tag{2.1.22}$$

In what follows we shall use the Euler Γ-function that is given by

$$\Gamma(s) = \int_0^\infty x^{s-1} e^{-x}\, dx, \quad s > 0. \tag{2.1.23}$$

For $t < 1$, the following equality holds:

$$E\left\{e^{t(X-\mu)/\sigma}\right\} = E(e^{tX_0}) = E(Z^{-t}) \tag{2.1.24}$$

$$= \int_0^\infty x^{-t} e^{-x}\, dx = \Gamma(1-t). \tag{2.1.25}$$

From the last equality it follows that the moment generating function of random variable $X = \mu + \sigma X_0$ is given by

$$E(e^{tX}) = e^{t\mu}\Gamma(1 - \sigma t), \quad \sigma t < 1, \tag{2.1.26}$$

while the generating function of the cumulants (semi-invariants) of this random variable is

$$\widetilde{G}(t) = \mu t - \ln \Gamma(1 - \sigma t). \tag{2.1.27}$$

The mathematical expectation and variance of random variable X with Gumbel distribution function $G_{0,\mu,\sigma}$ are given by

$$E(X) = \mu + \gamma_0\sigma \approx \mu + 0.57722\sigma, \tag{2.1.28}$$

$$\mathrm{Var}(X) = \frac{\pi^2\sigma^2}{6} \approx 1.64493\sigma^2, \tag{2.1.29}$$

where $\gamma_0 = 0.57722\ldots$ is the Euler constant. Therefore $E(X) = 0$ and $\mathrm{Var}(X) = 1$ if and only if

$$\sigma = \frac{\sqrt{6}}{\pi}, \quad \mu = -\gamma_0\sigma = -\frac{\gamma_0\sqrt{6}}{\pi}, \tag{2.1.30}$$

Table 2.1 Quantiles of Gumbel distributions

p	x'_p	x''_p
0.0005	−2.0325	−2.0283
0.0001	−1.9569	−2.2203
0.005	−1.7501	−1.6674
0.01	−1.6408	−1.5272
0.05	−1.3055	−1.0972
0.1	−1.1004	0.8340
0.9	1.3046	2.2504
0.95	1.8658	2.9702
0.99	3.1367	4.6001
0.9975	4.2205	5.9902
0.999	4.9355	6.9073

i.e., $\sigma \approx 0.77970$ and $\mu \approx -0.45006$. The probability distribution of random variable X is unimodal with mode μ. The folding points of probability density function $g_{0,\mu,\sigma}(x)$ are given by

$$x_1 = \mu - \sigma \ln \frac{3+\sqrt{5}}{2}, \quad x_2 = \mu + \sigma \ln \frac{3+\sqrt{5}}{2}. \tag{2.1.31}$$

The quantile of order $p \in (0, 1)$, i.e., real number x_p such that $G_{0,\mu,\sigma}(x_p) = p$, is given by

$$x_p = \mu - \sigma \ln(-\ln p). \tag{2.1.32}$$

In Table 2.1 we provide quantiles of Gumbel distributions: x'_p is the p-quantile of the Gumbel distribution with zero mean and unit variance ($\sigma = \sqrt{6}/\pi$, $\mu = -\sqrt{6}\gamma_0/\pi$), and x''_p is the p-quantile of the Gumbel distribution $G_{0,0,1}$ with parameters $\mu = 0$ and $\sigma = 1$.

MOMENTS OF THE FRECHET AND WEIBULL DISTRIBUTION Let X_0, $X_{1,\alpha}$, and $X_{2,\alpha}$ ($\alpha > 0$) be random variables with distribution functions G_0, $G_{1,\alpha}$, and $G_{2,\alpha}$, respectively. Let X_γ, $\gamma \neq 0$, be a random variable with the Fréchet or Weibull distribution function G_γ (in γ-parametrization) (Figs. 2.2 and 2.3).

For the Fréchet random variable $X_{1,\alpha}$ and $k > \alpha$, we obtain (after the substitution $t = x^{-\alpha}$) that the k-th moment is given by

$$\begin{aligned} E(X_{1,\alpha}^k) &= \int_0^{+\infty} x^k g_{1,\alpha}(x)\, dx \\ &= \int_0^{+\infty} t^{-k/\alpha} e^{-t}\, dt = \Gamma\left(1 - \frac{k}{\alpha}\right). \end{aligned} \tag{2.1.33}$$

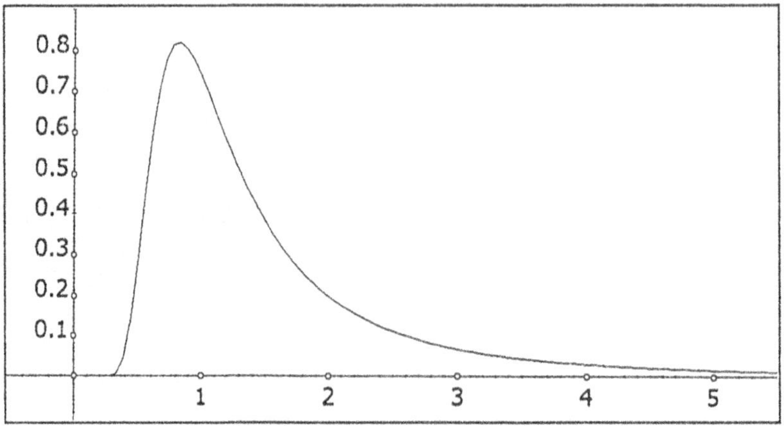

Fig. 2.2 Density of the Fréchet distribution: $\alpha = 2$, $\mu = 0$, and $\sigma = 1$

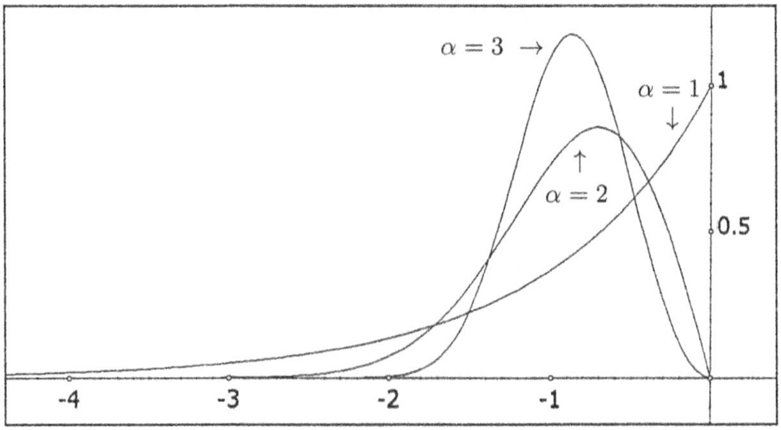

Fig. 2.3 Densities of the Weibull distributions: $\alpha \in \{1, 2, 3\}$, $\mu = 0$, and $\sigma = 1$

Consequently the mean and variance of $X_{1,\alpha}$ are given by

$$E(X_{1,\alpha}) = \Gamma\left(1 - \frac{1}{\alpha}\right), \quad \alpha > 1; \tag{2.1.34}$$

$$\text{Var}(X_{1,\alpha}) = \Gamma\left(1 - \frac{2}{\alpha}\right) - \Gamma^2\left(1 - \frac{1}{\alpha}\right), \quad \alpha > 2. \tag{2.1.35}$$

Note also that

$$E(X_{1,\alpha}) = +\infty, \quad \text{if } 0 < \alpha \leqslant 1; \tag{2.1.36}$$

$$D(X_{1,\alpha}) = +\infty, \quad \text{if } 0 < \alpha \leqslant 2. \tag{2.1.37}$$

For random variable $X_{2,\alpha}$ with the Weibull distribution, we get

$$E(X_{2,\alpha}^k) = \int_{-\infty}^{0} x^k g_{2,\alpha}(x). \tag{2.1.38}$$

By the substitution $t = (-x)^\alpha$, it follows that $x = -t^{1/\alpha}$, $dx = -\frac{1}{\alpha}t^{1/\alpha-1}\,dt$, and using (2.1.38) we obtain for $\alpha > 0$ and $1 + k/\alpha > 0$,

$$E(X_{2,\alpha}^k) = (-1)^k \int_0^\infty t^{k/\alpha} e^{-t}\,dt = (-1)^k \Gamma\left(1 + \frac{k}{\alpha}\right). \tag{2.1.39}$$

Now (2.1.39) implies that

$$D(X_{2,\alpha}) = \Gamma\left(1 + \frac{2}{\alpha}\right) - \Gamma^2\left(1 + \frac{1}{\alpha}\right), \quad \alpha > 0. \tag{2.1.40}$$

For Fréchet and Weibull distributions in γ-parametrization, we obtain that

$$E(X_\gamma) = \frac{\Gamma(1-\gamma) - 1}{\gamma}, \quad \gamma < 1; \tag{2.1.41}$$

$$\mathrm{Var}(X_\gamma) = \frac{\Gamma(1-2\gamma) - \Gamma^2(1-\gamma)}{\gamma^2}, \quad \gamma < 1/2. \tag{2.1.42}$$

The mean and variance of the standard Gumbel distribution can be obtained from (2.1.41) and (2.1.42) as follows:

$$E(X_0) = \lim_{\gamma\to 0} E(X_\gamma) = -\int_0^{+\infty} \ln x\, e^{-x}\,dx = 0.577216\ldots \tag{2.1.43}$$

$$\mathrm{Var}(X_0) = \lim_{\gamma\to 0} D(X_\gamma) = \frac{\pi^2}{6}. \tag{2.1.44}$$

2.1.3 Auxiliary Results

The following two theorems are related to the Cauchy equations and will be used in what follows.

Theorem 2.1.2 *Let $F : \mathbb{R} \to \mathbb{R}$ be a measurable function such that for every $t, s \in \mathbb{R}$ the following equality holds:*

$$F(t+s) = F(t) \cdot F(s). \tag{2.1.45}$$

Then $F(t) = 0$ for every $t \in \mathbb{R}$, or there exists a real number ρ such that $F(t) = e^{\rho t}$ for every $t \in \mathbb{R}$.

Proof Equality (2.1.45) implies that $F(t) = F^2(t/2) \geqslant 0$ for every $t \in \mathbb{R}$. If $F(t) = 0$ for some $t \in \mathbb{R}$, then for every $s \in \mathbb{R}$ we have $F(t+s) = 0$, i.e., $F(x) = 0$ for every real number x.

If $F(t) > 0$ for every $t \in \mathbb{R}$, then the functional equation (2.1.45) is equivalent to

$$\ln F(t+s) = \ln F(t) + \ln F(s), \quad \text{for every } t, s \in \mathbb{R}. \tag{2.1.46}$$

All solutions of Eq. (2.1.46) have the form $\ln F(t) = \rho t$, where ρ is a real constant. Therefore, in this case, all solutions of Eq. (2.1.45) have the form $F(t) = e^{\rho t}$, $t \in \mathbb{R}$, where ρ is a real constant. □

Theorem 2.1.3 *Let $F : \mathbb{R} \to \mathbb{R}$ be a function such that for every $t, s \in \mathbb{R}$ the following equality holds:*

$$F(t \cdot s) = F(t) + F(s). \tag{2.1.47}$$

Then $F(t) = 0$ for every $t \in \mathbb{R}$.

Proof If we put $s = 0$ in (2.1.47), we obtain that $F(0) = F(t) + F(0)$ for every $t \in \mathbb{R}$, and consequently $F(t) = 0$ for every $t \in \mathbb{R}$. □

THE ASYMPTOTIC BEHAVIOR OF PROBABILITY $P\{M_n \leqslant u_n\}$ Let us suppose that we are interested in the asymptotic behavior of maximum M_n given by equality (2.1.1). The crucial role belongs to events of the form $\{M_n \leqslant u_n\}$, where (u_n) is a sequence of real numbers that tends to $x_F = \sup\{t : F(t) < 1\}$. The limiting distribution of M_n is determined by the asymptotic behavior of the tail $1 - F(x)$ as $x \to x_F$. In what follows we shall use the following two simple theorems whose proofs can be found in Leadbetter et al. [143], Theorems 1.5.1 and 1.7.3.

Theorem 2.1.4 *Let $(X_n)_{n \geqslant 1}$ be a sequence of i.i.d. random variables with common distribution function F, $M_n = \max\{X_1, \ldots, X_n\}$, $(u_n)_{n \geqslant 1}$ be a sequence of real numbers, and $0 \leqslant \tau \leqslant +\infty$. Then the equality*

$$\lim_{n \to \infty} n(1 - F(u_n)) = \tau \tag{2.1.48}$$

holds if and only if

$$\lim_{n \to \infty} P\{M_n \leqslant u_n\} = e^{-\tau}. \tag{2.1.49}$$

Theorem 2.1.5 *Let $(X_n)_{n \geqslant 1}$ be a sequence of i.i.d. random variables with common distribution function F, $M_n = \max\{X_1, \ldots, X_n\}$, and $0 < \tau < +\infty$. Then a* *sequence* $(u_n)_{n\geqslant 1}$ *that satisfies condition* (2.1.48) *exists if and only if*

$$\lim_{x\to x_F} \frac{F(x)-F(x-0)}{1-F(x-0)} = 0. \tag{2.1.50}$$

Remark 2.1.6 Using Theorems 2.1.4 and 2.1.5 it is easy to prove the following statements:

(a) $M_n \to x_F$ as $n \to \infty$ with probability 1.
(b) If $x_F < +\infty$ and $F(x_F - 0) < 1$ and for some sequence (u_n) equality (2.1.48) holds, then $\tau = 0$ or $\tau = +\infty$. Moreover, for $\tau = 0$ or $\tau = +\infty$, a sequence (u_n) that satisfies (2.1.48) always exists.
(c) A sequence (u_n) that satisfies (2.1.48) exists if and only if

$$\lim_{x\to x_F} \frac{1-F(x)}{1-F(x-0)} = 1. \tag{2.1.51}$$

2.1.4 *Maximum Stable Distributions*

In this subsection we introduce the notion of M-stability and give examples of M-stable distributions.

Definition 2.1.7 A nondegenerate distribution function G is *maximum stable* (or shortly, M-*stable*), if for every positive integer $n \geqslant 2$ there exist constants $a_n > 0$ and b_n, such that for every real number x the following equality holds:

$$G^n(a_n x + b_n) = G(x). \tag{2.1.52}$$

Definition 2.1.8 Distribution functions G_1 and G_2 are of the *same type*, if there exist constants $a > 0$ and b, such that $G_2(x) = G_1(ax + b)$ for every real number x.

Obviously, a nondegenerate distribution function G is M-stable if and only if for every positive integer n the distribution functions G^n and G are of the same type. We shall also use the expression *a distribution is M-stable* if the corresponding distribution function is M-stable.

Example 2.1.9

(a) For $a_n = 1$, $b_n = \ln n$, and every $x \in \mathbb{R}$, we obtain

$$\left\{\exp\left(-e^{-(a_n x+b_n)}\right)\right\}^n = \left\{\exp\left(-e^{-(x+\ln n)}\right)\right\}^n$$
$$= \exp\left(-ne^{-\ln n}e^{-x}\right) = \exp(-e^{-x}),$$

and hence the standard Gumbel distribution function is M-stable.

(b) For $x > 0$, $a_n = n^{1/\alpha}$, where $\alpha > 0$, and $b_n = 0$, we obtain that

$$\left\{\exp\left(-(n^{1/\alpha}x)^{-\alpha}\right)\right\}^n = \exp\left(-n \cdot n^{-1} \cdot x^{-\alpha}\right)$$
$$= \exp(-x^{-\alpha}).$$

Hence, the Fréchet distribution function $G_{1,\alpha}$, $\alpha > 0$, is M-stable.

(c) For $x < 0$, $a_n = n^{-1/\alpha}$, where $\alpha > 0$, and $b_n = 0$, we obtain that

$$\left\{\exp\left(-(-n^{-1/\alpha}x)^{\alpha}\right)\right\}^n = \exp\left(-n \cdot n^{-1} \cdot (-x)^{\alpha}\right)$$
$$= \exp(-(-x)^{\alpha}).$$

Hence, the Weibull distribution function $G_{2,\alpha}$, $\alpha > 0$, is M-stable.

It is easy to check that for every location and scale parameters μ and $\sigma > 0$, the extreme value distribution functions $G_{0,\mu,\sigma}$, $G_{1,\alpha,\mu,\sigma}$, and $G_{2,\alpha,\mu,\sigma}$ are all M-stable. △

Notation For distribution functions F_n, $n \geqslant 1$, and F, the notation $F_n \Longrightarrow F$ means that $F_n(x) \to F(x)$ as $n \to \infty$ for every $x \in C(F)$, where $C(F)$ is the set of the continuity points of F. If F is continuous, then $F_n \Longrightarrow F$ means that $F_n(x) \to F(x)$ for every $x \in \mathbb{R}$.

The conditions that are equivalent to M-stability are provided by the following theorems.

Theorem 2.1.10 *A nondegenerate distribution function G is M-stable if and only if there exist a sequence of distribution functions $(F_n)_{n\geqslant 1}$ and the sequences $a_n > 0$ and $b_n \in \mathbb{R}$, $n \in \mathbb{N}$, such that for every positive integer k,*

$$F_n(a_{nk}x + b_{nk}) \Longrightarrow G^{1/k}(x), \quad n \to \infty, \tag{2.1.53}$$

where subscript nk is the product of n and k.

Proof

(a) Let us suppose that there exist sequences $a_n > 0$ and $b_n \in \mathbb{R}$, $n \in \mathbb{N}$, such that relation (2.1.53) holds. Since G is a nondegenerate distribution function, it follows that for every positive integer $k \geqslant 2$, $G^{1/k}$ is also a nondegenerate distribution function. Now, using condition (2.1.53) and the Khinchin Theorem 1.4.10, we conclude that there exist sequences $\alpha_k > 0$ and $\beta_k \in \mathbb{R}$, such that for every $x \in \mathbb{R}$ and every positive integer $k \geqslant 2$, equality $G(\alpha_k x + \beta_k) = G^{1/k}(x)$ holds. Therefore,

$$G^k(\alpha_k x + \beta_k) = G(x)$$

for every positive integer $k \geqslant 2$, i.e., distribution function G is M-stable.

(b) Let G be an M-stable distribution function, and suppose that there exist constants $a_n > 0$ and $b_n \in \mathbb{R}$ such that for every positive integer $n \geqslant 2$ the equality $G^n(a_n x + b_n) = G(x)$ holds for every $x \in \mathbb{R}$. Let us denote $F_n = G^n$. Then we have

$$F_n(a_{nk}x + b_{nk}) = G^n(a_{nk}x + b_{nk})$$
$$= (G^{nk}(a_{nk}x + b_{nk}))^{1/k} = G^{1/k}(x),$$

and this completes the proof of the theorem. □

Theorem 2.1.11 *A nondegenerate distribution function G is M-stable if and only if there exist functions $a : \mathbb{R}_+ \to \mathbb{R}_+$ and $b : \mathbb{R}_+ \to \mathbb{R}$, such that for every $x \in \mathbb{R}$ and every $t > 0$,*

$$G^t(a(t)x + b(t)) = G(x). \tag{2.1.54}$$

Proof

(a) Let us suppose that there exist functions $a(t) > 0$ and $b(t)$, such that equality (2.1.54) holds. For $t = n$, $a_n = a(n)$, and $b_n = b(n)$, we obtain that (2.1.52) holds, i.e., G is an M-stable distribution function.

(b) Let us suppose that G is M-stable. Then there exist constants $a_n > 0$ and $b_n \in \mathbb{R}$ for $n \geqslant 2$, such that (2.1.52) holds. For a fixed $t > 0$ and sufficiently large n ($[nt] \geqslant 2$), we have $G^{[nt]}(a_{[nt]}x + b_{[nt]}) = G(x)$, and

$$\lim_{n\to\infty} G^n(a_{[nt]}x + b_{[nt]}) = G^{1/t}(x). \tag{2.1.55}$$

Using (2.1.52) and (2.1.55) and the Khinchin Theorem 1.4.10, we conclude that there exist constants $a(t) > 0$ and $b(t) \in \mathbb{R}$, such that $G(a(t)x + b(t)) = G^{1/t}(x)$, i.e., $G^t(a(t)x + b(t)) = G(x)$. □

Theorem 2.1.12 *A nondegenerate distribution function G is M-stable if and only if $DA(G) \neq \varnothing$. In this case we have $G \in DA(G)$.*

Proof

(a) Let G be an M-stable distribution function. Then for every $n \geqslant 2$, there exist constants $a_n > 0$ and $b_n \in \mathbb{R}$ such that $G^n(a_n x + b_n) = G(x)$. Obviously $\lim_{n\to\infty} G^n(a_n x + b_n) = G(x)$, and hence $G \in DA(G)$.

(b) Let us suppose that G is a nondegenerate distribution function such that $DA(G) \neq \varnothing$. There exists a distribution function $F \in DA(G)$, i.e., there exist constants $a_n > 0$ and $b_n \in \mathbb{R}$, such that $F^n(a_n x + b_n) \Longrightarrow G(x)$ as $n \to \infty$. Consequently we obtain that $F^{nk}(a_{nk}x + b_{nk}) \Longrightarrow G(x)$ and finally $F^n(a_{nk}x + b_{nk}) \Longrightarrow G^{1/k}(x)$. Since F^n is a distribution function, it follows from Theorem 2.1.10 that G is an M-stable distribution function. □

2.2 Extremal Type Theorem

Theorem 2.2.1 (Gnedenko [81] and de Haan [87]) *Let $(X_n)_{n\geqslant 1}$ be a sequence of i.i.d. random variables with the common distribution function F, and $M_n = \max\{X_1, \dots, X_n\}$. Let us suppose that there exist sequences of constants $a_n > 0$ and $b_n \in \mathbb{R}$, $n \in \mathbb{N}$, such that*

$$P\left\{\frac{M_n - b_n}{a_n} \leqslant x\right\} = F^n(a_n x + b_n) \Longrightarrow G(x), \tag{2.2.1}$$

where G is a nondegenerate distribution function. Then function G is the same type as one of the extreme value distribution functions given by equalities (2.1.9)–(2.1.11)*:*

Proof Relation (2.2.1) implies that

$$F^{[nt]}(a_{[nt]}x + b_{[nt]}) \Longrightarrow G(x), \tag{2.2.2}$$

$$F^{[nt]}(a_n x + b_n) = (F^n(a_n x + b_n))^{[nt]/n} \Longrightarrow G^t(x). \tag{2.2.3}$$

Using the Khinchine theorem and relations (2.2.2) and (2.2.3), we obtain that there exist constants $u(t) > 0$ and $v(t) \in \mathbb{R}$, where $t > 0$, such that

$$\lim_{n\to\infty} \frac{a_n}{a_{[nt]}} = u(t), \qquad \lim_{n\to\infty} \frac{b_n - b_{[nt]}}{a_{[nt]}} = v(t), \tag{2.2.4}$$

$$G^t(x) = G(u(t)x + v(t)). \tag{2.2.5}$$

If all the terms of sequence $(a_n)_{n\geqslant 1}$ (respectively, $(b_n)_{n\geqslant 1}$) are mutually different, then we have

$$\{t : a_{[nt]} = a_k\} = [k/n, (k+1)/n),$$
$$\{t : b_{[nt]} = b_k\} = [k/n, (k+1)/n).$$

It is easy to see that functions $a_{[nt]}$ and $b_{[nt]}$ are measurable with respect to t, and consequently functions $u(t)$ and $v(t)$ are also measurable as the limit values of measurable functions. Similarly we can prove that functions $u(t)$ and $v(t)$ are measurable if sequences $(a_n)_{n\geqslant 1}$ and/or $(b_n)_{n\geqslant 1}$ contain equal terms.

Using equality (2.2.5), we obtain that for all $t > 0$ and $s > 0$ the following equalities hold:

$$G^{ts}(x) = G(u(ts)x + v(ts)), \tag{2.2.6}$$

$$\begin{aligned} G^{ts}(x) &= (G^s(x))^t = G^t(u(s)x + v(s)) \\ &= G(u(t)(u(s)x + v(s)) + v(t)) \end{aligned}$$

$$= G(u(t)u(s)x + u(t)v(s) + v(t)). \tag{2.2.7}$$

Since G is a nondegenerate distribution function, it follows that for all $t > 0$ and $s > 0$,

$$u(ts) = u(t)u(s), \tag{2.2.8}$$

$$v(ts) = u(t)v(s) + v(t). \tag{2.2.9}$$

All measurable and nonnegative functions that satisfy equation (2.2.8) are given by

$$u(t) = t^{-\rho}, \quad \rho \in \mathbb{R}. \tag{2.2.10}$$

Now we shall consider the cases $\rho = 0$ and $\rho \neq 0$ (with subcases $\rho > 0$ and $\rho < 0$).
Case $\rho = 0$. Then we have $u(t) = 1$ for all $t > 0$, and equality (2.2.9) implies that for all $t > 0$ and $s > 0$ the following equality holds:

$$v(ts) = v(t) + v(s). \tag{2.2.11}$$

All the solutions of Eq. (2.2.11) are given by

$$v(t) = -c \ln t, \quad t > 0, \quad c \in \mathbb{R}. \tag{2.2.12}$$

Equality (2.2.5) now gets the form

$$G^t(x) = G(x - c \ln t). \tag{2.2.13}$$

Since equality (2.2.13) holds for every real number x, it follows that constant c is not equal to 0 (in the opposite case distribution function G would be degenerate). Constant c has to be greater than 0, because, for a fixed x, function $G^t(x)$ is decreasing with respect to variable t.

Let us prove that $0 < G(x) < 1$ for every $x \in \mathbb{R}$. Let us suppose on the contrary that $G(x_0) = 0$ for some $x_0 \in \mathbb{R}$. Then, equality (2.2.13) implies that $G(x_0 - c \ln t) = 0$ for every $t > 0$, and hence $G(y) = 0$ for every $y \in \mathbb{R}$. The last conclusion contradicts the fact that G is a distribution function. Similarly, if $G(x_0) = 1$ for some $x_0 \in \mathbb{R}$, then equality (2.2.13) Implies that $G(x_0 - c \ln t) = 1$ for every $t > 0$, i.e., $G(y) = 1$ for every $y \in \mathbb{R}$. Again we obtain a contradiction.

If we put $x = 0$ in (2.2.13), then we obtain that for every $t > 0$,

$$G^t(0) = G(-c \ln t). \tag{2.2.14}$$

Let us denote $G(0) = \exp(-e^{-h})$, where $h \in \mathbb{R}$, and $y = -c \ln t$. When variable t passes through interval $(0, +\infty)$, then variable $y = -c \ln t$ passes through the whole set $\mathbb{R}$. Equality (2.2.14) implies that, for every $y \in \mathbb{R}$,

$$G(y) = \exp(-te^{-h}) = \exp\left(-e^{-(y/c+h)}\right) = G_0\left(\frac{y}{c} + h\right),$$

i.e., function G is the same type as the Gumbel distribution function G_0.
Case $\rho \neq 0$ It follows from (2.2.10) that $u(1) = 1$ and $u(t) \neq 1$ for every $t \in (0, 1) \cup (1, +\infty)$. Equality (2.2.9) implies that

$$u(t)v(s) + v(t) = u(s)v(t) + v(s),$$

and therefore for $t \neq 1$ and $s \neq 1$ we obtain

$$\frac{v(t)}{1 - u(t)} = \frac{v(s)}{1 - u(s)} = const = c. \tag{2.2.15}$$

For $t \neq 1$, equalities (2.2.15) and (2.2.10) imply that

$$v(t) = c(1 - u(t)) = c(1 - t^{-\rho}). \tag{2.2.16}$$

Using equalities (2.2.5) and (2.2.16) we obtain

$$G^t(x) = G(xt^{-\rho} + c(1 - t^{-\rho})) = G((x - c)t^{-\rho} + c),$$

and therefore

$$G^t(x + c) = G(xt^{-\rho} + c).$$

Let us denote $H(x) = G(x + c)$. Since functions G and H are the same type, it is sufficient to solve the following functional equation under the assumption that H is an unknown nondegenerate distribution function:

$$H^t(x) = H(xt^{-\rho}), \quad t > 0, \quad x \in \mathbb{R}. \tag{2.2.17}$$

For $x = 0$, equality (2.2.17) implies that $H^t(0) = H(0)$ for every $t > 0$. Therefore, it follows that $H(0) = 0$ or $H(0) = 1$.
Subcase $\rho > 0$ Let us suppose that $H(0) = 1$. Then, there exists $x_0 < 0$, such that $0 < H(x_0) < 1$, and

$$\lim_{x \to 0-} H(x) = \lim_{t \to +\infty} H(x_0 t^{-\rho}) = \lim_{t \to +\infty} H^t(x_0) = 0. \tag{2.2.18}$$

Equality (2.2.18) contradicts the assumption that H is a nondegenerate distribution function. Hence, $H(0) = 0$. Note that (2.2.17) implies that

$$H^t(1) = H(t^{-\rho}), \quad t > 0. \tag{2.2.19}$$

If $H(1) = 0$, then $H(t^{-\rho}) = 0$ for every $t > 0$. If $H(1) = 1$, then $H(t^{-\rho}) = 1$ for every $t > 0$. Since $\{t^{-\rho} : t > 0\} = (0, +\infty)$, we obtain a contradiction in both cases. Hence, $0 < H(1) < 1$.

Let us denote $\alpha = \rho^{-1}$, $H(1) = \exp(-h^{-\alpha})$ for some $h > 0$, and $y = t^{-\rho}$ $(t = y^{-\alpha})$. For $y > 0$, equality (2.2.19) implies that

$$H(y) = \exp(-h^{-\alpha}t) = \exp(-(hy)^{-\alpha}) = G_{1,\alpha}(hy),$$

i.e., function H is the same type as the Fréchet distribution function $G_{1,\alpha}$, where $\alpha > 0$. Finally, G is the same type as $G_{1,\alpha}$.

Subcase $\rho < 0$ Let us suppose that $H(0) = 0$. Then there exists $x_0 > 0$, such that $0 < H(x_0) < 1$, and

$$\lim_{x\to+\infty} H(x) = \lim_{t\to+\infty} H(x_0 t^{-\rho}) = \lim_{t\to+\infty} H^t(x_0) = 0.$$

The last equality contradicts the assumption that H is a distribution function. Hence, $H(0) = 1$. Equality (2.2.17) implies that

$$H^t(-1) = H(-t^{-\rho}), \quad t > 0. \tag{2.2.20}$$

If $H(-1) = 0$, then $H(-t^{-\rho}) = 0$ for every $t > 0$. If $H(-1) = 1$, then $H(-t^{-\rho}) = 1$ for every $t > 0$. In both cases we obtain a contradiction. Since $\{-t^{-\rho} : t > 0\} = (-\infty, 0)$, it follows that we have reached a contradiction in both cases. Hence, $0 < H(-1) < 1$.

Let us denote $\alpha = -\rho^{-1}$, $H(-1) = \exp(-(-h)^{\alpha})$ for some $h < 0$, and $y = -t^{-\rho}$ $(t = (-y)^{-1/\rho} = (-y)^{\alpha})$. For $y < 0$, equality (2.2.20) implies that

$$H(y) = e^{-t(-h)^{\alpha}} = e^{-(-y)^{\alpha}(-h)^{\alpha}} = e^{-(-(-h)y)^{\alpha}} = G_{2,\alpha}((-h)y),$$

i.e., function H is the same type as the Weibull distribution function $G_{2,\alpha}$, where $\alpha > 0$, and hence G is the same type as $G_{2,\alpha}$. □

Remark 2.2.2 Fréchet [70] obtained a class of possible limiting distributions for maximum M_n. Fisher and Tippett [66] obtained all three extreme value distributions as possible limit distributions for maxima. The strict foundation of extreme value theory and the necessary and sufficient conditions for weak convergence of maximal order statistics were given by Gnedenko [81]. Further development of extreme value theory based on regularly varying functions was given by Mejzler [150], Marcus, Pinsky [146], and de Haan [87, 88]. Extreme value theory is presented in several books, for example, Leadbetter et al. [143], Galambos [71], and Resnick [192].

2.3 Domains of Attraction of EV Distributions

2.3.1 Domain of Attraction of the Fréchet Distribution

We shall first formulate a theorem that provides sufficient conditions for an absolutely continuous distribution function to belong to the domain of attraction of the Fréchet extreme value distribution function.

Theorem 2.3.1 (von Mises [153]) *Let F be an absolutely continuous distribution function with the corresponding density function f. If the following two conditions are satisfied:*

(a) $f(x) > 0$ *for all* $x \in (a, x_F)$ *and*
(b) $\lim_{t\to\infty} \frac{t\, f(t)}{1 - F(t)} = \alpha > 0$

then $F \in DA(G_{1,\alpha})$.

The following theorem gives the necessary and sufficient conditions in general case. This theorem also shows how to determine the norming constants.

Theorem 2.3.2 (Gnedenko [81]) *A distribution function F belongs to the domain of attraction $DA\,(G_{1,\alpha})$ if and only if $x_F = \sup\{t : F(t) < 1\} = +\infty$ and $1 - F \in RV_{-\alpha}$, i.e., for every $x > 0$, the following equality holds:*

$$\lim_{t\to\infty} \frac{1 - F(tx)}{1 - F(t)} = x^{-\alpha}, \quad \text{where } \alpha > 0. \tag{2.3.1}$$

In this case $F^n(a_n x) \to G_{1,\alpha}(x)$ as $n \to \infty$, where

$$a_n = \left(\frac{1}{1-F}\right)^{-1}(n) = \inf\left\{x : F(x) \geqslant 1 - \frac{1}{n}\right\}. \tag{2.3.2}$$

Proof

(a) Let us suppose that the conditions of the theorem are satisfied, i.e., $x_F = +\infty$ and $1 - F \in RV_{-\alpha}$. If constant a_n is given by (2.3.2), then $1 - F(a_n) \sim \frac{1}{n}$ and $a_n \to \infty$, as $n \to \infty$, and hence, for every $x > 0$,

$$n(1 - F(a_n x)) \sim \frac{1 - F(a_n x)}{1 - F(a_n)} \to x^{-\alpha}.$$

If $x > 0$, then using Theorem 2.1.4 we obtain that $F^n(a_n x) \to \exp(-x^{-\alpha})$ as $n \to \infty$. If $x \leqslant 0$, then $F^n(a_n x) \leqslant F^n(0) \to 0$ as $n \to \infty$. This proves the fact that $F \in DA(G_{1,\alpha})$.

(b) Conversely, let us suppose that $F \in DA(G_{1,\alpha})$. Then there exist constants $a_n > 0$ and $b_n \in \mathbb{R}$, such that for every $x \in \mathbb{R}$,

$$F^n(a_n x + b_n) \to G_{1,\alpha}(x), \quad \text{as } n \to \infty. \tag{2.3.3}$$

From Theorem 2.1.4 we consequently obtain

$$n(1 - F(a_n x + b_n)) \to x^{-\alpha}, \quad \text{for } x > 0. \tag{2.3.4}$$

Let us denote $H(x) = \frac{1}{1-F(x)}$. Then equality (2.3.4) can be rewritten equivalently as

$$\frac{H(a_n x + b_n)}{n} \to x^{\alpha}, \quad x > 0. \tag{2.3.5}$$

Let us denote $y = x^{\alpha}$, for $x > 0$. Then $x = y^{1/\alpha}$, and (2.3.5) implies that

$$\frac{H^{-1}(ny) - b_n}{a_n} \to x = y^{1/\alpha}, \tag{2.3.6}$$

where H^{-1} is the generalized inverse of H. For $y = 1$, relation (2.3.6) implies that $(H^{-1}(n) - b_n)/a_n \to 1$, and consequently we obtain

$$\begin{aligned}\frac{H^{-1}(ny) - H^{-1}(n)}{a_n} &= \frac{H^{-1}(ny) - b_n}{a_n} - \frac{H^{-1}(n) - b_n}{a_n} \\ &\to y^{1/\alpha} - 1, \quad n \to \infty. \end{aligned} \tag{2.3.7}$$

Let us denote $a(t) = a_{[t]}$, where $t \geqslant 1$, and choose $\varepsilon > 0$. Then, for sufficiently large t, we have $[t] + 1 \leqslant [t](1 + \varepsilon)$, and since H^{-1} is an increasing function, for these large values of t we obtain

$$\begin{aligned}\frac{H^{-1}(ty) - H^{-1}(t)}{a(t)} &\geqslant \frac{H^{-1}([t]y) - H^{-1}([t] + 1)}{a(t)} \\ &\geqslant \frac{H^{-1}([t]y) - H^{-1}([t])}{a(t)} \\ &\quad - \frac{H^{-1}([t](1+\varepsilon)) - H^{-1}([t])}{a(t)}; \end{aligned} \tag{2.3.8}$$

$$\begin{aligned}\frac{H^{-1}(ty) - H^{-1}(t)}{a(t)} &\leqslant \frac{H^{-1}(([t] + 1)y) - H^{-1}([t])}{a(t)} \\ &\leqslant \frac{H^{-1}([t]y(1+\varepsilon)) - H^{-1}([t])}{a(t)}. \end{aligned} \tag{2.3.9}$$

Using relations (2.3.8) and (2.3.9), we obtain as $t \to \infty$

$$\begin{aligned} y^{1/\alpha} - 1 - ((1+\varepsilon)^{1/\alpha} - 1) &\leqslant \liminf_{t\to\infty} \frac{H^{-1}(ty) - H^{-1}(t)}{a(t)} \\ &\leqslant \limsup_{t\to\infty} \frac{H^{-1}(ty) - H^{-1}(t)}{a(t)} \\ &\leqslant (y(1+\varepsilon))^{1/\alpha} - 1. \end{aligned} \tag{2.3.10}$$

Since $\varepsilon > 0$ can be chosen arbitrarily, relation (2.3.10) implies that for every positive y,

$$\lim_{t\to\infty} \frac{H^{-1}(ty) - H^{-1}(t)}{a(t)} = y^{1/\alpha} - 1. \tag{2.3.11}$$

We shall prove that function $a(t)$ is regularly varying as $t \to \infty$. Indeed, for every $x > 0$, we obtain

$$\begin{aligned} \frac{a(tx)}{a(t)} &= \frac{H^{-1}(tx) - H^{-1}(t)}{a(t)} \cdot \frac{-a(tx)}{H^{-1}(txx^{-1}) - H^{-1}(tx)} \\ &\to \frac{x^{1/\alpha} - 1}{-(x^{-1})^{1/\alpha} + 1} = \frac{x^{1/\alpha} - 1}{1 - x^{-1/\alpha}} = x^{1/\alpha}, \quad t \to \infty, \end{aligned}$$

and hence $a(t) \in RV_{1/\alpha}$. For every $y > 0$, let us denote $H_y(t) = H^{-1}(ty) - H^{-1}(t)$. Relation (2.3.11) can be rewritten in the form $H_y(t)/a(t) \to y^{1/\alpha} - 1$ as $t \to \infty$, and hence, for every $x > 0$, we obtain that

$$\begin{aligned} \frac{H_y(tx)}{H_y(t)} &= \frac{H_y(tx)}{a(tx)} \cdot \frac{a(tx)}{a(t)} \cdot \frac{a(t)}{H_y(t)} \\ &\to \frac{(y^{1/\alpha} - 1)x^{1/\alpha}}{y^{1/\alpha} - 1} = x^{1/\alpha}, \quad t \to \infty, \end{aligned}$$

i.e., $H_y \in RV_{1/\alpha}$. By Karamata's Theorem 1.2.1(a), we obtain that

$$\lim_{t\to\infty} \frac{t(H^{-1}(ty) - H^{-1}(t))}{\int_2^t (H^{-1}(sy) - H^{-1}(s))\, ds} = \frac{1}{\alpha} + 1,$$

i.e.,

$$\lim_{t\to\infty} \frac{\int_2^t (H^{-1}(sy) - H^{-1}(s))\, ds}{t(H^{-1}(ty) - H^{-1}(t))} = \frac{\alpha}{\alpha + 1}. \tag{2.3.12}$$

Relations (2.3.11) and (2.3.12) imply that

$$\lim_{t\to\infty} \frac{\int_2^t (H^{-1}(sy) - H^{-1}(s))\,ds}{ta(t)} = \frac{\alpha}{\alpha+1}(y^{1/\alpha} - 1). \tag{2.3.13}$$

Let $y \in (0, 1)$. Note that

$$\int_2^t (H^{-1}(sy) - H^{-1}(s))\,ds = \frac{1}{y}\int_{2y}^{2} H^{-1}(s)\,ds + \frac{1}{y}\int_2^{ty} H^{-1}(s)\,ds$$
$$- \int_2^t H^{-1}(s)\,ds. \tag{2.3.14}$$

Since $t(H^{-1}(ty) - H^{-1}(t)) \to \infty$ as $t \to \infty$, it follows that

$$\lim_{t\to\infty} \int_{2y}^{2} \frac{H^{-1}(s)\,ds}{t(H^{-1}(ty) - H^{-1}(t))} = 0. \tag{2.3.15}$$

Relations (2.3.13), (2.3.14), and (2.3.15) imply that

$$\lim_{t\to\infty} \frac{\frac{1}{ty}\int_2^{ty} H^{-1}(s)\,ds - \frac{1}{t}\int_2^t H^{-1}(s)\,ds}{a(t)} = \frac{\alpha}{\alpha+1}(y^{1/\alpha} - 1). \tag{2.3.16}$$

Note that $g(y) = (H^{-1}(ty) - H^{-1}(t))/a(t)$ is a monotone function of y. Since limiting function $y^{1/\alpha} - 1$ is continuous, it follows that the convergence in (2.3.11) is uniform on every closed interval of the form $[\varepsilon, 1]$, where $0 < \varepsilon < 1$. Therefore

$$\lim_{t\to\infty} \int_\varepsilon^1 \frac{H^{-1}(t) - H^{-1}(ty)}{a(t)}\,dt$$
$$= \int_\varepsilon^1 (1 - y^{1/\alpha})\,dy = \frac{1}{\alpha+1} - \varepsilon + \frac{\alpha}{\alpha+1}\varepsilon^{1/\alpha+1}. \tag{2.3.17}$$

For $t\varepsilon > 2$, the integral on the left-hand side of (2.3.17) can be represented in the following form:

$$\frac{1-\varepsilon}{a(t)}H^{-1}(t) - \frac{1}{ta(t)}\int_{t\varepsilon}^{t} H^{-1}(s)\,ds = \frac{1-\varepsilon}{a(t)}\left\{H^{-1}(t) - \frac{1}{t}\int_2^t H^{-1}(s)\,ds\right\}$$
$$+ \frac{\varepsilon}{a(t)}\left\{\frac{1}{t\varepsilon}\int_2^{t\varepsilon} H^{-1}(s)\,ds - \frac{1}{t}\int_2^t H^{-1}(s)\,ds\right\}. \tag{2.3.18}$$

Using relations (2.3.16), (2.3.17), and (2.3.18), we obtain that

$$\lim_{t\to\infty}\frac{1-\varepsilon}{a(t)}\left\{H^{-1}(t)-\frac{1}{t}\int_2^t H^{-1}(s)\,ds\right\}$$

$$=\frac{1}{\alpha+1}-\varepsilon+\frac{\alpha}{\alpha+1}\varepsilon^{1/\alpha+1}-\frac{\varepsilon\alpha}{\alpha+1}(\varepsilon^{1/\alpha}-1)=\frac{1-\varepsilon}{\alpha+1},$$

i.e.,

$$\lim_{t\to\infty}\frac{1}{a(t)}\left\{H^{-1}(t)-\frac{1}{t}\int_2^t H^{-1}(s)\,ds\right\}=\frac{1}{\alpha+1}. \tag{2.3.19}$$

Let us denote

$$B(t)=H^{-1}(t)-\frac{1}{t}\int_2^t H^{-1}(s)\,ds. \tag{2.3.20}$$

Then, $B(t)\sim a(t)/(1+\alpha)$ as $t\to\infty$, and since $a(t)\in RV_{1/\alpha}$, it follows that $B(t)\in RV_{1/\alpha}$. Using (2.3.20) we obtain that

$$\begin{aligned}
\int_2^y \frac{B(t)}{t}\,dt &= \int_2^y \frac{H^{-1}(t)}{t}\,dt-\int_2^y \frac{1}{t^2}\int_2^t H^{-1}(s)\,ds\,dt\\
&=\int_2^y \frac{H^{-1}(t)}{t}\,dt-\int_2^y H^{-1}(s)\left(\int_s^y \frac{dt}{t^2}\right)ds\\
&=\int_2^y \frac{H^{-1}(t)}{t}\,dt-\int_2^y\left(\frac{1}{s}-\frac{1}{y}\right)H^{-1}(s)\,ds\\
&=\frac{1}{y}\int_2^y H^{-1}(s)\,ds=H^{-1}(y)-B(y),
\end{aligned}$$

and consequently

$$H^{-1}(y)=B(y)+\int_2^y \frac{B(t)}{t}\,dt. \tag{2.3.21}$$

Using the fact that $B(t)$ is regularly varying with exponent $1/\alpha$, relation (2.3.21), and Karamata's Theorem 1.2.1(a), we obtain that

$$\lim_{y\to\infty}\frac{H^{-1}(y)}{B(y)}=1+\lim_{y\to\infty}\frac{1}{B(y)}\int_2^y \frac{B(t)}{t}\,dt=1+\frac{1}{1/\alpha}=1+\alpha.$$

Hence,

$$H^{-1}(t)\sim(1+\alpha)B(t)\sim a(t)\in RV_{1/\alpha}.$$

Note that $\frac{1}{1-F(t)} = H(t) \sim (H^{-1})^{-1}(t)$, as $t \to \infty$. Using Theorems 1.1.6 and 1.4.7, we obtain that $\frac{1}{1-F} \in RV_\alpha$ and $1 - F \in RV_{-\alpha}$. □

Example 2.3.3 *The Fréchet distribution.* Function $G_{1,\alpha}$ belongs to the domain of attraction $DA(G_{1,\alpha})$. It follows from Example 2.1.9 (b) that for every positive integer n and every real number x the following equality holds:

$$G_{1,\alpha}^n(n^{1/\alpha}x) = G_{1,\alpha}(x).$$

The norming constants are $a_n = n^{1/\alpha}$ and $b_n = 0$. △

Example 2.3.4 *The Pareto distribution.* Let us consider the distribution function given by

$$F(x) = 1 - kx^{-\alpha}, \quad \alpha > 0, \ k > 0, \ x \geqslant k^{1/\alpha}, \tag{2.3.22}$$

and $F(x) = 0$ for $x < k^{1/\alpha}$. Then we obtain

$$\lim_{t\to\infty} \frac{1 - F(tx)}{1 - F(t)} = \lim_{t\to\infty} \frac{k(tx)^{-\alpha}}{kt^{-\alpha}} = x^{-\alpha},$$

and from Theorem 2.3.2 relation $F \in DA(G_{1,\alpha})$ follows. Moreover,

$$\begin{aligned} P\{M_n \leqslant (kn)^{1/\alpha}x\} &= F^n((kn)^{1/\alpha}x) = (1 - k(kn)^{-1}x^{-\alpha})^n \\ &= \left(1 - \frac{x^{-\alpha}}{n}\right)^n \to \exp(-x^{-\alpha}), \quad n \to \infty, \end{aligned}$$

i.e., the norming constants are given by $a_n = (kn)^{1/\alpha}$ and $b_n = 0$. △

Example 2.3.5 *The Cauchy distribution.* Let us consider the distribution function given by

$$F(x) = \frac{1}{2} + \frac{1}{\pi} \operatorname{arctg} x. \tag{2.3.23}$$

Then for every $x > 0$,

$$\frac{1 - F(tx)}{1 - F(t)} = \frac{\frac{\pi}{2} - \operatorname{arctg}(tx)}{\frac{\pi}{2} - \operatorname{arctg} t} \to \frac{1}{x}, \quad t \to +\infty. \tag{2.3.24}$$

Indeed, if we introduce the substitution $\operatorname{arctg} t = \pi/2 - \vartheta$, we obtain that $t = \operatorname{tg}(\pi/2 - \vartheta) = \operatorname{ctg} \vartheta$, and consequently

$$\lim_{t\to\infty} \left(\frac{\pi}{2} - \operatorname{arctg} t\right) \cdot t = \lim_{\vartheta\to 0} \vartheta \operatorname{ctg} \vartheta = 1.$$

Therefore

$$\lim_{t\to\infty}\left(\frac{\pi}{2}-\operatorname{arctg}(tx)\right)tx=1,$$

and (2.3.24) follows easily, i.e., $F \in DA(G_{1,1})$. The norming constants are a_n and $b_n = 0$, where constant a_n satisfies the condition

$$1-F(a_n)=\frac{1}{2}-\frac{1}{\pi}\operatorname{arctg} a_n=\frac{1}{n}.$$

Thus, $a_n = \operatorname{tg}\left(\frac{\pi}{2}-\frac{\pi}{n}\right) = \operatorname{ctg}\frac{\pi}{n}$. For every $x > 0$, we have

$$\lim_{n\to\infty} P\left\{M_n \leqslant x\cdot\operatorname{ctg}\frac{\pi}{n}\right\}=\lim_{n\to\infty}F^n\left(x\operatorname{ctg}\frac{\pi}{n}\right)=e^{-x^{-1}}.$$

△

2.3.2 Domain of Attraction of the Weibull Distribution

The following theorem gives the sufficient conditions for an absolutely continuous distribution function to belong to the domain of attraction of the Weibull extreme value distribution function.

Theorem 2.3.6 (von Mises [153]) *Let F be an absolutely continuous distribution function with the corresponding density function f. If the following three conditions are satisfied:*

(a) $x_F := \sup\{t : F(t) < 1\} < +\infty$
(b) $f(x) > 0$ *for all* $x \in (a, x_F)$
(c) $\lim_{t\uparrow x_F}\frac{(x_F-t)f(t)}{1-F(t)}=\alpha>0$

then $F \in DA(G_{2,\alpha})$.

The general case was considered by Gnedenko who proved the following theorem.

Theorem 2.3.7 (Gnedenko [81]) *A distribution function F belongs to the domain of attraction $DA\,(G_{2,\alpha})$ if and only if the following conditions are satisfied:*

(a) $x_F := \sup\{t : F(t) < 1\} < +\infty$.
(b) $1-F\left(x_F-\frac{1}{x}\right)\in RV_{-\alpha}$ *as* $x\to\infty$.
In that case for every $x < 0$ *we have*

$$\lim_{n\to\infty}F^n((x_F-\gamma_n)x+x_F)=G_{2,\alpha}(x), \tag{2.3.25}$$

where constant γ_n is given by

$$\gamma_n = \inf\left\{s : \frac{1}{1-F(s)} \geqslant n\right\} = \left(\frac{1}{1-F}\right)^{-1}(n). \tag{2.3.26}$$

Remark 2.3.8 Condition (b) can be rewritten equivalently as follows:

$$\lim_{h\downarrow 0} \frac{1-F(x_F - hx)}{1-F(x_F - h)} = x^{\alpha}, \quad \text{for all } x > 0. \tag{2.3.27}$$

Proof of Theorem 2.3.7 Let us suppose that the conditions of the Theorem are satisfied, i.e., $x_F < +\infty$ and $1 - F\left(x_F - \frac{1}{x}\right) \in RV_{-\alpha}$. Function F_1 given by

$$F_1(x) = \begin{cases} 0, & \text{if } x \leqslant 0, \\ F\left(x_F - \frac{1}{x}\right), & \text{if } x > 0 \end{cases} \tag{2.3.28}$$

is regularly varying at infinity with exponent $-\alpha$. From Theorem 2.3.2 it follows that $F_1 \in DA(G_{1,\alpha})$. Let us denote

$$a_n^* = \left(\frac{1}{1-F_1}\right)^{-1}(n). \tag{2.3.29}$$

Then we obtain

$$\begin{aligned} a_n^* &= \inf\left\{x : \frac{1}{1-F_1(x)} \geqslant n\right\} = \inf\left\{x : \frac{1}{1-F(x_F - 1/x)} \geqslant n\right\} \\ &= \inf\left\{\frac{1}{x_F - s} : \frac{1}{1-F(s)} \geqslant n\right\} = \frac{1}{x_F - \gamma_n}. \end{aligned} \tag{2.3.30}$$

Since $F_1 \in DA(G_{1,\alpha})$, it follows that, for every $x > 0$,

$$\lim_{n\to\infty} F_1^n(a_n^* x) = \exp\{-x^{-\alpha}\}. \tag{2.3.31}$$

If we put F_1 from (2.3.28) into (2.3.31), we obtain that

$$\lim_{n\to\infty} F^n\left(x_F - \frac{1}{a_n^* x}\right) = \exp\{-x^{-\alpha}\}. \tag{2.3.32}$$

Let us denote $t = -\frac{1}{x}$. Using equalities (2.3.30) and (2.3.32), we obtain that for every $t < 0$ the following relation holds:

$$\lim_{n\to\infty} F^n((x_F-\gamma_n)t+x_F)=\lim_{n\to\infty} F^n\left(x_F+\frac{t}{a_n^*}\right)=\exp\{-(-t)^\alpha\}.$$

Therefore, $F\in DA(G_{2,\alpha})$.

Conversely, let us suppose that $F\in DA(G_{2,\alpha})$. Then there exist sequences of constants $a_n>0$ and $b_n\in\mathbb{R}$, where $n\in\mathbb{N}$, such that for every real number x,

$$\lim_{n\to\infty} F^n(a_nx+b_n)=G_{2,\alpha}(x). \tag{2.3.33}$$

Relation (2.3.33) implies that

$$\lim_{n\to\infty} F^{2n}(a_nx+b_n)=G_{2,\alpha}^2(x)=G_{2,\alpha}(cx), \tag{2.3.34}$$

where $c=2^{1/\alpha}$. Relation (2.3.34) implies that

$$\lim_{n\to\infty} F^{2n}\left(\frac{a_n}{c}x+b_n\right)=G_{2,\alpha}(x), \tag{2.3.35}$$

while from (2.3.33) we obtain that

$$\lim_{n\to\infty} F^{2n}(a_{2n}x+b_{2n})=G_{2,\alpha}(x). \tag{2.3.36}$$

Using relations (2.3.35) and (2.3.36) and Khinchin's Theorem 1.4.10, we obtain that constants a_n and b_n can be chosen such that

$$a_{2n}=\frac{a_n}{c},\qquad b_{2n}=b_n, \tag{2.3.37}$$

for every positive integer n. For $x=0$, relation (2.3.33) implies that

$$\lim_{n\to\infty} F^n(b_n)=G_{2,\alpha}(0)=1. \tag{2.3.38}$$

If $x_F=\sup\{t:F(t)<1\}=+\infty$, then we have $\lim\limits_{n\to\infty} b_n=0$, and this contradicts (2.3.37). Hence, $x_F<+\infty$, i.e., condition (a) is satisfied. Relation (2.3.38) implies that $b_n\to x_F$, and using equalities (2.3.37), we obtain that $b_n=x_F$ for every positive integer n. Now, using (2.3.33) and Theorem 2.1.4, we obtain that, for every $x<0$,

$$\lim_{n\to\infty} n(1-F(a_nx+x_F))=(-x)^\alpha, \tag{2.3.39}$$

and consequently $\lim\limits_{n\to\infty} a_n=0$. Let us suppose that $t\downarrow 0$. For sufficiently small t, there exists sufficiently large positive integer n, such that $a_{n+1}\leqslant t\leqslant a_n$ or $a_n\leqslant t\leqslant a_{n+1}$. In the first case we have

$$1 - F(x_F - a_{n+1}) \leqslant 1 - F(x_F - t) \leqslant 1 - F(x_F - a_n), \tag{2.3.40}$$

and for every $x > 0$,

$$1 - F(x_F - a_{n+1}x) \leqslant 1 - F(x_F - tx) \leqslant 1 - F(x_F - a_n x). \tag{2.3.41}$$

Using inequalities (2.3.40) and (2.3.41), we obtain that

$$\frac{1 - F(x_F - a_{n+1}x)}{1 - F(x_F - a_n)} \leqslant \frac{1 - F(x_F - tx)}{1 - F(x_F - t)} \cdot \frac{1 - F(x_F - a_n x)}{1 - F(x_F - a_{n+1})}.$$

In the second case we obtain that

$$\frac{1 - F(x_F - a_n x)}{1 - F(x_F - a_{n+1})} \leqslant \frac{1 - F(x_F - tx)}{1 - F(x_F - t)} \cdot \frac{1 - F(x_F - a_{n+1}x)}{1 - F(x_F - a_n)}.$$

In both cases, taking into account (2.3.39), we obtain that for every $x > 0$,

$$\lim_{t \downarrow 0} \frac{1 - F(x_F - tx)}{1 - F(x_F - t)} = x^\alpha. \qquad \square$$

Example 2.3.9 *The Weibull distribution function* $G_{2,\alpha}$ belongs to the domain of attraction $DA(G_{2,\alpha})$. Indeed, it follows from Example 2.1.9 (c) that for every positive integer n and every $x \in \mathbb{R}$ the following equality holds:

$$G_{2,\alpha}^n(n^{-1/\alpha}x) = G_{2,\alpha}(x).$$

The norming constants are given by $a_n = n^{-1/\alpha}$ and $b_n = 0$. △

Example 2.3.10 *The beta distribution.* A random variable X has the beta distribution with parameters (p, q), notation beta(p, q), where $p > 0$ and $q > 0$, if its probability density function is given by

$$f(x) = \frac{\Gamma(p+q)}{\Gamma(p)\Gamma(q)} x^{p-1}(1-x)^{q-1}, \quad x \in [0, 1], \tag{2.3.42}$$

and $f(x) = 0$ for $x \notin [0, 1]$. For the right tail $1 - F(x) = \int_x^1 f(t)\,dt$ of the beta(p, q) distribution, we obtain the following asymptotic behavior:

$$1 - F(x) \sim \frac{1-x}{q} f(x) = \frac{1}{q} \cdot \frac{\Gamma(p+q)}{\Gamma(p)\Gamma(q)} x^{p-1}(1-x)^q, \quad \text{as } x \uparrow 1. \tag{2.3.43}$$

Relation (2.3.43) implies that for every $x > 0$,

$$\lim_{t \downarrow 0} \frac{1 - F(1 - tx)}{1 - F(1 - t)} = \lim_{t \downarrow 0} \frac{(1 - tx)^{p-1}(tx)^q}{(1 - t)^{p-1}t^q} = x^q. \tag{2.3.44}$$

Therefore, the beta (p, q) distribution belongs to the domain of attraction $DA(G_{2,q})$. The relation $\gamma_n = (1/(1-F))^{-1}(n)$ implies that

$$1-\gamma_n \sim \left(\frac{q\Gamma(p)\Gamma(q)}{\Gamma(p+q)} \cdot \frac{1}{n}\right)^{1/q}, \quad \text{as } n \to \infty. \tag{2.3.45}$$

Then, using the asymptotic (2.3.43), it is easy to check that for γ_n satisfying (2.3.45), and for each $x < 0$, we have

$$\lim_{n\to\infty} n\{1 - F((1-\gamma_n)x + 1)\} = (-x)^q. \tag{2.3.46}$$

Now we are interesting in probability $P\{M_n \leqslant (1-\gamma_n)x + 1\}$ for $x < 0$. Using (2.3.45), (2.3.46), and Theorems 2.1.4 and 2.3.7, we obtain the following proposition.

Let $(X_n)_{n\geqslant 1}$ be a sequence of independent random variables having the beta (p, q) distribution, and $M_n = \max\{X_1, \ldots, X_n\}$. Then for every $x < 0$, the following equality holds:

$$\lim_{n\to\infty} P\left\{\left(\frac{\Gamma(p+q)n}{q\Gamma(p)\Gamma(q)}\right)^{1/q}(M_n - 1) \leqslant x\right\} = e^{-(-x)^q}. \tag{2.3.47}$$

Note that the second parameter q of the beta(p, q) distribution is the shape parameter of the limiting Weibull extreme value distribution for M_n. The result for uniform $\mathcal{U}[0, 1] = \text{beta}[1, 1]$ distribution follows if we put $p = 1$ and $q = 1$.

Beta distributions are usually considered to be a four-parameter family of distributions whose densities are given by

$$f(x) = \frac{\Gamma(p+q)}{\Gamma(p)\Gamma(q)} \frac{(x-a)^{p-1}(b-x)^{q-1}}{(b-a)^{p+q-1}}, \quad x \in [a, b], \tag{2.3.48}$$

where $p > 0$, $q > 0$, $a < b$, notation beta$(p, q; a, b)$. Note that if $(Y_i)_{i\geqslant 1}$ is a sequence of i.i.d. random variables having the common beta$(p, q; a, b)$ distribution and $M_n = \max\{Y_1, \ldots, Y_n\}$, then equality (2.3.47) holds with $(M_n - a)/(b-a)$ instead of M_n, i.e., for every $x < 0$ we have

$$\lim_{n\to\infty} P\left\{\left(\frac{\Gamma(p+q)n}{q\Gamma(p)\Gamma(q)}\right)^{1/q} \frac{M_n - b}{b-a} \leqslant x\right\} = e^{-(-x)^q}. \tag{2.3.49}$$

Suppose now that we are interested in the minimum $m_n = \min\{X_1, \ldots, X_n\}$. First note that $-m_n = \max\{-X_1, \ldots, -X_n\}$, and the support of the distribution of $-X_1$ is interval $[-b, -a]$. Analogously as in the previous case it is easy now to obtain the limiting distribution for m_n in the following form:

$$\lim_{n\to\infty} P\left\{\left(\frac{\Gamma(p+q)n}{p\Gamma(p)\Gamma(q)}\right)^{1/p}\frac{a-m_n}{b-a}\leqslant x\right\} = e^{-(-x)^p}, \quad x<0. \tag{2.3.50}$$

△

2.3.3 Domain of Attraction of the Gumbel Distribution

Let F be a distribution function and $x_F = \sup\{t : F(t) < 1\}$. In the previous two sections it was proved that for every $F \in DA(G_{1,\alpha})$ condition $x_F = +\infty$ holds, while for every $F \in DA(G_{2,\alpha})$ condition $x_F < +\infty$ holds. The domain of attraction of the Gumbel distribution function $DA(G_0)$ contains both the distribution functions F with $x_F = +\infty$ and the distribution functions F with $x_F < +\infty$. Similarly as in the previous cases we shall formulate a result that gives the sufficient conditions for an absolutely continuous distribution function to belong to $DA(G_0)$ and then the results that give necessary and sufficient conditions for $F \in DA(G_0)$ in the general case.

Theorem 2.3.11 (von Mises [153]) *Let F be an absolutely continuous distribution function and $x_F := \sup\{t : F(t) < 1\}$. If the following conditions are satisfied:*

(a) $F''(x) < 0$ *for all* $x \in (a, x_F)$, *where* $x_F \leqslant \infty$
(b) $F'(x) = 0$ *for* $x \geqslant x_F$
(c) $\displaystyle\lim_{t\uparrow x_F} \frac{F''(t)(1-F(t))}{(F'(t))^2} = -1$

then $F \in DA(G_0)$.

Proof For $x \in (a, x_F)$, let us denote $f(x) = \dfrac{1-F(x)}{F'(x)}$. Then we have $f(x) > 0$, and the following relation holds:

$$f'(x) = -1 - \frac{F''(x)(1-F(x))}{(F'(x))^2}. \tag{2.3.51}$$

From condition (c) it follows that $\lim_{x\uparrow x_0} f'(x) = 0$. Let us note also that for $x \in (a, x_F)$ the following inequality holds:

$$1-F(x) = \int_x^{x_F} F'(t)\,dt \leqslant F'(x)(x_F - x). \tag{2.3.52}$$

If $x_F < +\infty$, then using (2.3.52) we obtain that

$$\lim_{x\uparrow x_F} \frac{f(x)}{x_F - x} = \lim_{x\uparrow x_F} \frac{1-F(x)}{(x_F-x)F'(x)} \leqslant \lim_{x\uparrow x_F} \frac{F'(x)(x_F-x)}{(x_F-x)F'(x)} = 1,$$

and consequently $f(x_F) := \lim_{x\uparrow x_F} f(x) = 0$. Next, let us note that

$$\lim_{x\uparrow x_F} \frac{f(x)}{x_F - x} = \lim_{x\uparrow x_F} \left\{ - \int_x^{x_F} \frac{f'(t)}{x_F - x}\, dt \right\}. \tag{2.3.53}$$

Using the substitutions $t = x_F - u$ and $v = x_F - x$ and the fact that $\lim_{x\uparrow x_F} f'(x) = 0$ implies the equality $\lim_{u\downarrow 0} f'(x_F - u) = 0$, we obtain that (2.3.53) implies

$$\lim_{x\uparrow x_F} \frac{f(x)}{x_F - x} = \lim_{v\downarrow 0} \left\{ -\frac{1}{v} \int_0^{v} f'(x_F - u)\, du \right\} = 0. \tag{2.3.54}$$

Let us prove that the equality

$$\lim_{x\uparrow x_F} \frac{f(t + xf(t))}{f(t)} = 1 \tag{2.3.55}$$

holds uniformly for x in every bounded interval. First, note that

$$f(t + xf(t)) = f(t) + xf(t)\, f'(t + \vartheta \cdot xf(t)), \tag{2.3.56}$$

where $\vartheta = \vartheta(x, t) \in [0, 1]$. If $x_F = +\infty$, then equality $\lim_{t\to\infty} f'(t) = 0$ implies that $\lim_{t\to\infty} \frac{f(t)}{t} = 0$. Consequently, we obtain that for every bounded interval $I \subset \mathbb{R}$ there exists $t_0 \geqslant 0$, such that for every $t \geqslant t_0$ and all $x \in I$,

$$1 + \vartheta(x, t) \cdot x \frac{f(t)}{t} \geqslant \frac{1}{2}. \tag{2.3.57}$$

Using (2.3.57) and equality $\lim_{t\to\infty} f'(t) = 0$, we obtain that on the bounded interval I the following equality uniformly holds:

$$\lim_{t\to\infty} f'(t + \vartheta(x, t) \cdot xf(t)) = 0. \tag{2.3.58}$$

Relations (2.3.56) and (2.3.58) imply that (2.3.55) uniformly holds on I. If $x_F < +\infty$, we obtain that uniformly on the bounded interval I

$$\lim_{t\uparrow x_F} f'(t + \vartheta(x, t) \cdot xf(t)) =$$

$$= \lim_{t\uparrow x_F} f' \left\{ x_F - (x_F - t)\left(1 + \vartheta(x, t) \cdot x \frac{f(t)}{x_F - t} \right) \right\} = 0,$$

and consequently (2.3.55) also holds in this case.

If $a < b_n < x < x_F$, then we obtain that

$$1 - F(x) = (1 - F(b_n)) \exp\left\{-\int_a^x \frac{F'(t)\,dt}{1 - F(t)}\right\},$$
$$= (1 - F(b_n)) \exp\left\{-\int_a^x \frac{dt}{f(t)}\right\}. \tag{2.3.59}$$

Let b_n be a constant such that $1 - F(b_n) = 1/n$. Then we have $\lim_{n\to\infty} b_n = x_F$, and for large n the inequalities $a < b_n < x_F$ hold. Note also that for a given real number x and sufficiently large n the following inequality holds:

$$b_n < b_n + xf(b_n) < x_F.$$

Relation (2.3.59) implies that

$$1 - F(b_n + xf(b_n)) = (1 - F(b_n)) \exp\left\{-\int_{b_n}^{b_n + xf(b_n)} \frac{dt}{f(t)}\right\}. \tag{2.3.60}$$

If we introduce the substitution $t = b_n + sf(b_n)$, then relations (2.3.60) and (2.3.55) imply that

$$n\{1 - F(b_n + xf(b_n))\} = \exp\left\{-\int_0^x \frac{f(b_n)\,ds}{f(b_n + sf(b_n))}\right\}$$
$$\to e^{-x}, \quad n \to \infty. \tag{2.3.61}$$

Finally, using Theorem 2.1.4 and relation (2.3.61) we obtain that for every $x \in \mathbb{R}$ the following equality holds:

$$\lim_{n\to\infty} F^n(b_n + xf(b_n)) = \exp\{-e^{-x}\}. \qquad \square$$

The first characterization of the Gumbel domain of attraction was given by Gnedenko [81]. This characterization includes an auxiliary function and was not final and suitable for application, so the problem was revisited by several authors. Mejzler [150] gave the characterization of the domain $DA(G_0)$ in terms of the generalized inverse of distribution function F. Papers by Marcus and Pinsky [146] and Balkema and de Haan [17] are related to this problem. De Haan [87, 88, 90] gave the characterizations of $DA(G_0)$ in terms of distribution function F. Theorem 1.5.14 gives the necessary and sufficient conditions under which $F \in DA(G_0)$. Combining the results and notation of Gnedenko, Mejzler, and de Haan, we can now reformulate this theorem as follows.

Theorem 2.3.12 (Gnedenko [81], Mejzler [150], and de Haan [87]) *Let F be a distribution function, $x_F = \sup\{t : F(t) < 1\}$, and*

$$H(x) = \frac{1}{1 - F(x)} \quad \text{for} \quad x < x_F.$$

Then the following conditions are equivalent:

(a) *$F \in DA(G_0)$, i.e., there exist sequences of constants $a_n > 0$ and $b_n \in \mathbb{R}$, $n \in \mathbb{N}$, such that for every $x \in \mathbb{R}$ the following equality holds:*

$$\lim_{n\to\infty} F^n(a_n x + b_n) = e^{-e^{-x}}.$$

(b) *$H \in \Gamma$, i.e., there exists function $g : (c, x_F) \to \mathbb{R}_+$, such that for every $x \in \mathbb{R}$,*

$$\lim_{t\uparrow x_F} \frac{1 - F(t + xg(t))}{1 - F(t)} = e^{-x}.$$

(c) *$H^{-1} \in \Pi$, i.e., there exists function $a : (c, +\infty) \to \mathbb{R}_+$, such that, for every $x > 0$,*

$$\lim_{t\to\infty} \frac{H^{-1}(tx) - H^{-1}(t)}{a(t)} = \ln x.$$

We shall now formulate two more theorems that give the necessary and sufficient conditions under which distribution function F belongs to the domain of attraction of the Gumbel distribution. The conditions are formulated only in terms of distribution function F, and the proofs can be found in de Haan [87] and Resnick [192]. As usual $x_F := \sup\{t : F(t) < 1\}$ is the notation for the right endpoint of the support of the distribution determined by F.

Theorem 2.3.13 (de Haan [87]) *$F \in DA(G_0)$ if and only if*

$$\lim_{x\uparrow x_F} \frac{(1 - F(x)) \int_x^{x_F} \int_y^{x_F} (1 - F(t))\, dt\, dy}{\left\{\int_x^{x_F} (1 - F(t))\, dt\right\}^2} = 1, \tag{2.3.62}$$

and all integrals in (2.3.62) are finite. Moreover, $\frac{1}{1-F} \in \Gamma$, an auxiliary function g from Theorem 2.3.12(b) can be determined such that

$$g(t) = \frac{\int_t^{x_F} \int_y^{x_F} (1 - F(s))\, ds\, dy}{\int_t^{x_F} (1 - F(s))\, ds}, \quad t < x_F, \tag{2.3.63}$$

or

$$g(t) = \frac{\int_t^{x_F} (1 - F(s))\, ds}{1 - F(t)}, \quad t < x_F, \tag{2.3.64}$$

and the norming constants a_n and b_n can be chosen such that

$$a_n = g(b_n), \quad b_n = \left(\frac{1}{1 - F}\right)^{-1}(n). \tag{2.3.65}$$

Theorem 2.3.14 (de Haan [87]) *$F \in DA(G_0)$ if and only if*

$$\lim_{x \uparrow x_F} \frac{\int_x^{x_F} (1 - F(t))^{\alpha}\, dt}{(1 - F(x)) \int_x^{x_F} (1 - F(t))^{\alpha-1}\, dt} = \frac{\alpha - 1}{\alpha}, \tag{2.3.66}$$

for some $\alpha > 1$. In that case equality (2.3.66) *holds for every $\alpha > 1$.*

Example 2.3.15 (The Gumbel Distribution) We shall prove that $G_0 \in DA(G_0)$. Indeed, since

$$G_0^n(x) = \exp(-e^{-x+\ln n}),$$

it follows that $G_0^n(x + \ln n) = G_0(x)$. Hence, for $a_n = 1$ and $b_n = \ln n$,

$$\lim_{n\to\infty} P\{M_n \leqslant a_n x + b_n\} = \lim_{n\to\infty} G_0^n(a_n x + b_n) = G_0(x).$$

It is easy to check that $g(t) = 1$ is the auxiliary function from Theorem 2.3.12(b). △

Example 2.3.16 (The Gamma Distribution) A random variable X has the *gamma* distribution with parameters $\lambda > 0$ and $\alpha > 0$, denoted by $\Gamma(\lambda, \alpha)$, if its probability density function is given by

$$f(x) = \frac{\lambda^{\alpha}}{\Gamma(\alpha)} x^{\alpha-1} e^{-\lambda x}, \quad x > 0, \tag{2.3.67}$$

and $f(x) = 0$ for $x \leqslant 0$. For the tail of the distribution, we have

$$1 - F(x) = \int_x^{\infty} \frac{\lambda^{\alpha}}{\Gamma(\alpha)} t^{\alpha-1} e^{-\lambda t}\, dt \sim \frac{1}{\lambda} f(x), \quad \text{as } n \to \infty. \tag{2.3.68}$$

Let us determine $u_n = u_n(x) = a_n x + b_n$, $a_n > 0$, $b_n \in \mathbb{R}$, from the condition

$$1 - F(u_n) \sim \frac{\lambda^{\alpha-1}}{\Gamma(\alpha)} u_n^{\alpha-1} e^{-\lambda u_n} \sim \frac{e^{-x}}{n}, \quad \text{as } n \to \infty. \tag{2.3.69}$$

It follows from (2.3.69) that

$$(\alpha-1)\ln\lambda-\ln\Gamma(\alpha)+(\alpha-1)\ln u_n-\lambda u_n\sim -x-\ln n,\quad \text{as } n\to\infty. \tag{2.3.70}$$

Relation (2.3.70) implies that

$$\lim_{n\to\infty}\frac{\lambda u_n}{\ln n}=1, \tag{2.3.71}$$

and consequently $\ln u_n = \ln\ln n - \ln\lambda + o(1)$, as $n\to\infty$. By replacing $\ln u_n$ from (2.3.71) into (2.3.70), we obtain that

$$u_n=\frac{x}{\lambda}+\frac{\ln n+(\alpha-1)\ln\ln n-\ln\Gamma(\alpha)}{\lambda}+o(1),\quad \text{as } n\to\infty.$$

Hence, $a_n = 1/\lambda$, $b_n = (\ln n + (\alpha-1)\ln\ln n - \ln\Gamma(\alpha))/\lambda$, and $F\in DA(G_0)$. Particularly, for the standard exponential distribution ($\lambda = 1$, $\alpha = 1$), we have $a_n = 1$ and $b_n = \ln n$. △

Example 2.3.17 (Normal Distribution) The density and the distribution function of the standard normal distribution are given by

$$\varphi(x)=\frac{1}{\sqrt{2\pi}}e^{-x^2/2},\quad \Phi(x)=\int_{-\infty}^{x}\varphi(t)\,dt,\quad x\in\mathbb{R}.$$

The following asymptotic relation holds:

$$1-\Phi(x)\sim\frac{\varphi(x)}{x},\quad x\to\infty. \tag{2.3.72}$$

Using (2.3.72) we obtain that

$$\begin{aligned}\frac{1-\Phi(t+xg(t))}{1-\Phi(t)}&=\frac{e^{-(t+xg(t))^2/2}}{t+xg(t)}\cdot\frac{t}{e^{-t^2/2}}(1+o(1))\\&=\frac{t}{t+xg(t)}e^{-txg(t)}e^{-x^2g^2(t)/2}(1+o(1)).\end{aligned}$$

For $g(t) = 1/t$, $t>0$, the following relations hold:

$$\begin{aligned}\frac{1-\Phi(t+xg(t))}{1-\Phi(t)}&=\frac{t^2}{t^2+x}e^{-x}e^{-x^2/(2t^2)}(1+o(1))\\&\to e^{-x},\quad t\to\infty. \end{aligned}\tag{2.3.73}$$

Theorem 2.3.12 and relation (2.3.73) imply that $\Phi\in DA(G_0)$.

Let us now determine the norming constants $a_n > 0$ and b_n. Let x be a fixed real number. We shall determine the constant $u_n = a_n x + b_n$ from the condition:

$$1 - \Phi(u_n) \sim \frac{e^{-x}}{n}, \qquad n \to \infty. \tag{2.3.74}$$

Using asymptotic relations (2.3.72) and (2.3.74), we obtain that $\frac{e^{-x}}{n} \cdot \frac{u_n}{\varphi(u_n)} \to 1$, as $n \to \infty$, and therefore

$$-x - \ln n + \ln u_n + \frac{1}{2}\ln 2\pi + \frac{1}{2}u_n^2 \to 0. \tag{2.3.75}$$

It is clear that $u_n \to \infty$ as $n \to \infty$, and (2.3.75) implies that

$$\lim_{n\to\infty} \frac{u_n^2}{2\ln n} = 1. \tag{2.3.76}$$

It follows from (2.3.76) that $\ln u_n = \frac{1}{2}(\ln 2 + \ln\ln n) + o(1)$, as $n \to \infty$. Consequently (2.3.75) implies that

$$\frac{u_n^2}{2} = x + \ln n - \frac{1}{2}\ln 2 - \frac{1}{2}\ln\ln n - \frac{1}{2}\ln 2\pi + o(1),$$

and therefore

$$u_n^2 = 2\ln n \left\{1 + \frac{x - \frac{1}{2}(\ln 4\pi + \ln\ln n)}{\ln n} + o\left(\frac{1}{\ln n}\right)\right\}$$

$$u_n = \sqrt{2\ln n}\left\{1 + \frac{x - \frac{1}{2}(\ln 4\pi + \ln\ln n)}{2\ln n} + o\left(\frac{1}{\ln n}\right)\right\},$$

and finally

$$u_n = \frac{x}{\sqrt{2\ln n}} + \sqrt{2\ln n} - \frac{\ln\ln n + \ln 4\pi}{2\sqrt{2\ln n}} + o\left(\frac{1}{\ln n}\right). \tag{2.3.77}$$

If constant u_n is given by (2.3.77), then relation (2.3.74) holds. Now, using Theorem 2.1.4 we obtain that for

$$a_n = \frac{1}{\sqrt{2\ln n}}, \tag{2.3.78}$$

$$b_n = \sqrt{2\ln n} - \frac{\ln\ln n + \ln 4\pi}{2\sqrt{2\ln n}} \tag{2.3.79}$$

and each $x \in \mathbb{R}$, the relation $\lim_{n\to\infty} P\{M_n \leqslant a_n x + b_n\} = \exp(-e^{-x})$ holds. △

Example 2.3.18 Let us consider the distribution function given by

$$F(x) = \begin{cases} 1 - e^{1/x}, & \text{if } x < 0, \\ 1, & \text{if } x \geqslant 0. \end{cases} \tag{2.3.80}$$

We shall prove that $F \in DA(G_0)$. It is easy to check that function g from Theorem 2.3.12 can be chosen such that $g(t) = t^2$, for $-\infty < t < 0$. Constant u_n for which the relation $1 - F(u_n) \sim e^{-x}/n$, i.e., $e^{1/u_n} \sim e^{-x}/n$, holds as $n \to \infty$ can be chosen such that $1/u_n = -x - \ln n$, i.e.,

$$\begin{aligned} u_n &= -\frac{1}{x + \ln n} = -\frac{1}{\ln n}\left(1 + \frac{x}{\ln n}\right)^{-1} \\ &= \frac{x}{(\ln n)^2} - \frac{1}{\ln n} + o\left(\frac{1}{\ln n}\right)^2. \end{aligned}$$

Hence, $a_n = \dfrac{1}{(\ln n)^2}$ and $b_n = -\dfrac{1}{\ln n}$ are the norming constants. △

2.3.4 *Distributions Out of the Domains of Attraction*

For some distribution functions F, it is not possible to obtain a nondegenerate limit distribution for maximum M_n under any linear normalization. Regarding this statement the following theorem is important. Theorem 2.1.5 and equality (2.1.50) give the necessary and sufficient conditions under which the limit theorem is possible under the linear normalization of maxima.

Remark 2.3.19 Let us assume that F is the distribution function of a discrete random variable X with support $\mathbb{N}_0 = \{0, 1, 2, \dots\}$, and let us denote

$$p_n = P\{X = n\} = F(n) - F(n-0), \quad n \in \mathbb{N}_0.$$

Condition (2.1.50) is then equivalent to the following relation:

$$\lim_{n\to\infty} \frac{p_n}{\sum_{k=n}^{\infty} p_k} = 0. \tag{2.3.81}$$

Some very well-known discrete distributions (for example, geometric, negative binomial, and Poisson distributions) do not belong to any of the maximum domains of attraction of EV distributions. We shall prove this fact here for the negative binomial distribution. The corresponding result for the geometric distribution then follows as a special case.

Example 2.3.20 *The negative binomial distribution.* Let $r \geqslant 1$ be a positive integer, $0 < p < 1$, and

$$p_n = \binom{n-1}{r-1} p^r (1-p)^{n-r}, \quad n \in \{r, r+1, \dots\}. \tag{2.3.82}$$

If $q = 1 - p$, then for every $n \in \mathbb{N}$ we have

$$\begin{aligned} \frac{p_n}{\sum\limits_{k=n}^{\infty} p_k} &= \frac{(n-1)(n-2)\dots(n-r+1)q^{n-r}}{\sum\limits_{k=n}^{\infty}(k-1)(k-2)\dots(k-r+1)q^{k-r}} \\ &= \frac{(n-1)(n-2)\dots(n-r+1)q^{n-r}}{\left(\sum\limits_{k=n}^{\infty} q^{k-1}\right)^{(r-1)}} \\ &= \frac{(n-1)(n-2)\dots(n-r+1)q^{n-r}}{\left(\dfrac{q^{n-1}}{1-q}\right)^{(r-1)}}. \end{aligned} \tag{2.3.83}$$

Using the Leibniz formula we obtain that

$$\begin{aligned} \left(\frac{q^{n-1}}{1-q}\right)^{(r-1)} &= \sum_{s=0}^{r-1} \binom{r-1}{s} (q^{n-1})^{(r-1-s)} \left(\frac{1}{1-q}\right)^{(s)} \\ &= (n-1)(n-2)\dots(n-r+1)\frac{q^{n-r}}{1-q} \\ &\quad + \sum_{s=1}^{r-1} \binom{r-1}{s} (q^{n-1})^{(r-1-s)} \left(\frac{1}{1-q}\right)^{(s)}. \end{aligned} \tag{2.3.84}$$

Using (2.3.83) and (2.3.84) we obtain that

$$\lim_{n\to\infty} \frac{p_n}{\sum_{k=n}^{\infty} p_k} = p \neq 0,$$

and hence the negative binomial distribution does not belong to any of the maximum domains of attraction of EV distributions. △

2.3.5 Mixture of Distributions and Extreme Values

Mixtures of distributions often appear in statistical applications. Here we consider two simple examples regarding extreme values in a sequence of i.i.d. random variables with a common distribution that is a mixture. *See* also Exercises 2.2 and 2.3.

Let X_1 and X_2 be random variables with distribution functions $F_1(x)$ and $F_2(x)$, respectively, and let us define

$$X = \begin{cases} X_1, & \text{with probability } p, \\ X_2, & \text{with probability } q, \end{cases} \tag{2.3.85}$$

where $p + q = 1$, $p > 0$, $q > 0$. Then the distribution function of random variable X is given by

$$F(x) = P\{X \leqslant x\} = pP\{X_1 \leqslant x\} + qP\{X_2 \leqslant x\} \tag{2.3.86}$$

$$= pF_1(x) + qF_2(x). \tag{2.3.87}$$

The distribution determined by distribution function F is a mixture of the distributions determined by F_1 and F_2.

Mixture of Normal Distributions In this section we shall first consider a sequence $(X_n)_{n\geqslant 1}$ of i.i.d. random variables with common distribution function F that is a mixture of two distributions and formulate some results related to the asymptotic behavior of the maximum $M_n = \max\{X_1, \ldots, X_n\}$.

Theorem 2.3.21 (Mladenović [156]) *Let $(Z_n)_{n\geqslant 1}$ be a sequence of i.i.d. random variables such that, for every $n \in \mathbb{N}$,*

$$Z_n \sim \begin{cases} \mathcal{N}(m_1, \sigma_1^2), & \text{with probability } p, \\ \mathcal{N}(m_2, \sigma_2^2), & \text{with probability } q, \end{cases} \tag{2.3.88}$$

where $p + q = 1$, $p > 0$, $q > 0$, and $M_n^* = \max\{Z_1, \ldots, Z_n\}$. If

(a) $\sigma_1 > \sigma_2$, $m_1, m_2 \in \mathbb{R}$ *or*
(b) $\sigma_1 = \sigma_2$ *and* $m_1 > m_2$

then for every $x \in \mathbb{R}$, equality

$$\lim_{n\to\infty} P\{M_n^* \leqslant a_n^* x + b_n^*\} = e^{-e^{-x}} \tag{2.3.89}$$

holds with the constants a_n^* and b_n^* given by

$$a_n^* = \frac{\sigma_1}{\sqrt{2\ln n}}, \tag{2.3.90}$$

$$b_n^* = m_1 + \sigma_1\sqrt{2\ln n} - \frac{\sigma_1}{2\sqrt{2\ln n}}\left(\ln\ln n + \ln\frac{4\pi}{p^2}\right). \tag{2.3.91}$$

Remark 2.3.22 It follows from Theorem 2.3.21 that in the case of a mixture of two normal distributions maximum M_n^* asymptotically comes from one of them. This is F_1 under condition (a) or (b) of this theorem.

Mixture of Cauchy Distributions Let $(X_n)_{n\geqslant 1}$ be a sequence of i.i.d. random variables with the common Cauchy $\mathcal{C}(1,0)$ distribution that is given by the distribution function

$$F(x) = \frac{1}{2} + \frac{1}{\pi}\operatorname{arctg} x,$$

and $M_n = \max\{X_1, \ldots, X_n\}$. It was proved in Example 2.3.5 that $F \in D(G_{1,1})$ and that for every $x > 0$,

$$\lim_{n\to\infty} P\left\{M_n \leqslant x \cdot \operatorname{ctg}\frac{\pi}{n}\right\} = e^{-x^{-1}}.$$

Let us first note the following simple fact. Let $(Y_n)_{n\geqslant 1}$ be a sequence of i.i.d. random variables with a common Cauchy $\mathcal{C}(\lambda,0)$ distribution, where $\lambda > 0$ is a scale parameter. The corresponding distribution function is $F_Y(x) = \frac{1}{2} + \frac{1}{\pi}\operatorname{arctg}\frac{x}{\lambda}$. Let us denote $\widetilde{M}_n = \max\{Y_1, \ldots, Y_n\}$ and $Y_n/\lambda = X_n$. Then, $\widetilde{M}_n = \lambda M_n$, and random variable Y_n/λ has $\mathcal{C}(1,0)$ distribution. Since the inequality $M_n \leqslant x \cdot \operatorname{ctg}\frac{\pi}{n}$ is equivalent to $\frac{1}{\lambda}\widetilde{M}_n \leqslant x \operatorname{ctg}\frac{\pi}{n}$, it follows that the norming constants for $\widetilde{M}_n$ are given by $\widetilde{a}_n = \lambda \operatorname{ctg}\frac{\pi}{n}$ and $\widetilde{b}_n = 0$, and for every $x > 0$,

$$\lim_{n\to\infty} P\left\{\widetilde{M}_n \leqslant \lambda x \cdot \operatorname{ctg}\frac{\pi}{n}\right\} = e^{-x^{-1}}.$$

Theorem 2.3.23 (Mladenović [156]) *Let $\mathcal{C}(\lambda_i,0)$ be a class of random variables with distribution function $F_i(x) = \frac{1}{2} + \frac{1}{\pi}\operatorname{arctg}\frac{x}{\lambda_i}$, $i \in \{1,2\}$. Let $(Z_n)_{n\geqslant 1}$ be a sequence of i.i.d. random variables such that for every $n \in \mathbb{N}$,*

$$Z_n \in \begin{cases} \mathcal{C}(\lambda_1,0), \text{ with probability } p, \\ \mathcal{C}(\lambda_2,0), \text{ with probability } q, \end{cases} \tag{2.3.92}$$

where $p+q=1$, $p>0$, $q>0$, and λ_1 and λ_2 are different scale parameters. Let us denote $M_n^* = \max\{Z_1, \ldots, Z_n\}$. Then, for every $x > 0$,

$$\lim_{n\to\infty} P\left\{\frac{\pi}{n(p\lambda_1+q\lambda_2)}M_n^* \leqslant x\right\} = e^{-x^{-1}}. \tag{2.3.93}$$

Remark 2.3.24 For Cauchy random variables with different scale parameters, M_n^* will come from either of the two parent distributions, which explains why the extremal distribution is also a mixture. More specifically, for pure Cauchy variables, $\pi M_n/n\lambda \to U$ in distribution, with $F_U(x) = G_{1,1}(x) = e^{-x^{-1}}$. In a sample of size n from the mixture of two Cauchy distributions, one observes approximately np and nq of the two types, and hence $M_n^* \approx \max\{M_{np}^{(1)}, M_{np}^{(2)})\}$ in distribution. Since F_U is a max-stable distribution, this explains the result.

2.4 Power Normalization

In Sects. 2.2–2.3 we considered limit theorems under the linear normalization of maxima in i.i.d. sequences. Other types of normalization have also been considered in the literature. Pancheva [176] and Mohan and Ravi [167] used the power normalization of maxima.

Definition 2.4.1 A distribution function F belongs to the domain of attraction of a distribution function H under power normalization (notation: $F \in D_p(H)$) if there exist constants $\alpha_n > 0$ and $\beta_n > 0$, $n \in \mathbb{N}$, such that

$$F^n(\alpha_n|x|^{\beta_n}\operatorname{sign}(x)) \Longrightarrow H(x), \quad \text{as } n \to \infty. \tag{2.4.1}$$

If $(X_n)_{n\geqslant 1}$ is a sequence of i.i.d. random variables with common distribution function F and $M_n = \max\{X_1, \dots, X_n\}$, then equality (2.4.1) is equivalent to the following asymptotic relation:

$$(|M_n|/\alpha_n)^{1/\beta_n} \cdot \operatorname{sign} M_n \xrightarrow{D} Z, \quad \text{as } n \to \infty,$$

where Z is a random variable with distribution function H. The following theorem was proved in Pancheva [176] and Mohan and Ravi [167].

Theorem 2.4.2 *The set of all nondegenerate distribution functions with a nonempty domain of attraction under the power normalization of maxima consists of the following functions:*

$$H_{1,\gamma} = \begin{cases} 0, & \text{if } x < 1, \\ \exp\{-(\ln x)^{-\gamma}\}, & \text{if } x \geqslant 1, \end{cases}$$

$$H_{2,\gamma} = \begin{cases} 0, & \text{if } x < 0, \\ \exp\{-(-\ln x)^{\gamma}\}, & \text{if } 0 \leqslant x < 1, \\ 1, & \text{if } x \geqslant 1, \end{cases}$$

$$H_3 = \begin{cases} 0, & \text{if } x < 0, \\ e^{-1/x}, & \text{if } x \geqslant 0, \end{cases}$$

$$H_{4,\gamma} = \begin{cases} 0, & \text{if } x < -1, \\ \exp\{-(-\ln(-x))^{-\gamma}\}, & \text{if } -1 \leqslant x < 0, \\ 1, & \text{if } x \geqslant 0, \end{cases}$$

$$H_{5,\gamma} = \begin{cases} \exp\{-(\ln(-x))^{\gamma}\}, & \text{if } x < -1, \\ 1, & \text{if } x \geqslant -1, \end{cases}$$

$$H_6 = \begin{cases} e^x, & \text{if } x < 0, \\ 1, & \text{if } x \geqslant 0, \end{cases}$$

where $\gamma > 0$, and all functions of the form $\widetilde{H}(x) = H(\alpha|x|^{\beta} \cdot \operatorname{sign} x)$, where H is a function from the above list, $\alpha > 0$ and $\beta > 0$.

2.5 Max-semistable Distributions

It was already mentioned in Sect. 2.3.4 that many integer-valued distributions do not belong to the domain of attraction of any extreme value distribution. The asymptotic behavior of the maximum in i.i.d. sequences with a common distribution function of this type cannot be described by the classical asymptotic results of extreme value theory. Anderson [10] considered a class of discrete distributions with infinite right endpoint that do not belong to the maximum domains of attraction and proved the following result that may be useful.

Theorem 2.5.1 (Anderson [10]) *Let $\mathcal{D}$ be a class of distribution functions whose support consists of all sufficiently large positive integers. If $F \in \mathcal{D}$, then the following two inequalities*

$$\limsup_{n\to\infty} F^n(x + b_n) \leqslant e^{-e^{-\alpha x}}, \tag{2.5.1}$$

$$\liminf_{n\to\infty} F^n(x + b_n) \geqslant e^{-e^{-\alpha(x-1)}} \tag{2.5.2}$$

hold for some $\alpha > 0$, for all $x \in \mathbb{R}$, and for some sequence of constants $(b_n)_{n\geqslant 1}$, if and only if

$$\lim_{n\to\infty} \frac{1 - F(n-1)}{1 - F(n)} = e^{\alpha}. \tag{2.5.3}$$

It can be proved (*see* Hall and Temido [98] and the references therein) that (2.5.3) holds for some $\alpha > 0$ if and only if there exists a nondecreasing sequence of positive integers $(k_n)_{n\geqslant 1}$ satisfying

$$\lim_{n\to\infty} \frac{k_{n+1}}{k_n} = r \in (1, +\infty), \tag{2.5.4}$$

such that

$$\lim_{n\to\infty} k_n\{1 - F(u_n)\} = \tau > 0, \tag{2.5.5}$$

for some sequence of real numbers $(u_n)_{n\geqslant 1}$, and $\tau > 0$.

Definition 2.5.2 (Pancheva [177]) A distribution function G is *max-semi-stable* if there exist real numbers $r > 1$, $a = a(r) > 0$, and $b = b(r)$ such that

$$G(x) = G^r(ax + b) \quad \text{for every } x \in \mathbb{R}, \tag{2.5.6}$$

or equivalently, if there exists a sequence of i.i.d. random variables with common distribution function F and there exist sequences $a_n > 0$, $b_n \in \mathbb{R}$ and $n \in \mathbb{N}$, such that

$$\lim_{n\to\infty} F^{k_n}(a_n x + b_n) = G(x), \tag{2.5.7}$$

for every continuity point of G, and where sequence $(k_n)_{n\geqslant 1}$ satisfies (2.5.4). In this case we say that distribution function F belongs to the *max-semistable domain of attraction* of G.

The difference between max-stable and max-semistable distributions is that, under certain conditions, for the max-stable distributions, relation (2.5.6) holds for every $r > 1$, while for max-semistable distributions this relation may hold only for some real numbers $r > 1$. Additionally, if a distribution function G is max-semistable, but not max-stable, then equality (2.5.6) holds only for some real $r > 1$ and all its natural powers. Max-semistable laws in extremes of stationary random sequences were studied in Canto e Castro et al. [37] and Temido and Canto e Castro [219].

2.6 Peaks Over Thresholds

2.6.1 Excesses Over a High Threshold

Excesses (peaks) over a high threshold have a crucial role in extreme value theory. Hence, it is important to build a mathematical model for such excesses when the threshold tends to the right endpoint of the support of the distribution considered.

Definition 2.6.1 Let X be a random variable with distribution function F, $x_0 = x_0(F) = \inf\{x : F(x) > 0\}$, $x_F = \sup\{t : F(t) < 1\}$, and let $u \in (x_0, x_F)$ be a given threshold. Function $F^{(u)}(x)$ that is given by

$$F^{(u)}(x) = P\{X - u \leqslant x | X > u\} = \frac{F(x+u) - F(u)}{1 - F(u)}, \quad x \geqslant 0, \tag{2.6.1}$$

is called the *conditional distribution function of excess* $X - u$ *under the condition that* $X > u$.

Definition 2.6.2 The conditional mathematical expectation of random variable $X - u$ under the condition that $X > u$, i.e.,

$$e_F(u) := E(X - u | X > u) = \int_{x_0}^{x_F} x \, dF^{(u)}(x), \quad x_0 < u < x_F, \tag{2.6.2}$$

is called the *mean excess function* for the given threshold.

Example 2.6.3

(a) For *exponential distribution* function $F(x) = 1 - e^{-\lambda x}$, $x \geqslant 0$, where $\lambda > 0$ is a parameter, we obtain that

$$F^{(u)}(x) = 1 - e^{-\lambda x} \text{ за } x \geqslant 0; \quad e_F(u) = 1/\lambda = const.$$

Note that the conditional distribution function of excess $X - u$ under the assumption that $X > u$ coincides in this case with the exponential distribution function.

(b) For the Pareto distribution function

$$F(x) = 1 - x_0^{\alpha}(x_0 + x)^{-\alpha}, \quad x \geqslant 0,$$

where $x_0 > 0$ and $\alpha > 0$ are the parameters, we obtain that

$$F^{(u)}(x) = 1 - \left(\frac{x_0 + u}{x_0 + u + x}\right)^{\alpha}, \quad x \geqslant 0; \quad e_F(u) = \frac{x_0 + u}{\alpha - 1}, \quad u > 0.$$

(c) For the *Weibull distribution* function $F(x) = 1 - e^{-cx^{\tau}}$, $x \geqslant 0$, where $c > 0$ and $\tau > 0$ are the parameters, we obtain that

$$F^{(u)}(x) = 1 - e^{-c(u^{\tau} - (u+x)^{\tau})}; \quad e_F(u) \sim \frac{u^{1-\tau}}{c\tau}, \quad u \to \infty.$$

For the exponential distribution the mean excess function is a constant, and for the Weibull distribution with parameter $\tau \geqslant 1$, it tends to a constant as $u \to \infty$. These two distributions belong to the class of light-tailed distributions. In the cases of the Pareto distribution and the Weibull distribution with parameter $\tau \in (0, 1)$, the mean excess functions tend to infinity as $u \to \infty$. These two distributions belong to the class of heavy-tailed distributions. For more theory on these classes of distributions and possible applications in actuarial and financial mathematics, *see* Embrechts et al. [58]. △

2.6.2 Generalized Pareto Distributions

In this subsection we consider the class of generalized Pareto distributions that are appropriate for modeling excesses over a high threshold.

Definition 2.6.4 Standard generalized Pareto (GP) distributions in α-parametrization are determined by the following distribution functions:

- The exponential distribution:

$$W_0(x) = \begin{cases} 0, & \text{if } x < 0, \\ 1 - e^{-x}, & \text{if } x \geqslant 0. \end{cases} \tag{2.6.3}$$

- The Pareto distribution with parameter $\alpha > 0$:

$$W_{1,\alpha}(x) = \begin{cases} 0, & \text{if } x < 1, \\ 1 - x^{-\alpha}, & \text{if } x \geqslant 1. \end{cases} \tag{2.6.4}$$

- The beta distribution with parameter $\alpha > 0$:

$$W_{2,\alpha}(x) = \begin{cases} 0, & \text{if } x < -1, \\ 1 - (-x)^{\alpha}, & \text{if } -1 \leqslant x \leqslant 0, \\ 1, & \text{if } x \geqslant 0. \end{cases} \tag{2.6.5}$$

There is a very simple analytic relation between a GP distribution function $W(x)$ and an EV distribution function $G(x)$:

$$W(x) = 1 + \ln G(x) \quad \text{if} \quad \ln G(x) > -1. \tag{2.6.6}$$

Note also that for $\alpha = 1$, $W_{2,1}$ corresponds to the uniform distribution on the interval $[-1, 0]$ and that the class of distributions determined by (2.6.5) is a subclass of the two-parameter class of beta distributions usually considered in the statistical literature.

The densities of the standard GP distributions are given as follows:

- The exponential distribution:

$$w_0(x) = e^{-x}, \quad x \geqslant 0. \tag{2.6.7}$$

- The Pareto distribution with parameter $\alpha > 0$:

$$w_{1,\alpha}(x) = \alpha x^{-(\alpha+1)}, \quad x \geqslant 1. \tag{2.6.8}$$

- The beta distribution with parameter $\alpha > 0$:

$$w_{2,\alpha}(x) = \alpha(-x)^{(\alpha-1)}, \quad -1 \leqslant x \leqslant 0. \tag{2.6.9}$$

Location and Scale Parameters If $W_{i,\alpha}(x)$, where $i \in \{1, 2\}$, is the distribution function of random variable X, then the distribution function of random variable $\sigma X + \mu$, where $\sigma > 0$ and $\mu \in \mathbb{R}$, is given by

$$W_{i,\alpha,\mu,\sigma}(x) = W_{i,\alpha}\left(\frac{x-\mu}{\sigma}\right).$$

Here, μ is the location parameter, and σ is the scale parameter. The support of the distribution determined by $W_{1,\alpha,\mu,\sigma}$ is the interval $[\mu + \sigma, +\infty)$, and the support of $W_{2,\alpha,\mu,\sigma}$ is the interval $[\mu - \sigma, \mu]$. Analogously, the exponential distribution function $W_{0,\mu,\sigma}$ with location and scale parameters μ and $\sigma > 0$ is defined on the interval $[\mu, +\infty)$.

γ-parametrization of GP distributions follows from α-parametrization after the substitution $\gamma = 1/\alpha$ for the Pareto distribution and $\gamma = -1/\alpha$ for the beta distribution. The obtained analytic expressions for the distribution functions are the following:

$$W_\gamma(x) = \begin{cases} 1 - e^{-x}, & x \geqslant 0 \ (\gamma = 0), \\ 1 - (1 + \gamma x)^{-1/\gamma}, & x \geqslant 0 \ (\gamma > 0), \\ 1 - (1 + \gamma x)^{-1/\gamma}, & x \in [0, -1/\gamma] \ (\gamma < 0). \end{cases} \tag{2.6.10}$$

The values $\gamma = 0$, $\gamma > 0$, and $\gamma < 0$ are related, respectively, to the exponential, Pareto, and beta distributions, and the following equalities also hold:

$$W_\gamma(x) = \begin{cases} W_{1,\alpha,-\alpha,\alpha}(x), & \text{if } \gamma = \frac{1}{\alpha} > 0, \\ W_{2,\alpha,\alpha,\alpha}(x), & \text{if } \gamma = -\frac{1}{\alpha} < 0. \end{cases} \tag{2.6.11}$$

$$\lim_{\gamma \to 0} W_\gamma(x) = W_0(x). \tag{2.6.12}$$

Moments of GP Distributions The moments of random variable X with distribution function W_γ are given by the following equalities:

$$E(1 + \gamma X)^r = \frac{1}{1 - \gamma r}, \quad \text{if } 1 - \gamma r > 0. \tag{2.6.13}$$

$$E(X^r) = \frac{r!}{\gamma^{r+1}} \frac{\Gamma\left(\frac{1}{\gamma} - r\right)}{\Gamma\left(\frac{1}{\gamma} + 1\right)}, \quad \text{if } \gamma < \frac{1}{r}. \tag{2.6.14}$$

It follows from (2.6.14) that the mathematical expectation and variance of random variable X are given by

$$E(X) = \frac{1}{1-\gamma}, \quad \text{if } \gamma < 1, \tag{2.6.15}$$

$$\text{Var}(X) = \frac{1}{(1-\gamma)^2(1-2\gamma)}, \quad \text{if } \gamma < \frac{1}{2}. \tag{2.6.16}$$

2.6.3 POT Stability

Let X be a random variable with distribution function F, $x_0 = x_0(F) = \inf\{x : F(x) > 0\}$, $x_F = \sup\{t : F(t) < 1\}$, and let $u \in (x_0, x_F)$ be a given threshold. Let us define

$$F^{[u]}(x) = \frac{F(x) - F(u)}{1 - F(u)}, \quad x \geqslant u. \tag{2.6.17}$$

Then $F^{[u]}$ is the distribution function of the excess of X over threshold u. The analytic relation between functions $F^{(u)}$ and $F^{[u]}$ that are given by (2.6.1) and (2.6.17), respectively, is the following:

$$F^{(u)}(x) = F^{[u]}(x+u) = \frac{F(x+u) - F(u)}{1 - F(u)}, \quad x \geqslant 0.$$

Definition 2.6.5 Distribution function F has the property of *pot-stability* (peaks over threshold stability) if there exist constants $a_u > 0$ and b_u, such that

$$F^{(u)}(a_u x + b_u) = F(x). \tag{2.6.18}$$

Example 2.6.6 GP distributions are pot-stable. Indeed, using the γ-parametrization we obtain the following equalities for these distributions:

$$W_0^{(u)}(x) = W_0(x); \quad W_\gamma^{(u)}((1+\gamma u)x) = W_\gamma(x), \quad \text{for } \gamma \neq 0. \tag{2.6.19}$$

Moreover, it can be proved that GP distributions are the only distributions that have this property. △

2.6.4 Domains of Attraction for Excesses and Limit Distributions

Let F be a distribution function and $x_F = \sup\{x : F(x) < 1\}$, and let us consider the following question: What are the possible limit distributions of the excess of X over a threshold u as $u \to x_F$? Before providing results related to this question that are obtained in Balkema and de Haan [18], we give the following definition.

Definition 2.6.7 Regarding excesses of a threshold, distribution function F belongs to the domain of attraction of some distribution function W if there exist constants $a_u > 0$ and b_u, such that

$$\lim_{u \to x_F} F^{(u)}(a_u x + b_u) = W(x) \quad \text{for every } x \in C(W). \tag{2.6.20}$$

Let $D_p(W)$ be the excess domain of attraction of distribution function W, D_0 be the set of distribution functions F that satisfy the condition $x_F = \sup\{x : F(x) < 1\} = +\infty$, and $DA(G)$ be the maximum domain of attraction of distribution function G.

Theorem 2.6.8 (Balkema and de Haan [18]) *Let F be a distribution function. If for every $x \in \mathbb{R}$,*

$$\lim_{u \to x_F} F^{(u)}(a_u x + b_u) = W(x), \tag{2.6.21}$$

where W is a continuous distribution function, $a_u > 0$ and $b_u \in \mathbb{R}$, then distribution function W is of the same type as some of the GP distributions W_γ, $\gamma \in \mathbb{R}$.

If we do not impose the condition that W is a continuous distribution function, then certain discrete distribution functions can also appear as limit distributions for excesses over high thresholds. More precisely, the following theorem holds.

Theorem 2.6.9 (Balkema and de Haan [18]) *Let F be a distribution function such that $x_F = \sup\{x : F(x) < 1\} = +\infty$. If there exist norming constants $a_u > 0$ and $b_u \in \mathbb{R}$, such that for every $x \in C(W)$,*

$$\lim_{u \to x_F} F^{(u)}(a_u x + b_u) = W(x), \tag{2.6.22}$$

where W is a nondegenerate distribution function, then W is of the same type as one of the following distribution functions:

$$W_0(x) = 1 - e^{-x}, \quad x \geqslant 0,$$
$$W_{1,\alpha}(x) = 1 - x^{-\alpha}, \quad x \geqslant 1,$$
$$T_\gamma(x) = 1 - e^{-\gamma[1+x]}, \quad x \geqslant 0,$$

$$Q_{\gamma,\alpha}(x) = 1 - e^{-\gamma[1+\alpha \ln(1+x)]}, \quad x \geqslant 0,$$

where $\alpha > 0$, $\gamma > 0$, *and* $[x]$ *is the integer part of* x.

Theorem 2.6.10 (Balkema and de Haan [18]) *The following equalities hold:* $D_p(W_0) = DA(G_0) \cap D_0$ *and* $D_p(W_{1,\alpha}) = DA(G_{1,\alpha})$, *for every* $\alpha > 0$.

The proof of Theorems 2.6.8–2.6.10 can be found in Balkema and de Haan [18]. For more theory about GP distributions, *see* Embrechts et al. [58] and Reiss and Thomas [190]. For parameter estimation of GP distributions, *see* Zea Bermudez and Kotz [226].

2.7 Stationary Sequences

In this section we consider extreme values in stationary sequences of random variables. Let us first recall the definition of strict stationarity and give some examples of strictly stationary sequences.

Definition 2.7.1 A sequence $(X_n)_{n\geqslant 1}$ of random variables is *strictly stationary* if for all $k, m \in \mathbb{N}$ and for all indices $1 \leqslant i_1 < i_2 < \cdots < j_k$ the random vectors $(X_{i_1}, X_{i_2} \ldots, X_{i_k})$ and $(X_{i_1+m}, X_{i_2+m} \ldots, X_{i_k+m})$ have the same probability distribution.

As an immediate consequence of Definition 2.7.1, it follows that all terms of strictly stationary sequence $(X_n)_{n\geqslant 1}$ have a common distribution function. In consideration of different questions related to stationary sequences, including those related to extreme values, one should take into account the dependence structure between random variables. The simplest question is how the dependence between X_i and X_j changes when the difference $|i - j|$ increases.

Various measures of dependence between terms in a family of dependent random variables have been introduced. We shall restrict our consideration to stationary sequences.

Example 2.7.2 Let $\xi, \eta \in \mathcal{N}(0, \sigma^2)$ be independent random variables, $\lambda \in \mathbb{R}$, and $(X_n)_{n\geqslant 1}$ a sequence of random variables defined by

$$X_n = \xi \cos(\lambda n) + \eta \sin(\lambda n). \tag{2.7.1}$$

Random variable X_n given by (2.7.1) is normally distributed as a linear combination of normal random variables. It is easy to check that for every positive integer r and every nonnegative integer s, $E(X_r) = 0$ and

$$E(X_r X_{r+s}) = \sigma^2 \cos(\lambda s). \tag{2.7.2}$$

It follows from (2.7.2) that for all n and k the covariance matrices of the random vectors $(X_1, X_2, \ldots, X_n)$ and $(X_{1+k}, X_{2+k}, \ldots, X_{n+k})$ coincide. It follows that these random vectors have the same multivariate normal distribution, and hence (X_n) is strictly stationary. △

Example 2.7.3 (Chernick [41]) Let $r \geqslant 2$ be a positive integer and $(X_n)_{n\geqslant 0}$ be a sequence of random variables defined as follows: X_0 is a random variable with uniform distribution over interval [0,1], and

$$X_n = \frac{1}{r}X_{n-1} + \varepsilon_n, \quad \text{for } n \geqslant 1,$$

where $(\varepsilon_n)_{n\geqslant 1}$ is a sequence of independent random variables with uniform distribution over the set $\{0, 1/r, 2/r, \ldots (r-1)/r\}$, such that ε_n does not depend on X_{n-1}. Then $(X_n)_{n\geqslant 0}$ is a strictly stationary random sequence whose terms are all uniformly distributed over the unit interval [0,1]. We shall refer to it as the *positively correlated uniform autoregressive order* 1 (notation: AR(1)) stationary sequence.△

Example 2.7.4 (Chernick and Davis [42]) Let $r \geqslant 2$ be a positive integer and $(X_n)_{n\geqslant 0}$ be a sequence of random variables defined as follows: X_0 is a random variable with uniform distribution over interval [0,1], and

$$X_n = -\frac{1}{r}X_{n-1} + \varepsilon_n, \quad \text{for } n \geqslant 1,$$

where $(\varepsilon_n)_{n\geqslant 1}$ is a sequence of independent random variables with uniform distribution over the set $\{1/r, 2/r, \ldots (r-1)/r, 1\}$, such that ε_n does not depend on X_{n-1}. Then $(X_n)_{n\geqslant 0}$ is a strictly stationary random sequence whose terms are all uniformly distributed over the unit interval [0,1]. We shall refer to it as the *negatively correlated uniform* AR(1) *stationary sequence.* △

2.7.1 Conditions of Weak Dependence

Let $(X_n)_{n\geqslant 1}$ be a stationary random sequence, and let $\mathfrak{M}_a^b$ be the σ-algebra generated by the family of random variables $\{X_j \mid a \leqslant j \leqslant b\}$.

Definition 2.7.5 (Rosenblatt [197]) A *strong mixing condition* is satisfied for stationary sequence $(X_n)_{n\geqslant 1}$ if there exists a sequence of positive real numbers α_k, $k \in \mathbb{N}$, such that

$$\sup_{A \in \mathfrak{M}_1^n,\, B \in \mathfrak{M}_{n+k+1}^\infty,\, n, k \in \mathbb{N}} |P(AB) - P(A)P(B)| \leqslant \alpha_k, \tag{2.7.3}$$

and $\alpha_k \to 0$ as $k \to \infty$. The α_k is called the *strong mixing coefficient*.

Definition 2.7.6 (Ibragimov and Linnik [233]) A *uniform strong mixing condition* is satisfied for stationary sequence $(X_n)_{n\geqslant 1}$ if there exists a sequence $\beta_k, k \in \mathbb{N}$, such that

$$\sup_{A \in \mathfrak{M}_1^n,\, B \in \mathfrak{M}_{n+k+1}^{\infty},\, P(A)>0,\, n,k\in\mathbb{N}} \frac{|P(AB) - P(A)P(B)|}{P(A)} \leqslant \beta_k, \tag{2.7.4}$$

and $\beta_k \to 0$ as $k \to \infty$. The β_k is called the *uniform strong mixing coefficient.*

Strong mixing and uniform strong mixing conditions were used firstly in Rosenblatt [197] and Ibragimov and Linnik [233] in proving the central limit theorem for weak-dependent stationary sequences. Several other mixing conditions have been defined. A survey and relations between these conditions can be found in Bradley [31]. Usually, in investigations of extreme values in stationary sequences, it is not necessary to consider all events as stated in Definitions 2.7.5 and 2.7.6. Events of the form $\{X_n \leqslant u\}$ and $\{X_n > u\}$ are crucial there.

Definition 2.7.7 (Leadbetter [141]) Let $(X_n)_{n\geqslant 1}$ be a stationary random sequence and $(u_n)_{n\geqslant 1}$ be a sequence of real numbers. The condition $D(u_n)$ is satisfied if for all $n, k, l \in \mathbb{N}_n$, and positive integers $j_1, j_2, \ldots, j_{k+l}$ such that

$$1 \leqslant j_1 < \cdots < j_k < j_{k+1} < \cdots < j_{k+l} \leqslant n,$$

and $j_{k+1} - j_k \geqslant l_n$, the following inequality holds:

$$\left| P\left(\bigcap_{i=1}^{k+l}\{X_{j_i} \leqslant u_n\}\right) - P\left(\bigcap_{i=1}^{k}\{X_{j_i} \leqslant u_n\}\right) P\left(\bigcap_{i=k+1}^{k+l}\{X_{j_i} \leqslant u_n\}\right)\right| \leqslant \alpha_{n,l_n},$$

and $\alpha_{n,l_n} \to 0$ as $n \to \infty$ for some sequence $l_n = o(n)$.

Usually, threshold u_n is considered a high level that depends on some real number x. For example, $u_n = u_n(x) = a_n x + b_n$, where $a_n > 0$ and $b_n \in \mathbb{R}$ are the norming constants for maximum M_n in i.i.d. settings. For some purposes, for example, when we consider the maxima of random variables X_n, $n \in \mathbb{N}$, on two or more disjoint sets of indices, it is necessary to introduce a modification of condition $D(u_n)$ that includes more than one threshold. A condition of this type that depends on two thresholds, called $D(u_n, v_n)$, will be used in Chap. 3.

Definition 2.7.8 (Loynes [145]) Condition $D'(u_n)$ is satisfied for a stationary random sequence $(X_n)_{n\geqslant 1}$ and a sequence of real numbers $(u_n)_{n\geqslant 1}$ if the following equality holds:

$$\lim_{k\to\infty} \limsup_{n\to\infty} n \cdot \sum_{j=2}^{[n/k]} P\{X_1 > u_n, X_j > u_n\} = 0.$$

Condition $D'(u_n)$ is very strong and prevents the clustering of extremes in stationary sequences. In Chap. 3 we shall see very different asymptotic behavior of partial maxima depending on whether this condition is satisfied or not.

2.7.2 Extremal Type Theorem for Stationary Sequences

Theorem 2.7.9 (Leadbetter [141]) *Let $(X_n)_{n\geqslant 1}$ be a strictly stationary random sequence, $a_n > 0$ and $b_n \in \mathbb{R}$, where $n \in \mathbb{N}$, and let $G(x)$ be a nondegenerate distribution function, such that for every $x \in C(G)$ the following equality holds:*

$$\lim_{n\to\infty} P\{M_n \leqslant a_n x + b_n\} = G(x).$$

If for every $x \in \mathbb{R}$ and $u_n = a_n x + b_n$ condition $D(u_n)$ is satisfied, then $G(x)$ is an extreme value distribution function.

Definition 2.7.10 Let $(X_n)_{n\geqslant 1}$ be a strictly stationary random sequence, and $F(x) = P\{X_1 \leqslant x\}$. The i.i.d. sequence $(X_n^*)_{n\geqslant 1}$ with common distribution function $F(x)$ is called an *associated sequence of independent random variables.*

The following theorem is analogous to Theorem 2.1.4 that is related to maxima in the sequences of i.i.d. random variables.

Theorem 2.7.11 *Let $(X_n)_{n\geqslant 1}$ be a strictly stationary random sequence with marginal distribution function F, and let $(u_n)_{n\geqslant 1}$ be a sequence of real numbers such that conditions $D(u_n)$ and $D'(u_n)$ are satisfied, and $0 \leqslant \tau < +\infty$. Then equality*

$$\lim_{n\to\infty} P\{M_n \leqslant u_n\} = e^{-\tau}$$

holds if and only if

$$\lim_{n\to\infty} n(1 - F(u_n)) = \tau.$$

The next theorem is related to the case when $P\{M_n \leqslant u_n\} \to 0$ and $n(1 - F(u_n)) \to \infty$.

Theorem 2.7.12 *Let $(X_n)_{n\geqslant 1}$ be a strictly stationary random sequence, $F(x) = P\{X_1 \leqslant x\}$, and let $(u_n)_{n\geqslant 1}$ be a sequence of real numbers.*

Let us suppose that for every $\tau \in [\tau_0, +\infty)$, there exists a sequence of real numbers $(v_n)_{n\geqslant 1}$ such that $\lim_{n\to\infty} n(1 - F(v_n)) = \tau$ and that conditions $D(v_n)$ and $D'(v_n)$ are satisfied. Then, for the sequence $(u_n)_{n\geqslant 1}$, equality

$$\lim_{n\to\infty} P\{M_n \leqslant u_n\} = 0$$

holds if and only if

$$\lim_{n\to\infty} n(1 - F(u_n)) = +\infty.$$

Theorem 2.7.13 (Leadbetter [141]) *Let $(X_n)_{n\geqslant 1}$ be a strictly stationary random sequence, $a_n > 0$, $b_n \in \mathbb{R}$, and $u_n = a_n x + b_n$ be sequences of constants, and $M_n = \max\{X_1, \ldots, X_n\}$. Let us assume that for every real number x, conditions $D(u_n)$ and $D'(u_n)$ are satisfied. Let (X_n^*) be the associated sequence of independent random variables, and $M_n^* = \max\{X_1^*, \ldots, X_n^*\}$. Then equality*

$$\lim_{n\to\infty} P\{M_n \leqslant a_n x + b_n\} = G(x)$$

holds for some nondegenerate distribution function G and every real number $x \in C(G)$, if and only if

$$\lim_{n\to\infty} P\{M_n^* \leqslant a_n x + b_n\} = G(x).$$

2.7.3 Stationary Normal Sequences

It was proved in Example 2.3.17 that the normal distribution belongs to the domain of attraction of the Gumbel extreme value distribution, and the corresponding limit theorem was obtained. Berman [25] proved that the analogous result holds for stationary normal sequences under some weak conditions imposed on the covariance sequence. These conditions are related to the rate of convergence to zero of the covariance sequence. Moreover, the norming constants are the same as in the i.i.d. case. More precisely, the following theorem holds.

Theorem 2.7.14 (Berman [25]) *Let $(X_n)_{n\geqslant 1}$ be a stationary normal random sequence such that $EX_1 = 0$, $\operatorname{Var} X_1 = 1$, and let $r_n = E(X_k X_{k+n})$, $n \in \mathbb{N}$, be the covariance sequence. If one of the conditions*

$$\text{(a)} \quad \lim_{n\to\infty} r_n \ln n = 0 \quad \text{or} \quad \text{(b)} \quad \sum_{n=1}^{\infty} r_n^2 < +\infty$$

holds true, then for every $x \in \mathbb{R}$ the following equality holds:

$$\lim_{n\to\infty} P\{M_n \leqslant a_n x + b_n\} = \exp\{-e^{-x}\},$$

where a_n and b_n are constants given by

$$a_n = \frac{1}{\sqrt{2\ln n}}, \quad b_n = \sqrt{2\ln n} - \frac{\ln\ln n + \ln 4\pi}{2\sqrt{2\ln n}}.$$

Berman's proof of the above theorem is based on the fact that probability $P\{M_n \leqslant u_n\}$ can be approximated by $\{\Phi(u_n)\}^n$, where Φ is the standard normal distribution function. More precisely, the proof is based on the following *normal comparison lemma*.

Lemma 2.7.15 (Berman [25]) *Let $(X_n)_{n\geqslant 1}$ be a stationary normal random sequence such that $EX_1 = 0$, $\operatorname{Var} X_1 = 1$, and let $l_1, \ldots, l_k$ be positive integers such that $1 \leqslant l_1 \leqslant \cdots \leqslant l_k$. Then there exists a positive constant C, such that for every $x \in \mathbb{R}$ the following inequality holds:*

$$\left| P\left(\bigcap_{j=1}^{k} \{X_{l_j} \leqslant u\}\right) - \Phi^k(u) \right| \leqslant C \sum_{1\leqslant i<j\leqslant k} |\rho_{ij}| \exp\left\{-\frac{u^2}{1+|\rho_{ij}|}\right\},$$

where $\rho_{ij} = r_{l_i - l_j} = E(X_{l_i} X_{l_j})$.

The general extreme value theory for stationary sequences that is based on conditions $D(u_n)$ and $D'(u_n)$ is given in Leadbetter [141] and also in the book by Leadbetter et al. [143], where additional references related to this topic are provided. We formulate here only one more result that gives sufficient conditions on the covariance sequence of a stationary normal random sequence under which conditions $D(u_n)$ and $D'(u_n)$ are satisfied.

Theorem 2.7.16 (Leadbetter [141]) *Let $(X_n)_{n\geqslant 1}$ be a stationary normal random sequence such that $EX_1 = 0$, $\operatorname{Var} X_1 = 1$, and $r_n = E(X_k X_{k+n})$. Let us assume that one of the conditions $\lim_{n\to\infty} r_n \ln n = 0$ or $\sum_{n=1}^{\infty} r_n^2 < +\infty$ is satisfied, and let $(u_n)_{n\geqslant 1}$ be a sequence of real numbers such that $1-\Phi(u_n) = \tau/n$. Then conditions $D(u_n)$ and $D'(u_n)$ are satisfied.*

2.7.4 Extremal Index

As before we use the following notation: $(X_n)_{n\geqslant 1}$ is a stationary random sequence; $(X_n^*)_{n\geqslant 1}$ is the associated i.i.d. sequence; F is the common distribution function of the random variables X_n and X_n^* for all $n \geqslant 1$; $M_n = \max\{X_1, \ldots, X_n\}$; and $M_n^* = \max\{X_1^*, \ldots, X_n^*\}$.

If $u_n = u_n(\tau)$ is a sequence of real numbers such that $n(1-F(u_n)) \to \tau$ as $n \to \infty$, where $0 \leqslant \tau \leqslant +\infty$, then from Theorem 2.1.4 we have the following result for maximum M_n^* of the first n terms of the associated i.i.d. sequence: $P\{M_n^* \leqslant u_n(\tau)\} = e^{-\tau}$ as $n \to \infty$.

The related result for maximum M_n of the first n terms in the stationary sequence considered is more complicated. Functions of the form $e^{-\vartheta\tau}$, $\tau > 0$, where $0 \leqslant \vartheta \leqslant 1$, can appear as the limit of the sequence $P\{M_n \leqslant u_n(\tau)\}$. More precisely, the following theorem holds.

Theorem 2.7.17 *Let us suppose that for every $\tau > 0$ a threshold $u_n(\tau)$ is defined such that the following three conditions hold:*

(a) $\lim_{n\to\infty} n\{1 - F(u_n(\tau))\} = \tau$ *for every* $\tau > 0$.
(b) *Condition* $D(u_n(\tau))$ *holds for every* $\tau > 0$.
(c) *There exists* $\lim_{n\to\infty} P\{M_n \leqslant u_n(\tau)\}$ *for at least one real number* $\tau > 0$.
Then, there exists $\vartheta \in [0, 1]$, *such that, for every* $\tau > 0$,

$$\lim_{n\to\infty} P\{M_n \leqslant u_n(\tau)\} = e^{-\vartheta\tau}. \tag{2.7.5}$$

The proof of Theorem 2.7.17 can be found in Loynes [145], Chernick [41], and Leadbetter et al. [143]. This result leads to the following definition.

Definition 2.7.18 A stationary sequence $(X_n)_{n\geqslant 1}$ has *extremal index* $\vartheta \in [0, 1]$, if for every $\tau > 0$ there exists a sequence $u_n(\tau)$, $n \in \mathbb{N}$, such that

$$\lim_{n\to\infty} n\{1 - F(u_n(\tau))\} = \tau, \tag{2.7.6}$$

$$\lim_{n\to\infty} P\{M_n \leqslant u_n(\tau)\} = e^{-\vartheta\tau}. \tag{2.7.7}$$

In Examples 2.7.19–2.7.21 we assume that F is a distribution function such that for every $\tau > 0$ there exists a sequence $u_n(\tau)$, $n \in \mathbb{N}$, satisfying the condition $n\{1 - F(u_n(\tau))\} \to \tau$ as $n \to \infty$.

Example 2.7.19 Let $(X_n)_{n\geqslant 1}$ be an i.i.d. sequence with common distribution function F. Then from Theorem 2.1.4 it follows that $\vartheta = 1$ is the extremal index for $(X_n)_{n\geqslant 1}$. △

Example 2.7.20 Let $(X_n)_{n\geqslant 1}$ be a strictly stationary random sequence with marginal distribution function F. If conditions $D(u_n)$ and $D'(u_n)$ are satisfied for every $\tau > 0$, then from Theorem 2.7.11 it follows that $\vartheta = 1$ is the extremal index for $(X_n)_{n\geqslant 1}$. △

Example 2.7.21 Let X be a random variable with distribution function F, and let us suppose that $X_n = X$ for every $n \in \mathbb{N}$. Then for every $n \in \mathbb{N}$, $M_n = \max\{X_1, \ldots, X_n\} = X$, and

$$P\{M_n \leqslant u_n\} = P\{X \leqslant u_n\} \to 1 \quad \text{as} \quad n \to \infty.$$

Hence, $\vartheta = 0$ is the extremal index for random sequence $(X_n)_{n\geqslant 1}$. △

Example 2.7.22 Let $(\varepsilon_n)_{n\geqslant 1}$ be a sequence of i.i.d. random variables with common distribution function F, and $X_n = \max\{\varepsilon_n, \varepsilon_{n+1}, \ldots, \varepsilon_{n+k-1}\}$, $n \in \mathbb{N}$, where k is a fixed positive integer. Let us denote $M_n = \max\{X_1, \ldots, X_n\}$. Obviously, $(X_n)_{n\geqslant 1}$ is a strictly stationary random sequence. Let F_X be the marginal distribution function of this sequence, and $\widetilde{u}_n(\tau) = u_n(\tau/k)$, for every $\tau > 0$. According to the assumption related to function F, we have $F(u_n(\tau/k)) \sim 1 - \frac{\tau}{kn}$ as $n \to \infty$, and consequently we obtain that

$$F_X(\widetilde{u}_n(\tau)) = F^k\left(u_n\left(\frac{\tau}{k}\right)\right) \sim \left(1 - \frac{\tau}{kn}\right)^k \sim 1 - \frac{\tau}{n}, \quad n \to \infty.$$

Therefore, $n\{1 - F_X(\widetilde{u}_n(\tau))\} \to \tau$ as $n \to \infty$. Moreover, as $n \to \infty$ we have

$$\begin{aligned} P\{M_n \leqslant \widetilde{u}_n(\tau)\} &= P\left\{\max\{\varepsilon_1, \varepsilon_2, \ldots, \varepsilon_{n+k-1}\} \leqslant u_n\left(\frac{\tau}{k}\right)\right\} \\ &= F^{n+k-1}\left(u_n\left(\frac{\tau}{k}\right)\right) \sim \left(1 - \frac{\tau}{kn}\right)^{n+k-1} \sim e^{-\tau/k}. \end{aligned}$$

From these relations it follows that $(X_n)_{n\geqslant 1}$ has extremal index $1/k$. △

Example 2.7.23 Let $(X_n)_{n\geqslant 1}$ be uniform and positively correlated AR(1) random sequences with parameter $r \in \mathbb{N} \setminus \{1\}$ that is defined in Example 2.7.3, and let us denote $M_n = \max\{X_1, \ldots, X_n\}$. Chernick [41] proved that for every $x \geqslant 0$ and $u_n = 1 - \frac{x}{n}$, condition $D(u_n)$ holds, condition $D'(u_n)$ does not hold, and

$$\lim_{n\to\infty} P\left\{M_n \leqslant 1 - \frac{x}{n}\right\} = \exp\left(-\frac{r-1}{r}x\right). \tag{2.7.8}$$

Hence, $(r-1)/r$ is the extremal index for random sequence $(X_n)_{n\geqslant 1}$. △

Example 2.7.24 Let $(X_n)_{n\geqslant 1}$ be uniform and negatively correlated AR(1) random sequences with parameter $r \in \mathbb{N} \setminus \{1\}$ that is defined in Example 2.7.4, and let us denote $M_n = \max\{X_1, \ldots, X_n\}$. Chernick and Davis [42] proved that for every $x \geqslant 0$ and $u_n = 1 - \frac{x}{n}$, condition $D(u_n)$ holds, condition $D'(u_n)$ does not hold, and

$$\lim_{n\to\infty} P\left\{M_n \leqslant 1 - \frac{x}{n}\right\} = \exp\left(-\frac{r^2-1}{r^2}x\right). \tag{2.7.9}$$

Hence, $(r^2-1)/r^2$ is the extremal index for random sequence $(X_n)_{n\geqslant 1}$. △

Let $(X_n)_{n\geqslant 1}$ be a stationary sequence with extremal index ϑ. Note that extremal index ϑ is referred to as *the reciprocal of the expected extremal cluster above high thresholds* for the following reason. Let us consider a sample $X_1, \ldots, X_n$. For a given r_n, split the sample into $k_n = [n/r_n]$ blocks of length r_n, and ignore the last block of length less than r_n. We call a block an *extremal cluster* relative to a threshold u_n, if at least one term in the block exceeds this threshold. We are interested in the case $u_n \uparrow x_F$ as $n \to \infty$, and $r_n \to \infty$, $r_n = o(n)$ as $n \to \infty$. The *expected cluster size of a block* is defined by

$$E\left\{\sum_{i=1}^{r_n} I_{\{X_i > u_n\}} \,\Big|\, M_{r_n} > u_n\right\}.$$

Using the stationarity of sequence $(X_n)_{n\geqslant 1}$, it is easy to prove that this expected value is equal to ϑ_n^{-1} for some $\vartheta_n \in [0, 1]$. Under some regularity conditions the equality $\lim_{n\to\infty} \vartheta_n = \vartheta$ holds. The exact formulation of the statement and more about the extremal index can be found in Davis and Hsing [50] and Buriticá et al. [36]. The *clustering of extreme values* has been studied by several authors. *See*, for example, Ferro and Segers [65], Segers [207, 208], Markovich [147], Weissman and Cohen [224], and the references therein.

2.7.5 Point Processes and Extreme Values

Example 2.7.25 Let us consider a stationary sequence $(X_n)_{n\geqslant 1}$ with marginal distribution function F. In the previous subsections we often considered the tail of distribution $1 - F(u_n(\tau)) = P\{X_j > u_n(\tau)\}$, where $u_n(\tau)$ is a high threshold that depends on n and $\tau > 0$ and is chosen such that $n\{1 - F(u_n(\tau))\} \to \tau$ as $n \to \infty$. Note that event $\{X_j > u_n(\tau)\}$ can be interpreted as the excess of random variable X_j over threshold $u_n(\tau)$, and $n\{1 - F(u_n(\tau))\}$ is the mathematical expectation of the number of the excesses of level $u_n(\tau)$ among the first n terms of sequence $(X_n)_{n\geqslant 1}$. Points j at which the excesses occur are randomly dispersed on the real line and can be considered as a *point process* with $\mathbb{R}$ as the *state space*. Let $M_n = \max\{X_1, \ldots, X_n\}$, and let N_n be the number of excesses of random variables $X_1, \ldots, X_n$ over the threshold $u_n(\tau)$. It is obvious that $\{M_n \leqslant u_n(\tau)\} = \{N_n = 0\}$. This is the simplest case that shows how point processes can be used to calculate probabilities related to maxima in a random sequence. △

We also need to consider point processes with more general state spaces, for example, a subset of the compactified Euclidean space $\mathbb{R}^n$, metric or topological spaces. For the general theory of point processes, *see* Chapter 3 in Resnick [192], Chapter 7 in Mikosch [152], and the books by Kallenberg [120, 121]. We shall provide here only basic notions and notation and results that are necessary for further considerations in Chaps. 3 and 4. For our purposes it is sufficient to consider point processes with state space that is a subset of the compactified Euclidean space $\mathbb{R}^n$, where $n \in \mathbb{N}$.

Let $\mathbb{E}$ be the state space and $\mathcal{B}_{\mathbb{E}}$ be the Borel σ-algebra of the subsets of $\mathbb{E}$, i.e., the σ-algebra generated by the open subsets of $\mathbb{E}$. For $x \in \mathbb{E}$ and $B \in \mathcal{B}_{\mathbb{E}}$, let us denote $\varepsilon_x(B) = 1$ if $x \in B$ and $\varepsilon_x(B) = 0$ if $x \notin B$.

Definition 2.7.26 Let $x_1, x_2, \ldots$ be a sequence of the elements of set $\mathbb{E}$. Function $m : \mathcal{B}_{\mathbb{E}} \to \{0, 1, 2, \ldots\} \cup \{\infty\}$ defined by

$$m(B) = \sum_{i=1}^{\infty} \varepsilon_{x_i}(B), \quad B \in \mathcal{B}_{\mathbb{E}}, \tag{2.7.10}$$

is a *point measure* on $\mathbb{E}$ if $m(K) < \infty$ for every compact set $K \in \mathcal{B}_{\mathbb{E}}$ (i.e., m is a *Radon measure*).

Hence, $m(B)$ is the number of terms x_i that belong to B. Note that the terms x_i and x_j with different indices may coincide. The *multiplicity* of an element $x \in \mathbb{E}$ with $m(\{x\}) > 0$ is the number of terms of the sequence $x_1, x_2, \dots$ that are equal to x. A point measure m is *simple* if $m(\{x\}) \leqslant 1$ for every point $x \in \mathbb{E}$.

Let $M_p(\mathbb{E})$ be the set of all point measures on $\mathbb{E}$, and let us consider the family $\mathcal{F}$ of all sets of point measures of the form

$$\{m \in M_p(\mathbb{E}) \,|\, m(B_1) \in B_2\}, \quad B_1 \in \mathcal{B}_{\mathbb{E}}, \quad B_2 \in \mathcal{B}_{[0,\infty]}.$$

Finally, let $\mathfrak{M}_p(\mathbb{E})$ be the smallest σ-algebra containing $\mathcal{F}$.

Definition 2.7.27 Let us suppose that $(\Omega, \mathcal{A}, P)$ is a probability space. A function $N : (\Omega, \mathcal{A}, P) \to (M_p(\mathbb{E}), \mathfrak{M}_p(\mathbb{E}))$ is called a *point process* on $\mathbb{E}$ if it is measurable with respect to the σ-algebras $\mathcal{A}$ and $\mathfrak{M}_p(\mathbb{E})$.

The probability distribution P_N of the point process N is the function $P_N : \mathfrak{M}_p(\mathbb{E}) \to \mathbb{R}$ defined by $P_N(B) = P\{N \in B\}$ for $B \in \mathfrak{M}_p(\mathbb{E})$. Note that point processes are special *random measures, see* Kallenberg [120, 121].

Note that N is a random element of $M_p(\mathbb{E})$, i.e., for every $\omega \in \Omega$, $N(\omega) = N(\omega, \cdot)$ is a function defined on $\mathcal{B}_{\mathbb{E}}$, and we shall use notation $N(\omega, B)$ for the value of $N(\omega)$ at the set $B \in \mathcal{B}_{\mathbb{E}}$. Points $x_1, x_2, \dots \in \mathbb{E}$ that determine point measure $N(\omega)$ can be considered realizations of random variables $X_1, X_2, \dots$, i.e., $X_j(\omega) = x_j$, $j \geqslant 1$, and point process N can be represented as follows:

$$N(B) = \sum_{i=1}^{\infty} \varepsilon_{X_i}(B), \quad B \in \mathcal{B}_{\mathbb{E}}, \tag{2.7.11}$$

or simply $N = \sum_{i=1}^{\infty} \varepsilon_{X_i}$. Here $N(\omega, B)$ is the number of points $x_1, x_2, \dots$ that belong to set B, and similarly $N(B)$ is the number of random variables $X_1, X_2, \dots$ that take values in B.

It is obvious that a point process can be considered a family of random variables $\{N(B), B \in \mathcal{B}_{\mathbb{E}}\}$, i.e., a random process indexed by sets $B \in \mathcal{B}_{\mathbb{E}}$. The distributions of random vectors $(N(B_1), \dots, N(B_k))$, where $B_1, \dots, B_k \in \mathcal{B}_{\mathbb{E}}$, $k \geqslant 1$, are called *finite-dimensional distributions* of point process N.

Definition 2.7.28 Let μ be a Radon measure on $\mathcal{B}_{\mathbb{E}}$. A point process N is a *Poisson process* or a *Poisson random measure* with mean measure μ (notation: PRM(μ)) if the following conditions are satisfied:

(a) For every $B \in \mathcal{B}_{\mathbb{E}}$ and every $k \in \{0, 1, 2, \dots\}$,

$$P\{N(B) = k\} = \frac{(\mu(B))^k}{k!} e^{-\mu(B)}, \tag{2.7.12}$$

where the right-hand side of (2.7.12) is equal to 0 if $\mu(B) = \infty$.

(b) If $B_1, \ldots, B_k$, where $k \in \mathbb{N}$, are mutually disjoint sets in $\mathcal{B}_{\mathbb{E}}$, then $N(B_1), \ldots, N(B_k)$ are independent random variables.

Definition 2.7.29 The *Laplace functional* of point process N is given by

$$\Psi_N(f) = E\exp\left\{-\int_{\mathbb{E}} f\,dN\right\} = \int_{M_p(\mathbb{E})} \exp\left\{-\int_{\mathbb{E}} f\,dm\right\} P_N(dm), \qquad (2.7.13)$$

for every nonnegative bounded measurable function $f : \mathbb{E} \to \mathbb{R}$.

The Laplace functional is well defined, i.e., the integral in (2.7.13) is finite. Note also that the Laplace functional Ψ_N of point process N uniquely determines the corresponding distribution P_N, *see* Exercise 2.11. The Laplace functional of PRM(μ) is given in Exercise 2.12.

Let $C_K(\mathbb{E})$ be the set of all continuous functions $f : \mathbb{E} \to \mathbb{R}$, such that there exists a compact set $K \subset \mathbb{E}$ with $f(x) = 0$ for all $x \in K^c = \mathbb{E} \setminus K$. In that case we say that f has compact support. Let $C_K^+(\mathbb{E}) \subset C_K(\mathbb{E})$ be the set of all nonnegative functions $f : \mathbb{E} \to \mathbb{R}$ with compact support.

Let $M_+(\mathbb{E})$ be the set of all nonnegative Radon measures on $(\mathbb{E}, \mathcal{B}_{\mathbb{E}})$. Using function $f \in C_K^+(\mathbb{E})$ we define a map from $M_+(\mathbb{E}) \to \mathbb{R}$ as follows:

$$m(f) = \int_{\mathbb{E}} f\,dm, \quad m \in M_+(\mathbb{E}). \qquad (2.7.14)$$

Let us consider the family $\mathcal{F}_+$ consisting of all sets of the form

$$\{m \in M_+(\mathbb{E}) \,|\, m(f) \in B\}, \quad f \in C_K^+(\mathbb{E}), \quad B \in \mathcal{B}_{[0,\infty]},$$

and let $\mathfrak{M}_+(\mathbb{E})$ be the minimal σ-algebra containing family $\mathcal{F}_+$.

Definition 2.7.30 Function $\xi : (\Omega, \mathcal{A}, P) \to (M_+(\mathbb{E}), \mathfrak{M}_+(\mathbb{E}))$ is a *random measure* if it is measurable with respect to σ-algebras $\mathcal{A}$ and $\mathfrak{M}_+(\mathbb{E})$. Probability distribution P_ξ of point process ξ is the function $P_\xi : \mathfrak{M}_+(\mathbb{E}) \to \mathbb{R}$ defined by $P_\xi(B) = P\{\xi \in B\}$ for $B \in \mathfrak{M}_+(\mathbb{E})$.

If $\xi : \Omega \to M_+(\mathbb{E})$ is a random measure and if for every $\omega \in \Omega$ the value $\xi(\omega)$ belongs to $M_p(\mathbb{E})$, then ξ is a point process.

Definition 2.7.31 Let $\mu, \mu_1, \mu_2, \cdots \in M_+(\mathbb{E})$ be random measures. We say that μ_n *converges vaguely* to μ (notation: $\mu_n \stackrel{v}{\to} \mu$) if

$$\int_{\mathbb{E}} f\,d\mu_n \to \int_{\mathbb{E}} f\,d\mu$$

for every $f \in C_K^+(\mathbb{E})$.

For conditions that are equivalent to the vague convergence of a sequence of random measures, *see* Exercise 2.13.

We also need to introduce the important notion of weak convergence of sequences of random elements in metric spaces. For a detailed presentation of the related theory, *see* Billingsley [27].

Let S be a complete separable metric space with metric d, and let $\mathcal{S}$ be the σ-algebra generated by the open subsets of S. Billingsley's book [27] contains the general theory and an exhaustive study of cases $S = C[0, 1]$, and $S = D[0, \infty)$, where $C[0, 1]$ is the set of continuous real functions on $[0, 1]$, and $D[0, \infty)$ is the set of real functions on $[0, \infty)$ that are right-continuous and have limits from the left. Here we are interested in the cases $S = M_p(\mathbb{E})$ and $S = M_+(\mathbb{E})$.

Both spaces $M_p(\mathbb{E})$ and $M_+(\mathbb{E})$, where $M_p(\mathbb{E}) \subset M_+(\mathbb{E})$, become complete and separable metric spaces if a metric d on $M_+(\mathbb{E})$ is defined as follows: Let us suppose that $\{G_n \mid n \in \mathbb{N}\}$ is a countable base of relatively compact sets and that this family is closed under finite unions and intersections. It is easy to prove that for every n there exists a sequence of functions $f_{ni} \in C_K^+$, $i \geqslant 1$, such that $f_{n1} \leqslant f_{n2} \leqslant \dots$, and $\lim_{i\to\infty} f_{ni} = \mathbf{1}_{G_n}$, and there exists a sequence of functions $g_{ni} \in C_K^+$, $i \geqslant 1$, such that $g_{n1} \geqslant g_{n2} \geqslant \dots$, and $\lim_{i\to\infty} g_{ni} = \mathbf{1}_{G_n}$. Let $h_1, h_2, \dots$ be a sequence of functions such that

$$\{h_1, h_2, \dots\} = \{f_{ni} \mid n \geqslant 1, i \geqslant 1\} \cup \{g_{ni} \mid n \geqslant 1, i \geqslant 1\}.$$

For two random measures $\mu_1, \mu_2 \in M_+(\mathbb{E})$, the distance between μ_1 and μ_2 is then defined by

$$d(\mu_1, \mu_2) = \sum_{n=1}^{\infty} \frac{1}{2^n} \left\{1 - e^{-|\mu_1(h_n) - \mu_2(h_n)|}\right\}. \tag{2.7.15}$$

The following theorem gives the sufficient condition under which a continuous function defined on state space naturally induces a continuous function defined on the space of random measures.

Theorem 2.7.32 (Resnick [192]) *Let $\mathbb{E}_1$ and $\mathbb{E}_2$ be two spaces that are locally compact with countable bases. If $T : \mathbb{E}_1 \to \mathbb{E}_2$ is a continuous function and $K_1 := T^{-1}(K_2)$ is a compact set in $\mathbb{E}_1$ for every compact set K_2 in $\mathbb{E}_2$, then the transformation $\widetilde{T} : M_+(\mathbb{E}_1) \to M_+(\mathbb{E}_2)$ defined by $\widetilde{T}\mu = \mu \circ T^{-1}$ is also continuous. Moreover, the restriction of $\widetilde{T}$ to $M_p(\mathbb{E}_1)$ is of the following form:*

$$\widetilde{T}\left(\sum_{i=1}^{\infty} \varepsilon_{x_i}\right) = \sum_{i=1}^{\infty} \varepsilon_{T(x_i)}. \tag{2.7.16}$$

Now we are going to define the notion of weak convergence in metric spaces. We shall follow the notation from Billingsley [27]. Let $(\Omega, \mathcal{A}, P)$ be a probability space. Let S be a complete and separable metric space equipped with metric d, and let $\mathcal{S}$ be the Borel σ-algebra generated by open sets. A function $X : \Omega \to S$ is a *random element* of S if it is measurable with respect to σ-algebras $\mathcal{A}$ and $\mathcal{S}$, and the *distribution* of X (the *probability measure generated by* X) is function $P_X : \mathcal{S} \to [0, 1]$ given by $P_X = P \circ X^{-1}$.

Remark 2.7.33 Note that if $S = \mathbb{R}$, then random element X is in fact a random variable. If $S = \mathbb{R}^n$, then X is a random vector. If $S = C[0, 1]$, then X is a random process with continuous trajectories on $[0, 1]$. If $S = D[0, \infty)$, then X is a random process with the sample paths that are right-continuous on $[0, \infty)$ and have limits from the left on $(0, \infty)$. If $S = M_p(E)$, then X is a point process. If $S = M_+(E)$, then X is a random measure.

Definition 2.7.34 Let $X_n : \Omega \to S$, $n \geqslant 1$, and $X : \Omega \to S$ be random elements, and P_n, $n \geqslant 1$, and P the corresponding distributions.

The sequence of probability measures $(P_n)_{n\geqslant 1}$ *converges weakly* to P as $n \to \infty$ if the following equality

$$\lim_{n\to\infty} \int_S f(x)\, P_n(dx) = \int_S f(x)\, P(dx) \tag{2.7.17}$$

holds for every bounded and continuous function $f : S \to \mathbb{R}$.

The sequence of random elements $(X_n)_{n\geqslant 1}$ *converges in distribution* to X as $n \to \infty$ if the corresponding sequence of probability measures $(P_n)_{n\geqslant 1}$ converges weakly to P.

For the notions defined above we use the following notation: $P_n \Rightarrow P$ and $X_n \Rightarrow X$. The meaning of the following notation $P_n \Rightarrow X$ and $X_n \Rightarrow P$ is then obvious and not confusing. Sometimes we also say that the sequence of random elements $(X_n)_{n\geqslant 1}$ *converges weakly* to X.

Note that $\int_S f(x)\, P_n(dx) = Ef(X_n)$ and $\int_S f(x)\, P(dx) = Ef(X)$. The conditions that are equivalent to weak convergence are given in Exercise 2.16. For the general theory of weak convergence in metric spaces that is based on the Prohorov theorem, *see* Chapter 1 in Billingsley [27].

Laplace functionals can be used in proving the weak convergence of sequences of point processes or random measures in a similar way that characteristic functions are used in proving convergence in the distribution of sequences of random variables. The following theorem holds.

Theorem 2.7.35 *Let $(\xi_n)_{n\geqslant 0}$ be a sequence of random measures, i.e., random elements in $M_+(\mathbb{E})$. Then $\xi_n \Rightarrow \xi_0$ if and only if $\Psi_{\xi_n}(f) \to \Psi_{\xi_0}(f)$ for every $f \in C_K^+(\mathbb{E})$.*

Proof *See* Proposition 3.19 in Resnick [192].

The following two theorems that give sufficient conditions for the weak convergence of a sequence of point processes will be useful for our purposes in Chaps. 3 and 4.

Theorem 2.7.36 (Resnick [192]) *Let X_{nj}, $j \geqslant 1$, $n \geqslant 1$, be a double array of random elements of $(\mathbb{E}, \mathcal{B}_{\mathbb{E}})$, such that for every $n \geqslant 1$, $(X_{nj})_{j \geqslant 1}$ is a sequence of i.i.d. random elements, and let μ be a Radon measure on $(\mathbb{E}, \mathcal{B}_{\mathbb{E}})$. Let ξ be a Poisson random measure on $[0, \infty) \times \mathbb{E}$ with mean measure $dt \times d\mu$, and let us define a sequence of point processes $(\xi_n)_{n \geqslant 1}$ as follows:*

$$\xi_n = \sum_{j=1}^{\infty} \varepsilon_{(j/n, X_{nj})}, \qquad n \geqslant 1. \tag{2.7.18}$$

Then, $\xi_n \Rightarrow \xi$ in $M_p([0, \infty) \times \mathbb{E})$ if and only if $nP\{X_{n1} \in \cdot\} \xrightarrow{v} \mu$ on $\mathbb{E}$.

Theorem 2.7.37 (Kallenberg [120] and Leadbetter et al. [143]) *Let N, N_1, N_2, ... be point processes with state space $(a, b] \subset \mathbb{R}$, and let us assume that N is a simple point process. Let us suppose that the following two conditions are satisfied:*

(a) *For all intervals $(c, d]$ such that $[c, d] \subset (a, b]$, we have*

$$\lim_{n \to \infty} E(N_n(c, d])) = E(N(c, d])).$$

(b) *For all sets B of the form $B = \bigcup_{i=1}^{k} (c_i, d_i]$, where $[c_i, d_i] \subset (a, b]$ for $i \in \{1, 2, \ldots, k\}$ and $k \in \mathbb{N}$, we have $\lim_{n \to \infty} P\{N_n(B) = 0\} = P\{N(B) = 0\}$.*

Then, $N_n \Rightarrow N$ as $n \to \infty$.

At the end of this subsection it is worth mentioning the *continuous mapping theorem*. This theorem allows conclusions of the following type: Let (S_1, d_1) and (S_2, d_2) be two metric spaces equipped with the Borel σ-algebras $\mathcal{S}_1$ and $\mathcal{S}_2$, respectively. If a sequence $(X_n)_{n \geqslant 0}$ of random elements in S_1 converges weakly to X_0 and $h : S_1 \to S_2$ is a continuous function, then $h(X_n) \Rightarrow h(X_0)$. More generally, the following theorem holds.

Theorem 2.7.38 *Let us consider two metric spaces (S_1, d_1) and (S_2, d_2) equipped with the Borel σ-algebras $\mathcal{S}_1$ and $\mathcal{S}_2$, respectively. Let $(X_n)_{n \geqslant 0}$ be a sequence of random elements in S_1, and let $h : S_1 \to S_2$ be a function such that $P\{X_0 \in D_h\} = 0$, where $D_h \subset S_1$ is a set of discontinuity points of function h. If $X_n \Rightarrow X_0$ in S_1, then $h(X_n) \Rightarrow h(X_0)$ in S_2.*

Exercises and Supplements

Domains of Attraction of EV Distributions

2.1 Prove that the Poisson distribution does not belong to any of the maximum domains of attraction of EV distributions.

2.2 Let F_1 and F_2 be two distribution functions such that $F_1 \in DA(G_0)$ and $F_2 \in DA(G_{1,\alpha})$ for some $\alpha > 0$. Let us define $F = \lambda_1 F_1 + \lambda_2 F_2$, where $\lambda_1 > 0$, $\lambda_2 > 0$, $\lambda_1 + \lambda_2 = 1$. Prove that $F \in DA(G_{1,\alpha})$.

2.3 Let $(Z_k)_{k \geqslant 1}$ be a sequence of independent random variables such that $Z_k \sim \mathcal{N}(\mu_k, \sigma_k^2)$, where $0 < \sigma_1 < \sigma_2 < \ldots$, and $\sigma_k \to \sigma_0$ as $k \to \infty$. Let us suppose that $(p_k)_{k \geqslant 1}$ is a sequence of nonnegative real numbers such that $\sum_{k=1}^{\infty} p_k = 1$. Let $K \in \mathbb{N}$ be a random variable such that $P\{K = k\} = p_k$ for every $k \in \mathbb{N}$. The distribution function of random variable Z_K is given by

$$F(x) = \sum_{k=1}^{\infty} p_k \Phi\left(\frac{x - \mu_k}{\sigma_k}\right),$$

where $\Phi(x)$ is the standard normal distribution function. Prove that F belongs to the maximum domain of attraction of the Gumbel EV distribution.

Let $(X_n)_{n \geqslant 1}$ be a sequence of i.i.d. random variables with common distribution function F, and $M_n^* = \max\{X_1, \ldots, X_n\}$. The norming constants for maximum M_n^* in the related limit theorem were determined in Shneina and Božin [212] under assumption $\mu_k = \mu_0$ for all $k \in \mathbb{N}$.

2.4 The *log-normal distribution* with parameters $m \in \mathbb{R}$ and $\sigma > 0$ is determined by the following density:

$$f(x) = \frac{1}{x\sigma\sqrt{2\pi}} \exp\left\{-\frac{(\ln x - m)^2}{2\sigma^2}\right\} \quad \text{if } x > 0,$$

and $f(x) = 0$ if $x \leqslant 0$. Prove that the corresponding distribution function belongs to $DA(G_0)$.

On the Rates of Convergence

2.5 Let $(X_n)_{n \geqslant 1}$ be a sequence of i.i.d. random variables with common distribution function F, $(u_n)_{n \geqslant 1}$ is a sequence of real numbers such that $\lim_{n\to\infty} = n(1-F(u_n)) = \tau$, $\tau_n = n(1 - F(u_n))$, and $M_n = \max\{X_1, \ldots, X_n\}$. Then we have

$$P\{M_n \leqslant u_n\} - e^{-\tau} = \left(\left(1 - \frac{\tau_n}{n}\right)^n - e^{-\tau_n}\right) + (e^{-\tau_n} - e^{-\tau}) =: A_n + B_n.$$

(a) Prove that $-\dfrac{\tau_n^2 e^{-\tau_n}}{2(n-1)} \leqslant A_n \leqslant 0$.

(b) If $\tau - \tau_n \leqslant \ln 2$, prove that $B_n = e^{-\tau}\left\{(\tau - \tau_n) + \vartheta(\tau - \tau_n)^2\right\}$ for some $\vartheta \in (0, 1)$.

2.6 Let $(X_n)_{n \geqslant 1}$ be a sequence of i.i.d. random variables with normal $\mathcal{N}(0, 1)$ distribution, and $M_n = \max\{X_1, \ldots, X_n\}$.

(a) If constants a_n and b_n are given by (2.3.78) and (2.3.79), then the following asymptotic relation holds as $n \to \infty$:

$$P\left\{\frac{M_n - b_n}{a_n} \leqslant x\right\} - e^{-e^{-x}} \sim \frac{\exp(-e^{-x})e^{-x}}{16} \cdot \frac{(\ln \ln n)^2}{\ln n}.$$

(b) Hall [93]: If the norming constants a_n and b_n are chosen such that

$$2\pi b_n^2 \exp(b_n^2) = n^2, \quad \text{and} \quad a_n = \frac{1}{b_n},$$

then there exist positive constants C_1 and $C_2 \leqslant 3$, which do not depend on n, such that for all n,

$$\frac{C_1}{\ln n} < \sup_{x \in \mathbb{R}} \left| \Phi^n(a_n x + b_n) - e^{-e^{-x}} \right| < \frac{C_2}{\ln n}.$$

The k-th Maximum

2.7 Let $(X_n)_{n \in \mathbb{N}}$ be a sequence of i.i.d. random variables with common distribution function F, $M_n = \max\{X_1, \ldots, X_n\}$, and let $M_n^{(k)}$ be the k-th maximum among random variables $X_1, \ldots, X_n$, i.e.,

$$M_n^{(n)} \leqslant M_n^{(n-1)} \leqslant \cdots \leqslant M_n^{(2)} \leqslant M_n^{(1)} = M_n.$$

Prove that the distribution function of the k-th order statistics $X_{(k)} = M_n^{(n-k+1)}$ is given by

$$F_{X_{(k)}}(x) = P\{X_{(k)} \leqslant x\} = \sum_{j=k}^{n} \binom{n}{j} F^j(x)(1 - F(x))^{n-j}.$$

2.6 *Continuation.* Let us assume that $F \in DA(G)$, where G is an extreme value distribution function, and let $a_n > 0$ and $b_n \in \mathbb{R}$, where $n \in \mathbb{N}$, be two sequences of constants such that $\lim_{n\to\infty} P\{M_n \leqslant a_n x + b_n\} = G(x)$ for every $x \in \mathbb{R}$. Prove that the following statements hold:

(a) Let k be a fixed positive integer, and $x \in \mathbb{R}$. If $0 < G(x) < 1$, then

$$\lim_{n\to\infty} P\{M_n^{(k)} \leqslant a_n x + b_n\} = G(x) \sum_{i=0}^{k-1} \frac{(-\ln G(x))^i}{i!}.$$

(b) If $G(x) = 1$, then $\lim_{n\to\infty} P\{M_n^{(k)} \leqslant a_n x + b_n\} = 1$.

(c) If $G(x) = 0$, then $\lim_{n\to\infty} P\{M_n^{(k)} \leqslant a_n x + b_n\} = 0$.

Stationary Sequences

2.9 Leadbetter [141]: Let $(X_n)_{n\geqslant 1}$ be a strictly stationary random sequence, and $(u_n)_{n\geqslant 1}$ be a sequence of real numbers such that condition $D(u_n)$ is satisfied. For fixed positive integers n, k, and l, let $A_1, A_2, \ldots, A_k \subset \mathbb{N}_n$ be mutually disjoint sets, such that for all r and s, where $r \in A_i$, $s \in A_j$, and $i \neq j$, inequality $|r - s| \geqslant l$ holds. Then

$$\left| P\left(\bigcap_{i=1}^{k} \{M(A_j) \leqslant u_n\}\right) - \prod_{i=1}^{k} P\{M(A_j) \leqslant u_n\} \right| \leqslant (k-1)\alpha_{n,l}.$$

2.10 Leadbetter [141]: Let $(X_n)_{n\geqslant 1}$ be a stationary random sequence and $(u_n)_{n\geqslant 1}$ be a sequence of real numbers such that condition $D(u_n)$ is satisfied. Then for every positive integer $k \geqslant 2$, the following relation holds as $n \to \infty$:

$$P\{M_n \leqslant u_n\} - P^k\{M_{[n/k]} \leqslant u_n\} \to 0.$$

Hint. Let k be a fixed positive integer and $m = [n/k]$. For large values of n, we can choose a positive integer l, such that $k < l < m$ holds. Let

$$\mathbb{N}_{mk} = (I_1 \cup J_1) \cup (I_2 \cup J_2) \cup \cdots \cup (I_k \cup J_k)$$

be a partition of the set $N_{mk} = \{1, 2, \ldots, mk\}$ into disjoint subsets such that the following conditions hold:

- Each of the sets $I_1, J_1, I_2, J_2, \ldots, I_k, J_k$ consists of consecutive natural numbers;
- $|I_1| = |I_2| = \cdots = |I_k| = m - l$, $|J_1| = |J_2| = \cdots = |J_k| = l$.
- The sets are ordered $I_1, J_1, I_2, J_2, \ldots, I_k, J_k$ such that set I_1 consists of the first $m - l$ natural numbers, set J_1 consists of the next l natural numbers, set I_2 consists of the next $m - l$ natural numbers, set J_2 consists of the next l natural numbers, etc.

  Then consider the maxima of sequence $(X_n)_{n\geqslant 1}$ on sets $I_1, J_1, I_2, J_2, \ldots, I_k, J_k$, and $\mathbb{N}_n \setminus \mathbb{N}_{mk}$. As $n \to \infty$, choose l such that $l = l_n = o(n)$. Use the stationarity of sequence $(X_n)_{n\geqslant 1}$, and apply condition $D(u_n)$.

Point Processes and Random Measures

2.11 Prove that the Laplace functional $\Psi_N(f)$ of a point process N uniquely determines its distribution P_N.

2.12 Prove that the Laplace functional $\Psi_N(f)$ of PRM(μ) on state space $\mathbb{E} \subset \mathbb{R}^n$ is given by

$$\Psi_N(f) = \exp\left\{-\int_{\mathbb{E}} \left(1 - e^{-f(x)}\right) \mu(dx)\right\},$$

for every nonnegative measurable function $f : \mathbb{E} \to \mathbb{R}$. Conversely a point process with Laplace functional of the given form is PRM(μ).

2.13 Let $\mu, \mu_1, \mu_2, \cdots \in M_+(\mathbb{E})$ be random measures. Prove that the following conditions are equivalent:

(a) $\mu_n \xrightarrow{v} \mu$ as $n \to \infty$.
(b) $\lim\limits_{n\to\infty} \mu_n(B) = \mu(B)$ for all relatively compact sets $B \in \mathfrak{M}_+(\mathbb{E})$ such that equality $\mu(\partial B) = 0$ holds, where ∂B is the boundary of B.
(c) $\limsup\limits_{n\to\infty} \mu_n(K) \leqslant \mu(K)$ for all compact sets K, and $\liminf\limits_{n\to\infty} \mu_n(G) \geqslant \mu(G)$ for all open and relatively compact sets G.

2.14 A topology on $M_+(\mathbb{E})$ that determines the notion of the vague convergence of a sequence of random measures in $M_+(\mathbb{E})$ is defined as follows: Let us consider all sets of the form $\{\mu \in M_+(\mathbb{E}) \mid a < \mu(f) < b\}$ for some $f \in C_K^+$ and $a, b \in \mathbb{R}$, $a < b$. A basis is the collection of all finite intersections of such sets, and every open set is a union of basis sets. Let $\mathcal{B}(M_+(\mathbb{E}))$ be the Borel σ-algebra generated by open sets. Prove that $\mathfrak{M}_+(\mathbb{E}) = \mathcal{B}(M_+(\mathbb{E}))$.

2.15 Prove that function $d(\mu_1, \mu_2)$ given by (2.7.15) indeed determines a metric on $M_+(\mathbb{E})$.

2.16 Let P_n, $n \geqslant 1$, and P be probability measures on $(S, \mathcal{S})$. Prove that the following conditions are equivalent:

(a) $P_n \Rightarrow P$ as $n \to \infty$.
(b) $\lim\limits_{n\to\infty} \int_S f\,dP_n = \int_S f\,dP$ for all bounded and uniformly continuous functions $f : S \to \mathbb{R}$.
(c) $\limsup\limits_{n\to\infty} P_n(F) \leqslant P(F)$ for all closed sets $F \in \mathcal{S}$.
(d) $\liminf\limits_{n\to\infty} P_n(G) \geqslant P(G)$ for all open sets $G \in \mathcal{S}$.
(d) $\lim\limits_{n\to\infty} P_n(B) = P(B)$ for all sets $B \in \mathcal{S}$ such that $P(\partial B) = 0$.

2.17 Let us assume that N and N_n, $n \geqslant 1$, are point processes with state space $\mathbb{E} \subset \mathbb{R}^n$, where $\mathbb{E}$ is a rectangle. Prove that $N_n \Rightarrow N$ if and only if $(N_n(B_1), \ldots, N_n(B_k))$ converges in distribution to $(N(B_1), \ldots, N(B_k))$ as $n \to \infty$ for every $k \in \mathbb{N}$, and all Borel sets $B_1, \ldots, B_k \subset \mathbb{E}$ such that $P\{N(\partial B_i) = 0\} = 1$ for $i \in \{1, \ldots, k\}$.

Chapter 3
Time Series and Missing Observations

3.1 Introduction

The general formulation of the problem we are going to consider in this chapter is the following. Let $\mathcal{F}$ be a family of dependent random variables, and suppose we are interested in the maximum

$$M = \max_{X \in \mathcal{F}} X.$$

Additionally, we usually assume that all random variables $X \in \mathcal{F}$ have the same marginal distribution. In practice we find cases where only the random variables from some subset $\widetilde{\mathcal{F}} \subset \mathcal{F}$ are observed, and consequently only the maximum

$$\widetilde{M} = \max_{X \in \widetilde{\mathcal{F}}} X$$

is known. The basic question that arises here is the following: What are the relations between these two maxima? We shall be interested in the joint distribution of M and $\widetilde{M}$, or the distribution of random vector $(\widetilde{M}, M)$.

Our attention will be focused on the sequences of random variables $(X_n)_{n \geqslant 1}$ (or $(X_n)_{n \geqslant 0}$), and the growing family $\{X_1, \dots X_n\}$, $n \geqslant 1$. Let us denote $M_n = \max\{X_1, \dots, X_n\}$, and let $\widetilde{M}_n$ be the maximum of observed random variables among the first n terms $X_1, \dots, X_n$. The main question that interests us is formulated as follows: What is the limiting distribution of random vector $(\widetilde{M}_n, M_n)$ as $n \to \infty$?

P. Mladenović, *Extreme Values In Random Sequences*,
Springer Series in Operations Research and Financial Engineering,
https://doi.org/10.1007/978-3-031-57412-2_3

3.2 Gaussian Sequences

Maxima in incomplete samples were probably first considered for Gaussian sequences in Mittal [154]. We shall give here a brief overview of Mittal's results using the original notation.

Let $(X_n)_{n\geqslant 0}$ be a standard Gaussian sequence ($E(X_i) = 0$, $\mathrm{var}(X_i) = 1$) with $E(X_i X_j) = r_{ij}$ for all $i \geqslant 0$ and $j \geqslant 0$. Let k be a positive integer, and for any positive integer n we define $n' = [n/k]$, where $[x]$ is the notation of the integer part of x. Let $G_n = \{t_1, t_2, \ldots, t_{n'}\}$ be a subset of the set $\{0, 1, 2 \ldots, n\}$. We are interested in the maxima

$$M_n = \max\{X_0, X_1, \ldots, X_n\}, \quad m_{n,k} = \max\{X_i : i \in G_n\}, \tag{3.2.1}$$

and their asymptotic behavior as $n \to \infty$.

Case 1. *Asymptotically independent sequences*

Let us suppose that the following two conditions are satisfied:

$$\sup_{i\geqslant 0,\, j\geqslant 0} |r_{ij}| < \delta < 1, \tag{3.2.2}$$

$$|r_{ij}| \leqslant r_{i-j} \quad \text{for } |i-j| > N_0 \quad \text{such that } r_n \ln n = o(1). \tag{3.2.3}$$

The limiting distribution of properly normalized random vector $(M_n - m_{n,k}, m_{n,k})$ is given by the following theorem.

Theorem 3.2.1 (Mittal [154]) *Let $(X_n)_{n\geqslant 0}$ be a standard Gaussian sequence with $E(X_i X_j) = r_{ij}$ for all $i \geqslant 0$ and $j \geqslant 0$, M_n and $m_{n,k}$ be defined by* (3.2.1)*. If conditions* (3.2.2)–(3.2.3) *are satisfied, then*

$$\lim_{n\to\infty} P\left\{M_n - m_{n,k} \leqslant \frac{x}{c_n},\, m_{n,k} \leqslant b_{n,k} + \frac{z}{c_n}\right\} = \frac{\exp\{-e^{-z}(1+(k-1)e^{-x})\}}{1+(k-1)e^{-x}}, \tag{3.2.4}$$

where

$$c_n = (2\ln n)^{1/2}, \quad b_{n,k} = \left(2\ln\left(\frac{n}{k}\right)\right)^{1/2} - \frac{\ln(4\pi \ln n)}{2c_n}, \tag{3.2.5}$$

and $0 \leqslant x < +\infty$, $-\infty < z < +\infty$.

Remark 3.2.2 Under the assumption of Theorem 3.2.1 it is easy to obtain the limiting distribution of the difference $M_n - m_{n,k}$ using (3.2.4):

$$\lim_{n\to\infty} P\left\{M_n - m_{n,k} \leqslant \frac{x}{c_n}\right\} = \frac{1}{1+(k-1)e^{-x}}, \quad 0 \leqslant x < +\infty. \tag{3.2.6}$$

Case 2. *Strongly dependent sequences*

In this case we impose the conditions that sequence $(X_n)_{n\geqslant 0}$ is stationary and

$$E(X_iX_j) = r_{i-j}, \quad r_n \text{ convex for } n \geqslant 0 \text{ and } r_n = o(1), \tag{3.2.7}$$

$$(r_n \ln n)^{-1} \text{ is monotone and } o(1). \tag{3.2.8}$$

Theorem 3.2.3 (Mittal [154]) *Let $(X_n)_{n\geqslant 0}$ be a standard stationary Gaussian sequence with $E(X_iX_j) = r_{i-j}$ such that conditions* (3.2.7)–(3.2.8) *are satisfied. Then*

$$\lim_{n\to\infty} P\left\{M_n - m_{n,k} \leqslant \frac{x}{c_n}\right\} = \frac{1}{1+(k-1)e^{-x}}, \tag{3.2.9}$$

where $c_n = (2\ln n)^{1/2}$, for all $0 \leqslant x < +\infty$.

Remark 3.2.4 The expression on the right-hand side of equality (3.2.9) is the logistic distribution function on the positive part of the real line. The value of the limiting distribution function at point 0 is $1/k$. This is the limiting probability that $\max\{X_0, X_1, \ldots, X_n\}$ is equal to X_i for some $i \in G_n$. Note also that $1/k$ can be considered part of the corresponding logistic distribution on the negative part of the real line. It is interesting that the limiting distribution of properly normalized difference $M_n - m_{n,k}$ is the same both in the case of asymptotically independent sequences and strongly dependent sequences, *see* (3.2.6) and (3.2.9).

3.3 Weakly Dependent Stationary Sequences

In this section we shall consider stationary sequences that satisfy Leadbetter type conditions of weak dependence. These conditions, namely $D(u_n)$ and $D'(u_n)$, were defined by Definitions 2.7.7 and 2.7.8. We shall consider here two maxima, the maximum of observed and the maximum of nonobserved random variables. As a consequence condition $D(u_n)$ should be modified into two level condition $D(u_n, v_n)$. Such modifications of condition $D(u_n)$ into multiple level conditions have been used in the literature several times in order to consider the maxima and minima of stationary processes, the maxima on several disjoint intervals, etc. *See*, for example, Davis [49], Leadbetter [141], and Leadbetter, Lindgren and Rootzén [143]. For our purposes in what follows we do not need to modify condition $D'(u_n)$ that prohibits the clustering of extremes.

Using conditions $D(u_n)$ and $D'(u_n)$ Leadbetter [141] obtained results related to the asymptotic behavior of maxima in stationary sequences that satisfy these conditions. The technique developed by Leadbetter is sufficient to obtain results on the joint limiting distribution of maxima in complete and incomplete samples from stationary sequences. Of course, instead of condition $D(u_n)$ the following $D(u_n, v_n)$ condition will be imposed, *see* Davis [49].

Definition 3.3.1 Let $(X_n)_{n\geqslant 1}$ be a strictly stationary random sequence, (u_n) and (v_n) two sequences of real numbers, and $\mathbb{N}_n = \{1, 2, \ldots, n\}$. *Condition* $D(u_n, v_n)$ is satisfied if for all $A_1,\ A_2,\ B_1,\ B_2 \subset \mathbb{N}_n$, such that $b-a \geqslant l$, for all $a \in A_1 \cup A_2$, $b \in B_1 \cup B_2$, $A_1 \cap A_2 = \varnothing$, $B_1 \cap B_2 = \varnothing$, the following inequality holds

$$\Bigg| P\Bigg(\bigcap_{j\in A_1\cup B_1}\{X_j \leqslant u_n\}\cap \bigcap_{j\in A_2\cup B_2}\{X_j \leqslant v_n\}\Bigg) - \\ -P\Bigg(\bigcap_{j\in A_1}\{X_j \leqslant u_n\}\cap \bigcap_{j\in A_2}\{X_j \leqslant v_n\}\Bigg)\times \\ \times P\Bigg(\bigcap_{j\in B_1}\{X_j \leqslant u_n\}\cap \bigcap_{j\in B_2}\{X_j \leqslant v_n\}\Bigg)\Bigg| \leqslant \alpha_{n,l}, \tag{3.3.1}$$

and $\alpha_{n,l_n} \to 0$ as $n \to \infty$ for some $l_n = o(n)$.

Along with stationary sequence $(X_n)_{n\geqslant 1}$ we shall consider a sequence of indicators $(\varepsilon_n)_{n\geqslant 1}$ that is independent of $(X_n)_{n\geqslant 1}$. Indicator ε_i determines whether term X_i is observed or not as follows: X_i is observed if $\varepsilon_i = 1$, and X_i is not observed if $\varepsilon_i = 0$.

Theorem 3.3.2 (Mladenović and Piterbarg [161]) *Let us suppose that the following conditions are satisfied:*

(a) $F \in D(G)$, *where* G *is an extreme value distribution function, i.e., there exist constants* $a_n > 0$ *and* $b_n \in \mathbb{R}$, $n \in \mathbb{N}$, *such that for every real* x, $F^n(a_n x + b_n) \to G(x)$ *as* $n \to \infty$.
(b) $(X_n)_{n\geqslant 1}$ *is a strictly stationary random sequence, such that conditions* $D(u_n, v_n)$ *and* $D'(u_n)$ *are satisfied for* $u_n = a_n x + b_n$ *and* $v_n = a_n y + b_n$, *where* $x < y$, *and* $M_n = \max\{X_i \mid 1 \leqslant i \leqslant n\}$.
(c) $(\varepsilon_n)_{n\geqslant 1}$ *is a sequence of indicators that is independent of* $(X_n)_{n\geqslant 1}$, $\widetilde{M}_n = \max\{X_i \mid 1 \leqslant i \leqslant n, \varepsilon_i = 1\}$, *and*

$$\frac{1}{n}\sum_{i=1}^{n}\varepsilon_i \xrightarrow{P} p \in [0, 1] \quad as \quad n \to \infty. \tag{3.3.2}$$

Then the following equality holds for all real $x < y$:

$$\lim_{n\to\infty} P\{\widetilde{M}_n \leqslant a_n x + b_n, M_n \leqslant a_n y + b_n\} = G^p(x)G^{1-p}(y). \tag{3.3.3}$$

Remark 3.3.3 Theorem 3.3.2 states that events $\{(\widetilde{M}_n - b_n)/a_n \leqslant x\}$ and $\{(M_n - b_n)/a_n \leqslant y\}$, where $x < y$, are asymptotically independent for stationary sequences that satisfy conditions $D(u_n, v_n)$ and $D'(u_n)$. The statement of Theorem 3.3.2 remains valid if the observation of random variables in sequence $(X_n)_{n\geqslant 1}$ is determined by a non-random 0–1 sequence $(c_i)_{i\geqslant 1}$, i.e., X_i is observed if $c_i = 1$, and X_i is not observed if $c_i = 0$, under the assumption

$$\frac{1}{n}\sum_{i=1}^{n} c_i \longrightarrow p \in [0, 1] \text{ as } n \to \infty. \tag{3.3.4}$$

If conditions $D(u_n, v_n)$ and $D'(u_n)$ are not fulfilled simultaneously, then the limit distribution of the appropriately normalized random vector $(\widetilde{M}_n, M_n)$ may be much more complicated. Stationary sequences of this type will be considered in some of the next sections.

Remark 3.3.4 For every $x < y$ the asymptotic independence of events $\{(\widetilde{M}_n - b_n)/a_n \leqslant x\}$ and $\{(M_n - b_n)/a_n \leqslant y\}$ was obtained in Mladenović and Piterbarg [161] under condition (3.3.2). Under the conditions of Theorem 3.3.2 with (3.3.2) replaced by

$$\frac{1}{n}\sum_{i=1}^{n} \varepsilon_i \stackrel{P}{\longrightarrow} \lambda \in [0, 1] \text{ as } n \to \infty, \tag{3.3.5}$$

where λ is a random variable, Krajka [136] proved that for every $x < y$, the following asymptotic relation holds:

$$\lim_{n\to\infty} P\{\widetilde{M}_n \leqslant a_n x + b_n, M_n \leqslant a_n y + b_n\} = E\big\{G^{\lambda}(x)G^{1-\lambda}(y)\big\}. \tag{3.3.6}$$

3.4 A Storage Process in Discrete Time

Various types of storage processes have been investigated in probabilistic and statistical literature. The properties of a storage process depend on the input process. Noros [173] considered a storage model with self-similar input. Piterbarg [182] and Hüsler and Piterbarg [113] investigated a storage process with fractional Brownian motion as input.

Let $X(t), t \geqslant 0, X(0) = 0$ a.s., be a fractional Brownian motion, i.e., a Gaussian zero mean process with stationary increments such that

$$E(X(t) - X(s))^2 = |t - s|^{2H}, \quad 0 < H \leqslant 1.$$

Consider a random sequence

$$Y_k = \sup_{j\geqslant k} (X(j) - X(k) - c(j - k)),$$

where $c > 0$ and $k, l \in \mathbb{Z}$. Since X has stationary increments, Y_k is a stationary sequence and we call it a discrete time storage process with *fractional Brownian motion* as input, *see* for details Piterbarg [182], where the *storage process with continuous time*

$$Y(t) = \sup_{\sigma \geqslant t} (X(\sigma) - X(t) - c(\sigma - t))$$

was studied. We have,

$$\begin{aligned} &P\left\{\max_{1\leqslant k\leqslant K} Y_k \leqslant u\right\} \\ &= P\{X(u(s+\tau)) - X(su) \leqslant u + cu\tau, \\ &\qquad\qquad \text{for some } \tau \geqslant 0, s \in [0, u^{-1}K], su, \tau u \in \mathbb{Z}\} \\ &= P\left\{\sup_{\tau>0, s\in[0,u^{-1}K], su, \tau u\in\mathbb{Z}} \frac{X(u(s+\tau)) - X(su)}{\tau^H u^H v(\tau)} \leqslant u^{1-H}\right\}, \end{aligned} \tag{3.4.1}$$

where $v(\tau) = \tau^{-H} + c\tau^{1-H}$. The Gaussian random field

$$Z(s,\tau) := \frac{X(u(s+\tau)) - X(su)}{\tau^H u^H v(\tau)}, \quad \tau > 0, \ s \in \mathbb{R},$$

is studied in Piterbarg [182]. Its distribution does not depend on u, $Z(s,\tau)$ is stationary in s but not in τ, its variance $\sigma_Z^2(\tau)$ depends only on τ and has a single maximum point at $\tau_0 = H/(c(1-H))$. Moreover,

$$\sigma_Z(\tau) = v^{-1}(\tau) = \frac{1}{A} - \frac{B}{2A^2}(\tau - \tau_0)^2 + O((\tau - \tau_0)^3) \tag{3.4.2}$$

as $\tau \to \tau_0$, where

$$A := \frac{1}{1-H}\left(\frac{H}{c(1-H)}\right)^{-H} = v(\tau_0), \tag{3.4.3}$$

$$B := H\left(\frac{H}{c(1-H)}\right)^{-H-2} = v''(\tau_0). \tag{3.4.4}$$

For the correlation function $r(s,\tau; s', \tau')$ of Z we have

$$r(s,\tau;s',\tau') = 1 - \frac{1+o(1)}{2\tau_0^{2H}}\left(|s - s' + \tau - \tau'|^{2H} + |s - s'|^{2H}\right), \tag{3.4.5}$$

as $s - s' \to 0$, $\tau - \tau' \to 0$, and for sufficiently large $|s - s'|$ and an absolute constant C,

$$|r(s,\tau;s',\tau')| \leqslant C|s - s'|^{2H-2}, \tag{3.4.6}$$

if $2H \neq 1$. For $2H = 1$, we have $r(s, \tau; s', \tau') = 0$ for large $|s - s'|$ since the increments of Brownian motion on disjoint intervals are independent, *see* for details Piterbarg [182], Hüsler and Piterbarg [113]. Here we consider the Gaussian field $Z(s, \tau)$ on the grid $\mathcal{R}_u = u^{-1}(\mathbb{Z}_+ \times \mathbb{Z})$.

Lemma 3.4.1 *Assume that* $1/2 < H < 1$. *There exists* $\delta > 0$ *such that*

$$P\left\{\max_{1\leqslant k\leqslant K} Y_k > u\right\} \sim P\left\{\max_{t\in[0,K]} Y(t) > u\right\}$$
$$\sim P\{Y(0) > u\} \sim P\{Y_1 > u\}$$

as $u \to \infty$, *where* K *may tend to infinity such that* $K = O(e^{\delta u^2})$.

Proof The second relation is proved in Piterbarg [182]. To prove the first and third relations, we use (3.4.1) and the fact that the grid $\mathcal{R}_u$ is *dense* for Gaussian field Z and level u^{1-H}, in the sense of Piterbarg [183]. Indeed, for level u^{1-H}, in view of (3.4.2) and (3.4.5), we get that $(u^{1-H})^{-2/(2H)} = u^{1-1/H}$, and for $H > 1/2$, $u^{-1} = o(u^{1-1/H})$. □

Lemma 3.4.2 *Let* $u_n, v_n \to +\infty$ *as* $n \to \infty$, *such that* u_n/v_n *is bounded from zero and infinity. Then condition* $D(u_n, v_n)$ *is fulfilled for the sequence* (Y_k).

Proof By (3.4.1), Definition 3.3.1 reduces to Gaussian field Z and levels u_n^{1-H} and v_n^{1-H}. Then the assertion of the Lemma follows from Piterbarg [181]. In estimating α_{n,l_n} we use relations (3.4.2) and (3.4.6). □

Theorem 3.4.3 (Mladenović and Piterbarg [161]) *Assume that* $1/2 < H < 1$. *Let* $(\varepsilon_n)_{n\geqslant 1}$ *be a sequence of indicators that are independent of the sequence* $(Y_k)_{k\geqslant 1}$, *and let us denote* $S_n = \varepsilon_1 + \cdots + \varepsilon_n$,

$$\widetilde{M}_n = \max\{Y_j \mid 1 \leqslant j \leqslant n, \varepsilon_j = 1\},$$

$$M_n = \max\{Y_1, \ldots, Y_n\},$$

$$a_n = \frac{(2A^{-2})^{1/(2(1-H))}}{2(1-H)}(\ln n)^{-(1-2H)/(2(1-H))},$$

$$b_n = (2A^{-2}\ln n)^{1/(2(1-H))} + \left\{\frac{h(2A^{-2})^{1/(2(1-H))}\ln(2A^{-2}\ln n)}{4(1-H)^2}\right.$$
$$\left.+\frac{(2A^{-2})^{1/(2(1-H))}\ln c}{2(1-H)}\right\}(\ln n)^{-(1-2H)/(2(1-H))},$$

where A *is given by Eq. (3.4.3),* $c = a^{2/H}(2b)^{-1/2}\mathcal{H}_{2H}^2 A^{2/H-2}$, $a = 1/(2\tau_0^{2H})$, $b = B/(2A)$, $h = 2(1-H)^2/H - 1$, *and* $\mathcal{H}_{2H}$ *is the Pickands constant given by*

$$H_{2H} = \lim_{T\to\infty}\frac{1}{T}E\exp\left\{\max_{0\leqslant t\leqslant T}\left(\sqrt{2}\,X(t) - |t|^{2H}\right)\right\}.$$

If $P\{S_n = 0\} \to 0$ as $n \to \infty$, then the following equality holds

$$\lim_{n\to\infty} P\{\widetilde{M}_n \leqslant a_n x + b_n, M_n \leqslant a_n y + b_n\} = e^{-e^{-\min\{x,y\}}}. \tag{3.4.7}$$

Proof If $x \geqslant y$, equality (3.4.7) follows immediately from Theorem 1 in Hüsler and Piterbarg [113]. Let $x < y$ and $u_n(x) = a_n x + b_n$, $u_n(y) = a_n y + b_n$. Then equality (3.4.7) also follows from the above-mentioned Theorem 1 and the following relations:

$$\begin{aligned} P\{\widetilde{M}_n \leqslant u_n(x), M_n \leqslant u_n(y)\} &\sim P\{\widetilde{M}_n \leqslant u_n(x)\} \\ &\sim P\{M_n \leqslant u_n(x)\}. \end{aligned} \tag{3.4.8}$$

Note that the second asymptotic relation in (3.4.8) follows from Lemma 3.4.1 if $P\{S_n = 0\} \to 0$ as $n \to \infty$. The first relation in (3.4.8) is a consequence of the following relations:

$$\{\widetilde{M}_n \leqslant u_n(x)\} = \{\widetilde{M}_n \leqslant u_n(x), M_n \leqslant u_n(y)\} \cup \{\widetilde{M}_n \leqslant u_n(x), M_n > u_n(y)\},$$
$$\{\widetilde{M}_n \leqslant u_n(x), M_n > u_n(y)\} \subset \{\widetilde{M}_n \leqslant u_n(y)\} \setminus \{M_n \leqslant u_n(y)\},$$
$$\{\widetilde{M}_n \leqslant u_n(y)\} \sim \{M_n \leqslant u_n(y)\}.$$

□

Remark 3.4.4 Let $X = \{X(t)\}_{t\geqslant 0}$ be an infinitely divisible stochastic process, with no Gaussian component, that is self-similar with index $H > 0$. For given constants $c > 0$ and $\gamma > H$, the storage process with a *self-similar and infinitely divisible* input is defined by

$$Y(t) = \sup_{s\geqslant t}(X(s) - X(t) - c(s-t)^{\gamma}), \quad t \geqslant 0. \tag{3.4.9}$$

For an input process X with stationary increments, the storage process $\boldsymbol{Y}$ is stationary, if finite. Under some additional conditions on process $\boldsymbol{X}$ and function $t = t(u)$, Albin and Samorodnitsky [4] proved that

$$\lim_{u\to\infty} \frac{P\left\{\sup_{s\in[0,t]} Y(s) > u\right\}}{P\left\{\inf_{s\in[0,t]} Y(s) > u\right\}} = 1. \tag{3.4.10}$$

In some cases, function $t(u)$ may tend to infinity, *see* Corollary 2 in Albin and Samorodnitsky [4]. Let (Y_k) be a sequence defined by $Y_k = Y(k)$, $k \in \{1, 2, \dots\}$. We call it a discrete time storage process with a self-similar and infinitely divisible input. Apparently, using (3.4.10), it would be possible to get the result that the components of vector $(\widetilde{M}_n, M_n)$ are asymptotically perfectly dependent, where $\widetilde{M}_n$ and M_n are defined as in Theorem 3.4.3.

3.5 Random Failures or Censoring in the Observation Mechanism

In this section we shall consider the effects of the random failures or censorships in the observation mechanism on the extreme values in stationary sequences that satisfy certain weak dependence conditions. The results that will be presented are given in Robert [194]. First, we introduce the necessary notation. Let $(X_n)_{n\geqslant 1}$ be a strictly stationary random sequence with marginal distribution function F, $x_{\min} = \inf\{x \in \mathbb{R} : F(x) > 0\}$ and $x_{\max} = \sup\{x \in \mathbb{R} : F(x) < 1\}$. We assume that for every $\tau > 0$, there exists a sequence $(u_n(\tau))_{n\geqslant 1}$ such that

$$\lim_{n\to\infty} n\big(1 - F(u_n(\tau))\big) = \lim_{n\to\infty} nP\{X_1 > u_n(\tau))\} = \tau. \tag{3.5.1}$$

For a positive integer r and a threshold u, let us define $N_r(u)$ to be the number of exceedances of threshold u by random variables $X_1, \dots, X_r$, i.e.,

$$N_r(u) = \sum_{i=1}^{r} I_{\{X_i > u\}}, \tag{3.5.2}$$

and

$$\pi_{1,r}(k; u) = P\{N_r(u) = k \mid N_r(u) > 0\}, \quad k \in \{1, \dots, r\}, \tag{3.5.3}$$

$$\vartheta_r(u) = \big(E(N_r(u) \mid N_r(u) > 0)\big)^{-1} = \left(\sum_{k=1}^{r} k\pi_{1,r}(k; u)\right)^{-1}. \tag{3.5.4}$$

For two thresholds u and v, such that $u < v$, and for integers i and j, such that $i \geqslant 1, i \geqslant j \geqslant 0$, we also define

$$\pi_{2,r}(i, j; u, v) = P\{N_r(u) = i, N_r(v) = j \mid N_r(u) > 0\}. \tag{3.5.5}$$

3.5.1 *The Effects of Random Failures*

Let ε_n be the indicator of the event that X_n is observed. The number of observed random variables among the first n terms of sequence $(X_i)_{i\geqslant 1}$ is $S_n = \varepsilon_1 + \cdots + \varepsilon_n$. For $n \geqslant 1$ we define

$$W_n = \begin{cases} X_n, & \text{if } \varepsilon_n = 1, \\ x_{\min}, & \text{if } \varepsilon_n = 0, \end{cases} \tag{3.5.6}$$

and $\widetilde{M}_n = \max\{W_1, \ldots, W_n\}$, $M_n = \max\{X_1, \ldots, X_n\}$. Clearly $\widetilde{M}_n$ is the maximum of the observed random variables. Let us also assume that there exists a positive function $p(x)$, $x \in (x_{\min}, x_{\max})$, such that for each $k \geqslant 1$,

$$p(x) = P\{\varepsilon_k = 1 \mid X_k = x\}. \tag{3.5.7}$$

This assumption means that the failure in the observation mechanism depends on the value of the underlying random variable. We also assume that for all $m \geqslant 1$ and $1 \leqslant i_1 < i_2 < \cdots < i_m$,

$$\mathcal{L}\big((\varepsilon_{i_1}, \ldots, \varepsilon_{i_m}) \mid (X_n)_{n \geqslant 1}\big) = \mathcal{L}\big(\varepsilon_{i_1} | X_{i_1}\big) \times \cdots \times \mathcal{L}\big(\varepsilon_{i_m} | X_{i_m}\big), \tag{3.5.8}$$

i.e., the failures are conditionally independent and their conditional distribution depends only on the corresponding random variables. If $p(x) = p$ for some $p \in (0, 1]$, then $(\varepsilon_n)_{n \geqslant 1}$ is an i.i.d. sequence. Note also that the law of large numbers holds for the sequence of indicators $(\varepsilon_n)_{n \geqslant 1}$, i.e., $\widetilde{S}_n/n \xrightarrow{P} E\{p(X_1)\}$, as $n \to \infty$.

Let us define function $\widetilde{\Phi}$ as follows:

$$\widetilde{\Phi}(u) = \frac{P\{W_i > u\}}{P\{X_i > u\}}. \tag{3.5.9}$$

Finally, we assume that for every $\sigma > 0$, there exists a sequence of levels $(w_n(\sigma))_{n \geqslant 1}$, such that for every $i \geqslant 1$,

$$\lim_{n \to \infty} n P\{W_i > w_n(\sigma)\} = \sigma. \tag{3.5.10}$$

Theorem 3.5.1 (Robert [194]) *Suppose that the following conditions hold:*

(a) $\lim_{x \to x_{\max}} \widetilde{\Phi}(x) = 0.$

(b) *There exists a sequence of positive integers $(r_n)_{n \geqslant 1}$, and a positive constant ϑ, such that*

$$\lim_{n \to \infty} r_n P\{X_1 > u_n(\tau)\} = 0, \qquad \lim_{n \to \infty} \vartheta_{r_n}(u_n(\tau)) = \vartheta. \tag{3.5.11}$$

(c) *There exists a sequence of positive integers $(l_n)_{n \geqslant 1}$, such that $l_n = o(r_n)$ and $n r_n^{-1} \alpha_{l_n} \to 0$ as $n \to \infty$, where α_l is the Rosenblatt strong mixing coefficient determined by Definition 2.7.5.*

Then, the following asymptotic relation holds:

$$\lim_{n \to \infty} P\{\widetilde{M}_n \leqslant w_n(\sigma), M_n \leqslant u_n(\tau)\} = e^{-(\sigma + \vartheta\tau)}. \tag{3.5.12}$$

Note that the maxima $\widetilde{M}_n$ and M_n are asymptotically independent under the conditions of Theorem 3.5.1. It follows from (3.5.12) that the extremal index of sequence $(W_n)_{n\geqslant 1}$ is equal to 1, even if the extremal index of $(X_n)_{n\geqslant 1}$ is smaller than 1. Hence, there is no clustering of extreme values in the sequence of observed random variables $(W_n)_{n\geqslant 1}$.

Theorem 3.5.2 (Robert [194]) *Suppose that the following conditions hold:*

(a) $\lim_{x\to x_{\max}} \widetilde{\Phi}(x) = p \in (0, 1)$.

(b) *For* $\sigma > \tau p$, *there exists a sequence of positive integers* $(r_n)_{n\geqslant 1}$, *such that* $r_n P\{X_1 > u_n(\tau)\} = o(1)$, *for every* $i \geqslant 1$ *there exists a constant* $\pi_2^{(\sigma/(p\tau))}(i, 0)$ *such that*

$$\lim_{n\to\infty} \pi_{2,r_n}\big(i, 0; u_n(\sigma/p), u_n(\tau)\big) = \pi_2^{(\sigma/(p\tau))}(i, 0),$$

and there exists a positive constant ϑ *such that* $\lim_{n\to\infty} \vartheta_{r_n}(u_n(\tau)) = \vartheta$.

(c) *There exists a sequence of positive integers* $(l_n)_{n\geqslant 1}$, *such that* $l_n = o(r_n)$ *and* $nr_n^{-1}\alpha_{l_n} \to 0$ *as* $n \to \infty$, *where* α_l *is the Rosenblatt strong mixing coefficient determined by Definition 2.7.5.*

Then for $\sigma > \tau p$ *the following asymptotic relation holds:*

$$\lim_{n\to\infty} P\{\widetilde{M}_n \leqslant u_n(\sigma/p), M_n \leqslant u_n(\tau)\}$$
$$= \exp\left\{-\frac{\sigma\vartheta}{p}\left(1 - \sum_{k=1}^{\infty} \pi_2^{(\sigma/(p\tau))}(k, 0)(1-p)^k\right)\right\}.$$

Remark 3.5.3 If $\widetilde{\Phi}(x) \to p \in (0, 1)$ as $x \to x_{\max}$, then

$$\lim_{n\to\infty} nP\{X_1 > w_n(\sigma)\} = \lim_{n\to\infty} nP\left\{X_1 > u_n\left(\frac{\sigma}{p}\right)\right\} = \sigma/p,$$

and therefore $w_n(\sigma)$ was replaced by $u_n(\sigma/p)$ in Theorem 3.5.2.

Note that the maxima $\widetilde{M}_n$ and M_n are symptotically dependent under the conditions of Theorem 3.5.2. If $p(x) = p$ for $x \in [x_{min}, x_{max}]$, then $\widetilde{\Phi}(x) = p$ for $x \in [x_{min}, x_{max}]$. If $\vartheta = 1$, then

$$\pi_2^{\sigma/(p\tau)}(1, 0) = 1 - \pi_2^{\sigma/(p\tau)}(1, 1) = \frac{(\sigma/p) - \tau}{\sigma/p},$$

and for $\sigma > \tau p$, the following asymptotic relation holds

$$\lim_{n\to\infty} P\left\{\widetilde{M}_n \leqslant u_n\left(\frac{\sigma}{p}\right), M_n \leqslant u_n(\tau)\right\} = e^{-(1-p)\tau - p(\sigma/p)},$$

which is in fact relation (3.3.3).

3.5.2 Censorships and Extreme Values

Let $(Y_n)_{n\geqslant 1}$ be a sequence of random variables and suppose that X_n is observed if $X_n \leqslant Y_n$. For every $n \geqslant 1$ let us denote $\delta_n = I_{\{X_n \leqslant Y_n\}}$,

$$S_n^* = \delta_1 + \cdots + \delta_n, \quad V_n = \min\{X_n, Y_n\},$$
$$M_n^* = \max\{V_1, \dots, V_n\}, \quad M_n = \max\{X_1, \dots, X_n\}.$$

In other words, S_n^* is the number of uncensored observations and M_n^* is the maximum of the observed random variables.

We assume that there exists a positive function $\rho(x)$, $x \in (x_{\min}, x_{\max})$, such that for each $i \geqslant 1$,

$$\rho(x) = P\{\delta_i = 1 \mid X_i = x\} = P\{X_i \leqslant Y_i \mid x_i = x\}. \tag{3.5.13}$$

We also assume that for all $m \geqslant 1$ and $1 \leqslant i_1 < i_2 < \cdots < i_m$,

$$\mathcal{L}\big((Y_{i_1}, \dots, Y_{i_m}) \mid (X_n)_{n\geqslant 1}\big) = \mathcal{L}\big(Y_{i_1} | X_{i_1}\big) \times \cdots \times \mathcal{L}\big(Y_{i_m} | X_{i_m}\big), \tag{3.5.14}$$

i.e., given the stationary sequence, the censored random variables are conditionally independent. Note also that the law of large numbers holds for the sequence of indicators $(\delta_n)_{n\geqslant 1}$, *see* Robert [194].

Let us define $\Phi^*(u) := P\{V_i > u\}/P\{X_i > u\}$. Finally, we assume that for every $\sigma > 0$, there exists a sequence of levels $(v_n(\sigma))_{n\geqslant 1}$, such that

$$\lim_{n\to\infty} nP\{V_i > v_n(\sigma)\} = \sigma. \tag{3.5.15}$$

Theorem 3.5.4 (Robert [194]) *Suppose that the following conditions hold:*

(a) $\lim_{x\to x_{\max}} \Phi^*(x) = 0.$

(b) *There exists a sequence of positive integers $(r_n)_{n\geqslant 1}$, and a positive constant ϑ, such that*

$$\lim_{n\to\infty} r_n P\{X_1 > u_n(\tau)\} = 0, \quad \lim_{n\to\infty} \vartheta_{r_n}(u_n(\tau)) = \vartheta. \tag{3.5.16}$$

(c) *There exists a sequence of positive integers $(l_n)_{n\geqslant 1}$, such that $l_n = o(r_n)$ and $nr_n^{-1}\alpha_{l_n} \to 0$ as $n \to \infty$, where α_l is the Rosenblatt strong mixing coefficient.*

Then the following asymptotic relation holds:

$$\lim_{n\to\infty} P\{M_n^* \leqslant v_n(\sigma), M_n \leqslant u_n(\tau)\} = e^{-(\sigma+\vartheta\tau)}. \tag{3.5.17}$$

Note that the maxima M_n^* and M_n are symptotically independent under the conditions of Theorem 3.5.4.

Theorem 3.5.5 (Robert [194]) *Suppose that the following conditions hold:*

(a) $\lim_{x \to x_{\max}} \Phi^*(x) = p \in (0, 1)$.

(b) *For* $\sigma > \tau p$, *there exists a sequence of positive integers* $(r_n)_{n \geqslant 1}$, *such that* $r_n P\{X_1 > u_n(\tau)\} = o(1)$, *for every* $i \geqslant 1$ *there exists a constant* $\pi_2^{(\sigma/(p\tau))}(i, 0)$ *such that*

$$\lim_{n \to \infty} \pi_{2,r_n}\big(i, 0; u_n(\sigma/p), u_n(\tau)\big) = \pi_2^{(\sigma/(p\tau))}(i, 0),$$

and there exists a positive constant ϑ *such that* $\lim_{n \to \infty} \vartheta_{r_n}(u_n(\sigma/p)) = \vartheta$.

(c) *There exists a sequence of positive integers* $(l_n)_{n \geqslant 1}$, *such that* $l_n = o(r_n)$ *and* $n r_n^{-1} \alpha_{l_n} \to 0$ *as* $n \to \infty$, *where* α_l *is the Rosenblatt strong mixing coefficient.*

Then for $\sigma > \tau p$ *the following asymptotic relation holds:*

$$\lim_{n \to \infty} P\{M_n^* \leqslant u_n(\sigma/p), M_n \leqslant u_n(\tau)\}$$
$$= \exp\left\{ -\frac{\sigma\vartheta}{p}\left(1 - \sum_{k=1}^{\infty} \pi_2^{(\sigma/(p\tau))}(k, 0)\left(1 - \frac{\sigma - \tau\varphi^*(\sigma/(p\tau))}{\sigma/p - \tau}\right)^k\right)\right\},$$

where $\varphi^*(\sigma/(\tau p)) := \lim_{n \to \infty} P\big\{Y_i > u_n(\sigma/p) \,|\, X_i > u_n(\tau)\big\}$, $\sigma > \tau p$.

Remark 3.5.6 If $\Phi^*(x) \to p \in (0, 1)$ as $x \to x_{\max}$, then

$$\lim_{n \to \infty} n P\{X_1 > v_n(\sigma)\} = \lim_{n \to \infty} n P\{X_1 > u_n(\sigma/p)\} = \sigma/p,$$

and therefore $v_n(\sigma)$ was replaced by $u_n(\sigma/p)$ in Theorem 3.5.5.

Note that the maxima M_n^* and M_n are symptotically dependent under the conditions of Theorem 3.5.5. If $\vartheta = 1$, then

$$\pi_2^{\sigma/(p\tau)}(1, 0) = 1 - \pi_2^{\sigma/(p\tau)}(1, 1) = \frac{(\sigma/p) - \tau}{\sigma/p},$$

and for $\sigma > \tau p$, the following asymptotic relation holds

$$\lim_{n \to \infty} P\big\{M_n^* \leqslant v_n(\sigma), M_n \leqslant u_n(\tau)\big\} = \exp\left\{ -\tau + \sigma - \tau\varphi^*\left(\frac{\sigma}{p\tau}\right)\right\}.$$

3.6 Sub-sampled Models

Extreme values in sub-sampled models have been considered by several authors. *See*, for example, Robinson and Tawn [195], Hall and Scotto [95, 96], Hall, Scotto and Ferreira [97], Scotto and Turkman [204], Scotto, Turkman, and Anderson [205], Scotto [202], Hall and Temido [98], Sebastião et al. [206], Peng, Weng and Nadarajah [179], Tan and Wang [217].

The general formulation of the problem is the following: Suppose that a discrete random process $(X_n)_{n\geqslant 1}$ is sampled with some sampling frequency. The observed values form a new process $X_k, X_{2k}, X_{3k}, \ldots$, where $k \geqslant 2$ is a positive integer. What can we say about relations between the true maximum of the first n terms, i.e., $M_n = \max\{X_i \mid 1 \leqslant i \leqslant n\}$, and the observed maximum $\widetilde{M}_n = \max\{X_{ik} \mid 1 \leqslant ik \leqslant n\}$? Different stochastic processes and different aspects of relations between $\widetilde{M}_n$ and M_n were considered in the papers mentioned above.

Scotto,Turkman, and Anderson [205] considered a linear process $(X_k)_{k\geqslant 0}$ where X_k is of the form $X_k = \sum_{j=1}^{\infty} \psi_j Z_{k-j}$, $(Z_k)_{k\in\mathbb{Z}}$ are i.i.d. random variables with regularly varying balanced tails, and the coefficients ψ_j satisfy certain conditions. The limit theory for maxima in the sequence $(X_k)_{k\geqslant 0}$ was developed by Davis and Resnick [51], where the point process approach was used, *see* also Resnick [192]. Scotto, Turkman and Anderson [205] investigated the sub-sampled series $(Y_k)_{k\geqslant 1}$ that is defined by $Y_k = X_{kM}$, where $M > 1$ is a fixed sampling interval. *See* Exercises and Supplements at the end of this chapter. More details about linear processes and relations between maxima in complete and incomplete samples will be given below in Sect. 3.8.

Scotto and Turkman [204] investigated a stationary process $(X_k)_{k\geqslant 0}$ satisfying the stochastic difference equation

$$X_k = A_k X_{k-1} + B_k, \quad k \geqslant 1, \tag{3.6.1}$$

where (A_k, B_k), $k \geqslant 1$, are i.i.d. random vectors. Again, the sub-sampled series $(Y_k)_{k\geqslant 1}$ is defined by $Y_k = X_{kM}$, where $M > 1$ is a fixed sampling interval. In a subsequent paper Scotto [202] considered the same process $(X_k)_{k\geqslant 1}$ with a more complicated procedure of obtaining the sub-sampled series. Note that for every $n \geqslant 1$,

$$\begin{aligned} X_n &= B_n + A_n B_{n-1} + \cdots + A_n A_{n-1} \ldots A_2 B_1 + A_n A_{n-1} \ldots A_1 X_0 \\ &\overset{d}{=} B_1 + A_1 B_2 + \cdots + A_1 A_2 \ldots A_{n-1} B_n + A_1 A_2 \ldots A_n X_0, \end{aligned}$$

where $\overset{d}{=}$ means equality in distribution. Kesten [128], Vervaat [222], and Goldie [82] investigated Eq. (3.6.1), and proved that under certain conditions there exists a stationary solution whose distribution coincides with the distribution of the sum

$$\sum_{s=1}^{\infty} \prod_{r=1}^{s-1} A_r B_s, \tag{3.6.2}$$

where $\prod_{r=1}^{0} A_r B_s = B_1$. The extremal properties of sequence $(X_k)_{k\geqslant 1}$ were studied by de Haan et al. [92] and Perfekt [180]. The tail behavior of X_1 is determined by the tail behavior of B_1. We formulate here the theorem that gives the characterization of this tail behavior.

Theorem 3.6.1 (Grey [84]) *Let (A_k, B_k), $k \geqslant 1$, be i.i.d. random vectors such that the following conditions are satisfied:*

(a) $E \ln^+ |B_1| < +\infty$.
(b) $P\{A_1 \geqslant 0\} = 1$.
(c) *For some $\beta > \alpha > 0$, $EA_1^\alpha < 1$, and $EA_1^\beta < \infty$.*

Let $L(x)$ be a slowly varying function at infinity. Then as $x \to \infty$,

$$P\{B_1 > x\} \sim x^{-\alpha} L(x) \iff P\{X_1 > x\} \sim \frac{1}{1 - EA_1^\alpha} x^{-\alpha} L(x). \tag{3.6.3}$$

Scotto and Turkman [204] investigated the extremal properties of the sub-sampled series $(Y_k)_{k \geqslant 1}$. They also provide some results for the original series $(X_k)_{k \geqslant 1}$, *see* also Scotto [202]. We shall formulate here the result related to the extremal index of the sub-sampled series.

Theorem 3.6.2 (Scotto and Turkman [204]) *Under the conditions of Theorem 3.6.1 for every $M > 1$ the sub-sampled series $Y_k = X_{kM}$, $k \geqslant 1$, has an extremal index ϑ_M given by*

$$\vartheta_M = \int_1^\infty P\left\{ \max_{1 \leqslant r \leqslant \infty} \prod_{s=1}^{rM} A_s \leqslant \frac{1}{y} \right\} \alpha y^{-\alpha-1}\, dy,$$

and $$\lim_{n\to\infty} P\left\{ \max_{1 \leqslant k \leqslant [n/M]} Y_k \leqslant a_n x \right\} = \exp\left\{ -\frac{\vartheta_M}{M} \frac{1}{1 - EA_1^\alpha} x^{-\alpha} \right\}.$$

Peng, Weng and Nadarajah [179] and Tan and Wang [217] obtained almost sure limit theorems related to maxima and minima in complete and incomplete samples from stationary sequences.

The maxima of integer-valued stationary time series with missing values were studied in Hall and Temido [98]. In that paper the authors considered three models described as follows: Let $(X_n)_{n \geqslant 1}$ be a strictly stationary sequence of random variables with marginal distribution function F in the domain of attraction of a max-semistable distribution, and $M_n = \max\{X_1, \dots, X_n\}$. Suppose that $0 < x_F = \sup\{t \mid F(t) < 1\}$. Let us suppose that $(U_n)_{n \geqslant 1}$ is another stationary sequences having Bernoulli marginal distribution with parameter β, $0 \leqslant \beta \leqslant 1$, and assume that the sequences $(X_n)_{n \geqslant 1}$ and $(U_n)_{n \geqslant 1}$ are independent.

- Model with missing values (that are replaced by 0's): $Y_n = X_n U_n$, $n \geqslant 1$.
- Model with sub-sampling: $Z_n = X_{i_n}$, $n \geqslant 1$, where i_n are the indices for which $U_{i_n} = 1$.
- Model with replaced missing values: $W_n = U_n X_n + (1 - U_n) X_n^*$, where X_n^* is independent of X_n, and $X_n^* \stackrel{d}{=} X_n$.

Hall and Temido [98] investigated the effect of missing values on the distribution of the maximum in each of the models described above. They assumed that the pattern of occurrence of the missing values is i.i.d. or strongly mixing. They also assumed that stationary sequence $(X_n)_{n\geqslant 1}$ satisfies an asymptotic independence condition similar to condition $D(u_n)$, *see* Definition 2.7.7. Additionally, local dependence conditions similar to conditions $D^{(2)}(u_n)$ and $D^{(3)}(u_n)$ of Chernick, Hsing and McCormick [43] were imposed. The extremal index for the resulting sequence is obtained.

3.7 Uniform AR(1) Processes

3.7.1 Definition and Introductory Notes

In this section we consider an apparently simple stationary stochastic process with standard uniform marginals. The process is defined as a first order autoregressive ($AR(1)$) model

$$X_n = \alpha X_{n-1} + \varepsilon_n, \quad n \geqslant 1, \tag{3.7.1}$$

where $(\varepsilon_n)_{n\geqslant 1}$ is an i.i.d. sequence of innovations, independent of the initial value X_0. We will consider two different $AR(1)$ processes, each one parametrized by an integer $r \geqslant 2$. In both cases the initial state X_0 is taken to be a standard uniform random variable.

Positively correlated uniform $AR(1)$ processes are defined by (3.7.1) with $\alpha = 1/r$, where $r \geqslant 2$ is an integer. In this case a generic noise variable ε_n takes one of the r discrete values $\{0, 1/r, 2/r, \ldots, (r-1)/r\}$ with equal probabilities $1/r$. These processes were introduced by Chernick [41], *see* Example 2.7.3. *Negatively correlated uniform $AR(1)$ processes* are also defined by (3.7.1), but now $\alpha = -1/r$, with $r \geqslant 2$ still an integer. This time a generic noise variable ε_n takes one of the r discrete values $\{1/r, 2/r, \ldots, (r-1)/r, 1\}$ with equal probabilities $1/r$. These processes were introduced by Chernick and Davis [42], *see* Example 2.7.4.

The extreme values of positively and negatively correlated uniform $AR(1)$ processes are interesting and have attracted attention because their extremes cluster in a somewhat unusual way. Recall that marginally the standard uniform distribution is in the Weibull domain of attraction, and for $u_n = 1 - x/n$ with $x > 0$ we have $P\{\max_{1\leqslant i\leqslant n} U_i \leqslant u_n\} \to e^{-x}$ as $n \to \infty$, if $U_1, U_2, \ldots$ are i.i.d. standard uniform random variables. For this sequence (u_n) both positively and negatively correlated uniform $AR(1)$ processes satisfy the $D(u_n)$ condition of Leadbetter [141]. This was shown by Chernick [41] and Chernick and Davis [42]. If these processes also satisfied the $D'(u_n)$ condition of Leadbetter [141], then the extremes of these processes would not cluster, and the partial maxima of these processes would satisfy a limit theorem with the same normalization and the same limit as the i.i.d. standard uniform sequence. This is, however, not true, and condition $D'(u_n)$ fails for positively and negatively correlated uniform $AR(1)$ processes. In fact, it was shown by Chernick [41] that the partial maximum $M_n = \max\{X_1, \dots, X_n\}$ of positively correlated uniform $AR(1)$ processes satisfies $P\{M_n \leqslant 1 - x/n\} \to \exp\left\{-(1 - r^{-1})x\right\}$ as $n \to \infty$, while Chernick and Davis [42] showed that for negatively correlated uniform $AR(1)$ processes we have $P\{M_n \leqslant 1 - x/n\} \to \exp\left\{-(1 - r^{-2})x\right\}$ as $n \to \infty$. In particular, the extremes of the two processes cluster, and the extremal index of the two processes is equal to $1 - r^{-1}$ and $1 - r^{-2}$ in positively correlated and negatively correlated cases, correspondingly. *See* Leadbetter [142].

Even among processes whose extremes cluster, uniform $AR(1)$ processes may be unusual. Let us explain this point briefly. The stationary process $Y_n = (1 - X_n)^{-1}$, $n = 0, 1, \dots$ has standard Pareto (1) marginals, and, more generally, multivariate regularly varying distributions with exponent 1 of regular variation. Therefore, *the spectral tail process* is well defined; it can be obtained (extending first the $\boldsymbol{Y}$ process in law to a stationary process $(\dots, Y_{-1}, Y_0, Y_1, Y_2, \dots)$ indexed by $\mathbb{Z}$) by

$$P\left\{\left(\frac{Y_n}{Y_0}, n \in \mathbb{Z}\right) \in \cdot \,\middle|\, |Y_0| > y\right\} \Rightarrow P\left\{\left(T_n, n \in \mathbb{Z}\right) \in \cdot\right\}$$

as $y \to \infty$; *see* Basrak and Segers [21]. Since process $\boldsymbol{Y}$ is a Markov chain, we can expect, in accordance with the theory of Segers [209] and Janssen and Segers [117], that the tail process is itself a Markov chain of a particular type, the so-called back-and-forth tail chain. This is, indeed, the case if process $\boldsymbol{Y}$ corresponds to a positively correlated uniform $AR(1)$ process. It is not difficult to check that in this situation we have, in law,

$$T_n = \begin{cases} r^n & \text{if } n \leqslant 0, \\ \prod_{j=1}^{n} A_j^{(r)} & \text{if } n > 0, \end{cases}$$

where $A_1^{(r)}, A_2^{(r)}, \dots$ are i.i.d. random variables, such that $P\{A_1^{(r)} = r\} = 1 - P\{A_1^{(r)} = 0\} = 1/r$. On the other hand, if process $\boldsymbol{Y}$ corresponds to a negatively correlated uniform $AR(1)$ process, then the spectral tail process is not even a Markov chain. It is not difficult to check that in the above notation we can now write, in law,

$$T_n = \begin{cases} r^n & \text{if } n \leqslant 0 \text{ is even}, \\ \prod_{j=1}^{n/2} A_j^{(r^2)} & \text{if } n > 0 \text{ is even}, \\ 0 & \text{if } n \text{ is odd}. \end{cases}$$

This happens because if process $\boldsymbol{Z}$ is a negatively correlated uniform $AR(1)$ process, then process $\boldsymbol{Y}$ does not satisfy Condition 2.2 of Segers [209] or of Janssen and Segers [117].

We are interested in studying the extreme value theory of uniform $AR(1)$ processes in incomplete samples. This means that only some of the observations of the process of interest are registered, and the extremal behavior of the registered observations is studied. Questions of this type are of obvious interest in situations where more than one natural frequency of observation exists (for example, in finance), or where we may be interested in less frequent observations of the state of a physical system (e.g., annual) than the natural frequency (e.g., daily observations). The observation scheme could more generally be driven by a mechanism independent of the process of interest. Studying extreme value theory for incomplete samples has a venerable history, and most of the work has concentrated on cases where the registered observations were either equally spaced or were registered in a periodic manner. In this and the next sections, periodically registered observations provide one of the examples. The results that will be presented in this section were obtained by Glavaš, Mladenović, and Samorodnitsky [80].

In order to obtain a fuller picture of the extremes of a partially registered random sequence, it is useful to understand the joint behavior of the maxima of both fully and partially registered observations. Let $\boldsymbol{X} = (X_n)_{n\geqslant 1}$ be a stationary process, and let $c_n \in \{0, 1\}$, $n = 1, 2, \ldots$, be a deterministic sequence defining the registration: observation X_n of the process is registered if $c_n = 1$ and is not registered otherwise. Then

$$M_n = \max_{1\leqslant i\leqslant n} X_i \quad \text{and} \quad \widetilde{M}_n = \max_{1\leqslant i\leqslant n:\, c_i=1} X_i$$

are the two partial maxima of interest. Clearly, M_n is bigger than $\widetilde{M}_n$ and the partial maximum of the nonregistered observations, so, in particular, $M_n \geqslant \widetilde{M}_n$ for every $n \in \mathbb{N}$. We will always assume the existence of an asymptotic sampling frequency

$$p = \lim_{n\to\infty} \frac{1}{n} \sum_{j=1}^{n} c_j \in (0, 1]\,. \tag{3.7.2}$$

Under very general conditions it was shown in Mladenović and Piterbarg [161] that if the extremes of process $\boldsymbol{X}$ do not cluster, i.e., if the $D'(u_n)$ condition of Leadbetter [141] holds for process $\boldsymbol{X}$ and an appropriate sequence $(u_n)_{n\geqslant 1}$, then the partial maxima of registered and nonregistered observations are asymptotically independent, which fully determines the joint limiting distribution of appropriately normalized pair $\left(\widetilde{M}_n, M_n\right)$. In particular, this limiting distribution exists and is completely determined by the asymptotic frequency p in (3.7.2). This result, of course, does not apply in the case of uniform $AR(1)$ processes. In fact, it has been shown by examples in Mladenović [163], Mladenović and Živadinović [166] and Glavaš [75] that neither does the asymptotic independence of the partial maxima of registered and nonregistered observations hold nor is the limiting law of $\left(\widetilde{M}_n, M_n\right)$ determined by the asymptotic frequency p.

A general solution to the problem of the joint asymptotic behavior of properly normalized pair $\left(\widetilde{M}_n, M_n\right)$ will be given in what follows. We will provide sufficient conditions for the existence of the limiting distribution, explain what features of sampling sequence $(c_n)_{n\geqslant 1}$ determine the limiting distribution, and describe the form of the limiting distribution. The main results are stated in Sect. 3.7.2, which also provides a number of examples and a discussion. The proofs are given in Sect. 3.7.3. Section 3.7.4 contains certain estimates concerning the application of the $D(u_n, v_n)$ condition in our arguments.

3.7.2 Limit Theorems for Partial Maxima

The marginal distributions of uniform $AR(1)$ processes dictate the manner in which partial maxima should be normalized. Specifically, we will study the limiting behavior of the random vector

$$\left(n(1-\widetilde{M}_n), n(1-M_n)\right)$$

and try to determine the existence and shape of limit

$$G(x, y) = \lim_{n\to\infty} P\left\{\widetilde{M}_n \leqslant 1 - \frac{x}{n},\ M_n \leqslant 1 - \frac{y}{n}\right\}, \quad x, y > 0. \tag{3.7.3}$$

Notice that, if $0 < x \leqslant y$, then

$$\begin{aligned} P\left\{\widetilde{M}_n \leqslant 1 - \frac{x}{n},\ M_n \leqslant 1 - \frac{y}{n}\right\} &= P\left\{M_n \leqslant 1 - \frac{y}{n}\right\} \\ &\to \begin{cases} \exp\left\{-(1-r^{-1})y\right\} \text{ for positively dependent } \boldsymbol{X}, \\ \exp\left\{-(1-r^{-2})y\right\} \text{ for negatively dependent } \boldsymbol{X}, \end{cases} \end{aligned} \tag{3.7.4}$$

by Chernick [41] and Chernick and Davis [42], irrespectively of the behavior of the sampling sequence $(c_n)_{n\geqslant 1}$. Therefore, the only non-trivial case as far as the limit in (3.7.3) is concerned is the case $0 < y < x$.

We shall state and discuss below the limiting results for positively correlated uniform $AR(1)$ processes and negatively correlated uniform $AR(1)$ processes separately. Before doing so, we would like to draw attention to the difference in the mechanisms that make extremes cluster in the positively correlated case and in the negatively correlated case, and we shall do it in an informal manner. In the positively correlated case, a large value of the process will be followed by large values as long as the subsequent noise variables keep taking the value $(r-1)/r$. In the negatively correlated case a large value of the process cannot be immediately followed by another large value, but the value two units of time away can also be large as long as the pair of the two subsequent noise variables takes the value

$(1/r, 1)$. The importance of parity in the negatively correlated case will be visible both in the statement of the results and in their proofs.

A. *Positively correlated uniform $AR(1)$ processes.* For a sampling sequence $(c_n)_{n\geqslant 1}$ we denote for $n, j = 1, 2, \ldots$ the empirical frequencies of consecutive zeroes,

$$f_{n,j} = \sum_{i=1}^{n-j+1} \mathbf{1}_{\{c_i = c_{i+1} = \ldots = c_{i+j-1} = 0\}}, \tag{3.7.5}$$

if $j \leqslant n$ and $f_{n,j} = 0$ if $j > n$. We will use the notation

$$f_j = \lim_{n\to\infty} \frac{f_{n,j}}{n}, \quad j \in \{1, 2, \ldots\}, \tag{3.7.6}$$

if the limit exists. Note that $f_1 = 1 - p$, where p is defined in (3.7.2), and we always assume the existence of this limit. The following theorem is the main result for positively correlated uniform $AR(1)$ processes.

Theorem 3.7.1 (Glavaš, Mladenović, and Samorodnitsky [80]) *Let X be a positively correlated uniform $AR(1)$ process.*

(a) *If the limiting frequencies f_j in (3.7.6) exist for each $j \in \{1, 2, \ldots\}$, then the sequence $\big(n(1 - \widetilde{M}_n), n(1 - M_n)\big)$, $n = 1, 2, \ldots$ converges weakly as $n \to \infty$.*
(b) *Let j be a positive integer, and suppose that the limit f_k in (3.7.6) exists for all $k \in \{1, \ldots, j\}$. Then the limit $G(x, y)$ in (3.7.3) exists for all (x, y) in the range $0 < xr^{-j} \leqslant y < xr^{-(j-1)}$ and is given by*

$$G(x, y) = \exp\left\{ -(r-1)x \sum_{k=0}^{j-1} \frac{f_k - f_{k+1}}{r^{k+1}} - (r-1)y\frac{f_j}{r} \right\}, \tag{3.7.7}$$

with the convention $f_0 = 1$.

Remark 3.7.2 An immediate conclusion from Theorem 3.7.1 and the pointwise ergodic theorem is that if the sampling sequence $(c_n)_{n\geqslant 1}$ is the realization of a stationary 0–1 valued process, then for almost every such realization, the sequence $\big(n(1 - \widetilde{M}_n), n(1 - M_n)\big)$, $n = 1, 2, \ldots$ converges weakly. In particular, if the underlying stationary process is an i.i.d. Bernoulli sequence with probability p of appearing 1, then for almost every realization we have $f_j = (1-p)^j$, $j = 1, 2, \ldots$, and the expression (3.7.7) for the limiting distribution reduces to

$$G(x, y) = \exp\left\{ -\frac{(r-1)p}{r+p-1}\big(1-(1-p)^j r^{-j}\big)x - \frac{(r-1)}{r}(1-p)^j y \right\}$$

for (x, y) in the range $0 < xr^{-j} \leqslant y < xr^{-(j-1)}$, $j = 1, 2, \ldots$.

Remark 3.7.3 Suppose that the sampling sequence (c_n) is periodic with period $k \geqslant 1$. This sequence is one of the k possible realizations of the stationary process, consisting of taking the sequence (c_n) and erasing a random number N of its initial entries, the random number N having a discrete uniform distribution between 0 and $k-1$. Since each realization has a positive probability, Remark 3.7.2 applies, and the sequence $\big(n(1-\widetilde{M}_n), n(1-M_n)\big)$, $n = 1, 2, \ldots$ converges weakly. If l is the largest number of consecutive zeroes within a period, we have $f_j > 0$ for $1 \leqslant j \leqslant l$, and $f_j = 0$ for $j > l$. We conclude that the expression (3.7.7) for the limiting distribution G in the range $0 < xr^{-(l+1)} \leqslant y < xr^{-l}$ remains valid in the entire range $0 < y < xr^{-l}$. The results of Mladenović [163], Mladenović and Živadinović [166] and Glavaš [75] treat such periodic sampling sequences.

Remark 3.7.4 Instead of considering the joint distribution of the maxima of all observations and of the registered observations, we can consider the joint distribution of the maxima of registered and nonregistered observations. Let

$$\widehat{M}_n = \max_{1\leqslant i\leqslant n:\, c_i=0} X_i$$

be the partial maximum of nonregistered observations. Then

$$P\left\{\widetilde{M}_n \leqslant 1-\frac{x}{n},\ \widehat{M}_n \leqslant 1-\frac{y}{n}\right\}$$
$$= \begin{cases} P\left\{\widetilde{M}_n \leqslant 1-\frac{x}{n},\ M_n \leqslant 1-\frac{y}{n}\right\} & \text{if } 0 < y \leqslant x, \\ P\left\{M_n \leqslant 1-\frac{x}{n},\ \widehat{M}_n \leqslant 1-\frac{y}{n}\right\} & \text{if } 0 < x \leqslant y, \end{cases}$$

and so sequence $\big(n(1-\widetilde{M}_n), n(1-\widehat{M}_n)\big)$, $n = 1, 2, \ldots$, converges weakly if and only if both sequences $\big(n(1-\widetilde{M}_n), n(1-M_n)\big)$, $n = 1, 2, \ldots$, and $\big(n(1-\widehat{M}_n), n(1-M_n)\big)$, $n = 1, 2, \ldots$, converge weakly, sufficient conditions for which is the existence of the asymptotic frequency of any number of consecutive zeroes and the existence of the asymptotic frequency of any number of consecutive ones. Suppose that the asymptotic frequency of ones $p \in (0, 1)$, so that the marginal limits of both random variables $\big(n(1-\widehat{M}_n)\big)$ and $\big(n(1-\widetilde{M}_n)\big)$ are nondegenerate. When are the maxima of registered observations and nonregistered observations asymptotically independent? Theorem 3.7.1 says that a necessary and sufficient condition for such asymptotic independence is $f_k = f_1$ for all $k \geqslant 1$. That is, asymptotically, zeroes arrive in arbitrarily long groups and then, automatically, ones arrive in arbitrarily long groups as well. Such sampling sequences (c_n) exist. An example is the following sequence. Let $k_j = [j(1-p)/p]$ for $j = 1, 2, \ldots$, and construct a sampling sequence by alternating j consecutive ones with k_j consecutive zeroes, $j = 1, 2, \ldots$. Clearly, $f_1 = 1 - p$ in this case.

B. *Negatively correlated uniform* $AR(1)$ *processes.* In this case it is natural to consider for a sampling sequence $(c_n)_{n\geqslant 1}$ the empirical frequencies of zeroes at consecutive positions of the same parity, so we define

$$\tilde{f}_{n,j} = \sum_{i=1}^{n-2j+2} \mathbf{1}_{\{c_i = c_{i+2} = \cdots = c_{i+2j-2} = 0\}},$$

if $j \leqslant n/2$, and $\tilde{f}_{n,j} = 0$ if $j > n/2$. We will use the notation

$$\tilde{f}_j = \lim_{n\to\infty} \frac{\tilde{f}_{n,j}}{n}, \quad j \in \{1, 2, \dots\}, \tag{3.7.8}$$

if the limit exists. Note that $\tilde{f}_{n,1} = f_{n,1}$ defined by (3.7.5), and as before we assume that the limit $p = 1 - \tilde{f}_1 = 1 - f_1$ in (3.7.2) exists. The following theorem is our main result for negatively correlated uniform $AR(1)$ processes.

Theorem 3.7.5 (Glavaš, Mladenović, and Samorodnitsky [80]) *Let X be a negatively correlated uniform $AR(1)$ process.*

(a) *If the limiting frequencies $\tilde{f}_j$ in (3.7.8) exist for each $j \in \{1, 2, \dots\}$, then the sequence $\big(n(1 - \widetilde{M}_n), n(1 - M_n)\big)$, $n = 1, 2, \dots$ converges weakly as $n \to \infty$.*
(b) *Let j be a positive integer, and suppose that limit $\tilde{f}_k$ in (3.7.8) exists for all $k \in \{1, \dots, j\}$. Then limit $G(x, y)$ in (3.7.3) exists for all (x, y) in the range $0 < xr^{-2j} \leqslant y < xr^{-(2j-2)}$ and is given by*

$$G(x, y) = \exp\left\{ -(r^2 - 1)x \sum_{k=0}^{j-1} \frac{\tilde{f}_k - \tilde{f}_{k+1}}{r^{2k+2}} - (r^2 - 1)y \frac{\tilde{f}_j}{r^2} \right\}, \tag{3.7.9}$$

with the convention $\tilde{f}_0 = 1$.

Remarks regarding Theorem 3.7.1 that appear above have obvious counterparts in the negatively correlated case. We shall only address them briefly. If the sampling sequence $(c_n)_{n\geqslant 1}$ is the realization of a stationary 0–1 valued process, then for almost every realization the sequence $\big(n(1 - \widetilde{M}_n), n(1 - M_n)\big)$, $n = 1, 2, \dots$ still converges weakly, and in the special case of an i.i.d. Bernoulli sequence the limiting distribution of the process is

$$G(x, y) = \exp\left\{ -\frac{(r^2 - 1)p}{r^2 + p - 1}\big(1 - (1 - p)^j r^{-2j}\big)x - \frac{(r^2 - 1)}{r^2}(1 - p)^j y \right\}$$

for (x, y) in the range $0 < xr^{-2j} \leqslant y < xr^{-2(j-1)}$, $j = 1, 2, \dots$. Furthermore, we always have weak convergence of the sequence $\big(n(1-\widetilde{M}_n), n(1-M_n)\big)$ in the case of a periodic sampling scheme, and the requirement for the asymptotic independence of the maxima of registered observations and nonregistered observations is $\tilde{f}_k = \tilde{f}_1$ for all $k \geqslant 1$.

3.7.3 Proofs

The argument we will use depends, as do most of the arguments in related statements, on a version of the $D(u_n)$ condition. Since we have to deal with two types of observations, registered observations and nonregistered observations, and since the corresponding arguments in multivariate cumulative distribution functions take, in general, different values, we need a two-sequence version of the $D(u_n)$ condition. This condition, the so-called $D(u_n, v_n)$ condition (with $u_n = 1 - x/n$, $v_n = 1 - y/n$, $x, y > 0$ in our case), was introduced by Davis [49], and Chernick and Davis [42] explained that this condition holds for both positively and negatively correlated uniform $AR(1)$ processes. This implies the following fact. For a fixed positive integer j, let $x, y > 0$ be either in the range $0 < xr^{-j} \leqslant y < xr^{-(j-1)}$ or in the range $0 < xr^{-2j} \leqslant y < xr^{-(2j-2)}$, depending on whether the process is positively correlated or negatively correlated. In the positively correlated case assume that limit f_k in (3.7.6) exists for all $k \in \{1, \ldots, j\}$, and in the negatively correlated case assume that limit $\tilde{f}_k$ in (3.7.8) exists for all $k \in \{1, \ldots, j\}$. Let (m_n) be a sequence of positive integers such that $m_n \to \infty$, $m_n/n \to 0$ as $n \to \infty$. If limit

$$\lim_{n\to\infty} \frac{n}{m_n}\Big(1 - P\Big(\widetilde{M}_{m_n} \leqslant 1 - \frac{x}{n},\ M_{m_n} \leqslant 1 - \frac{y}{n}\Big)\Big) = H(x, y) \tag{3.7.10}$$

exists, then limit $G(x, y)$ in (3.7.3) exists, and

$$G(x, y) = \exp\big\{ - H(x, y)\big\}. \tag{3.7.11}$$

We shall explain why this is true in Sect. 3.7.4.

Proof of Theorem 3.7.1 Since part (a) of the theorem is an obvious consequence of part (b), we will prove part (b) of the theorem. Throughout the proof we will use the notation

$$z_i = \begin{cases} x, \text{ if } c_i = 1, \\ y, \text{ if } c_i = 0. \end{cases} \tag{3.7.12}$$

For a positive integer m we denote

$$\begin{aligned} a_m &= P\big\{\widetilde{M}_m \leqslant 1 - x/n,\ M_m \leqslant 1 - y/n\big\} \\ &= P\left\{X_1 \leqslant 1 - z_1/n,\ X_2 \leqslant 1 - z_2/n, \ldots, X_m \leqslant 1 - z_m/n\right\}. \end{aligned} \tag{3.7.13}$$

We will let $m = m_n \to \infty$ sufficiently slowly with n so that $r^m/n \to 0$. Note that

$$X_i = r^{-(i-1)}X_1 + \sum_{j=2}^{i} r^{-(i-j)}\varepsilon_j,\quad i = 2, 3, \ldots, m\,.$$

This implies the following fact that we will use repeatedly in the subsequent calculations:

- Inequality $X_i \leqslant 1 - z_i/n$ holds if and only if

$$X_1 \leqslant r^{i-1} - \sum_{j=2}^{i} r^{j-1}\varepsilon_j - r^{i-1}z_i/n.$$

Writing $\varepsilon_j = k_j/r$ with $k_j \in \{0, 1, \dots, r-1\}$, this condition translates into

$$X_1 \leqslant r^{i-1} - \sum_{j=2}^{i} r^{j-2}k_j - r^{i-1}z_i/n\,. \tag{3.7.14}$$

We will often use the following elementary fact about the right-hand side of (3.7.14):

$$r^{i-1} - \sum_{j=2}^{i} r^{j-2}k_j \begin{cases} = 1, \text{ if } k_j = r-1 \text{ for all } j = 2, 3, \dots, i, \\ \geqslant 2, \text{ in all other cases.} \end{cases} \tag{3.7.15}$$

Using inequality (3.7.14), the probability in (3.7.13) can be written in the form

$$\begin{aligned} a_m &= \\ &= \sum_{k_2=0}^{r-1}\sum_{k_3=0}^{r-1}\cdots\sum_{k_m=0}^{r-1} r^{-(m-1)}P\Bigg(\bigcap_{i=1}^{m}\Big\{X_1 \leqslant r^{i-1} - \sum_{j=2}^{i} r^{j-2}k_j - r^{i-1}z_i/n\Big\}\Bigg) \\ &= S_1 + S_2 + \dots + S_m, \end{aligned} \tag{3.7.16}$$

where we have decomposed the full $(m-1)$-tuple sum on the first line by grouping its terms into sums S_l, $1 \leqslant l \leqslant m$ according to the following rules:

- Sum S_m consists of the single term with $k_2 = \dots = k_m = r-1$.
- For $2 \leqslant l \leqslant m-1$, sum S_l runs over all $k_2, \dots, k_m \in \{0, 1, \dots, r-1\}$ such that $k_2 = \dots = k_l = r-1$, but $k_{l+1} \neq r-1$.
- Sum S_1 runs over all $k_2, \dots, k_m \in \{0, 1, \dots, r-1\}$ such that $k_2 \neq r-1$.

It is easy to compute the number of terms in each sum. We shall use these numbers as part of the calculations below. Recall that according to our constraints on the rate of growth of m we have $r^{i-1}z_i/n \in (0, 1)$ for all n large enough. For all such n we have

$$S_m = r^{-(m-1)}P\Bigg(\bigcap_{i=1}^{m}\Big\{X_1 \leqslant 1 - \frac{r^{i-1}z_i}{n}\Big\}\Bigg)$$

$$= r^{-(m-1)}\left(1 - \frac{\max_{1\leqslant i\leqslant m} r^{i-1}z_i}{n}\right), \tag{3.7.17}$$

while for $1 \leqslant l \leqslant m-1$, we have

$$S_l = \left[(r-1)r^{m-l-1}\right]r^{-(m-1)}P\left(\bigcap_{i=1}^{l}\left\{X_1 \leqslant 1 - \frac{r^{i-1}z_i}{n}\right\}\right)$$

$$= \frac{r-1}{r^l}\left(1 - \frac{\max_{1\leqslant i\leqslant l} r^{i-1}z_i}{n}\right)$$

$$= \frac{r-1}{r^l} - \frac{(r-1)\max_{1\leqslant i\leqslant l} r^i z_i}{r^{l+1}n} := A_l - B_l\,. \tag{3.7.18}$$

Let $j = 1, 2, \ldots$ and consider a pair (x, y) such that $0 < xr^{-j} \leqslant y < xr^{-(j-1)}$. Let $l \in \{1, 2, \ldots, m-1\}$. The following cases are possible in (3.7.18).

Case 1 For some $k \in \{0, 1, \ldots, j-1\}$, $c_{l-k} = 1$, while $c_{l-k+1} = \ldots = c_l = 0$. In this case

$$\max_{1\leqslant i\leqslant l} r^i z_i = r^{l-k}x\,,$$

so that

$$B_l = \frac{(r-1)x}{r^{k+1}n}\,.$$

Note that this scenario is feasible only if $l - k \geqslant 1$.

Case 2 $c_d = 0$ for all integer d such that $\max\{l-j+1, 1\} \leqslant d \leqslant l$. In this case

$$\max_{1\leqslant i\leqslant l} r^i z_i = r^l y\,,$$

so that

$$B_l = \frac{(r-1)y}{rn}\,.$$

We conclude that each occurrence of the pattern $10\ldots0$ (one followed by k zeroes), $k \in \{0, 1, \ldots, j-1\}$, in sequence $c_1, \ldots, c_{m-1}$ contributes exactly

$$\frac{(r-1)x}{r^{k+1}n}$$

to sum $B_1 + \ldots + B_{m-1}$, while each occurrence of pattern $0\ldots0$ (j zeroes) contributes exactly

$$\frac{(r-1)y}{rn}$$

to that sum. It is elementary that the number of times that pattern $10\ldots0$ (one followed by k zeroes) occurs in sequence $c_1, \ldots, c_{m-1}$ is equal to $f_{m,k} - f_{m,k+1} - \delta$, with $\delta \in \{0, 1\}$ ($\delta = 1$ if the number of zeroes at the initial positions of sequence (c_n) is at least k) and the convention that $f_{m,0} = m$. Therefore,

$$\begin{aligned} S_1 + \ldots + S_{m-1} =& \frac{1}{r} - \frac{1}{r^m} \\ &- \frac{(r-1)x}{n} \sum_{k=0}^{j-1} \frac{f_{m,k} - f_{m,k+1}}{r^{k+1}} - \frac{(r-1)y}{rn} f_{m,j} + O(1/n), \end{aligned}$$

where the $O(1/n)$ term comes both from the δ correction above and from the fact that sum $B_1 + \ldots + B_{m-1}$ can also contain some additional terms $(r-1)y/(rn)$ due to the possible presence of a string of initial zeroes in sequence $c_1, \ldots, c_{m-1}$ of length smaller than j. We conclude by (3.7.16) that

$$a_m = 1 - \frac{(r-1)x}{n} \sum_{k=0}^{j-1} \frac{f_{m,k} - f_{m,k+1}}{r^{k+1}} - \frac{(r-1)y}{rn} f_{m,j} + O(1/n).$$

Since $m = m_n \to \infty$ as $n \to \infty$, we conclude by (3.7.13) that

$$\begin{aligned} &\frac{n}{m_n}\left(1 - P\left(\widetilde{M}_{m_n} \leqslant 1 - x/n,\ M_{m_n} \leqslant 1 - y/n\right)\right) \\ &= (r-1)x \sum_{k=0}^{j-1} \left(\frac{f_{m_n,k}}{m_n} - \frac{f_{m_n,k+1}}{m_n}\right) r^{-(k+1)} + \frac{r-1}{r} y \cdot \frac{f_{m_n,j}}{m_n} + o(1) \\ &\to (r-1)x \sum_{k=0}^{j-1} \frac{f_k - f_{k+1}}{r^{k+1}} + (r-1)y \frac{f_j}{r} \end{aligned} \tag{3.7.19}$$

as $n \to \infty$. Therefore, (3.7.10) holds. Referent to (3.7.11) proves (3.7.7) and, hence, establishes part (b) of the theorem. □

Proof of Theorem 3.7.5 As in the proof of Theorem 3.7.1, it is enough to prove part (b). We use once again the notation (3.7.13), but in this case it is convenient for us to let sequence $m = m_n$ consist of even numbers, so for $0 < y < x$ and a positive even integer $2m$ we will consider

$$a_{2m} = P\left\{\widetilde{M}_{2m} \leqslant 1 - \frac{x}{n},\ M_{2m} \leqslant 1 - \frac{y}{n}\right\}$$

$$= P\left\{X_1 \leqslant 1 - \frac{z_1}{n}, X_2 \leqslant 1 - \frac{z_2}{n}, \dots, X_{2m} \leqslant 1 - \frac{z_{2m}}{n}\right\}, \tag{3.7.20}$$

where z_i is once again given by (3.7.12). It is appropriate now to let $m = m_n \to \infty$ sufficiently slowly with n so that $r^{2m}/n \to 0$. The parity of the time variables is important now, so we will use a more detailed representation of stationary sequence (X_n) than in the proof of Theorem 3.7.1. Let us write

$$X_{2i-1} = r^{-2i+2}X_1 + \sum_{j=2}^{i} r^{-2i+2j}\varepsilon_{2j-1} - \sum_{j=1}^{i-1} r^{-2i+2j+1}\varepsilon_{2j}, \quad i \geqslant 2,$$

$$X_{2i} = -r^{-2i+1}X_1 - \sum_{j=1}^{i-1} r^{-2i+2j+1}\varepsilon_{2j+1} + \sum_{j=1}^{i} r^{-2i+2j}\varepsilon_{2j}, \quad i \geqslant 1.$$

Certain simple facts follow from the above representations. We will list them separately as they refer to odd-numbered observations and even-numbered observations. We write for each j, $\varepsilon_j = k_j/r$ with $k_j \in \{1, \dots, r\}$.

The first simple fact is that for $i \geqslant 2$ inequality $X_{2i-1} \leqslant 1 - z_{2i-1}/n$ holds if and only if

$$X_1 \leqslant r^{2i-2} + \sum_{j=1}^{i-1} r^{2j-2}k_{2j} - \sum_{j=2}^{i} r^{2j-3}k_{2j-1} - \frac{r^{2i-2}z_{2i-1}}{n}. \tag{3.7.21}$$

We will use the following elementary fact about the right-hand side of (3.7.21):

$$r^{2i-2} + \sum_{j=1}^{i-1} r^{2j-2}k_{2j} - \sum_{j=2}^{i} r^{2j-3}k_{2j-1}$$

$$\begin{cases} = 1, \text{ if } k_2 = k_4 = \dots = k_{2i-2} = 1, \;\; k_3 = \dots = k_{2i-3} = r, \\ \geqslant 2, \text{ in all other cases.} \end{cases} \tag{3.7.22}$$

The second simple fact is that for $i \geqslant 1$ inequality $X_{2i} \leqslant 1 - z_{2i}/n$ holds if and only if

$$X_1 \geqslant -r^{2i-1} - \sum_{j=1}^{i-1} r^{2j-1}k_{2j+1} + \sum_{j=1}^{i} r^{2j-2}k_{2j} + r^{2i-1}z_{2i}/n. \tag{3.7.23}$$

Once again, we will use an elementary fact about the right-hand side of (3.7.23):

$$-r^{2i-1} - \sum_{j=1}^{i-1} r^{2j-1} k_{2j+1} + \sum_{j=1}^{i} r^{2j-2} k_{2j}$$

$$\begin{cases} = 0, & \text{if } k_2 = k_4 = \cdots = k_{2i} = r,\ k_3 = \cdots = k_{2i-1} = 1, \\ \leqslant -1, & \text{in all other cases.} \end{cases} \tag{3.7.24}$$

We proceed with decomposition of the probability in (3.7.20) parallel to the decomposition in (3.7.16), but now we have to be careful about parity of the time stamp. We write

$$\begin{aligned} a_{2m} &= \sum_{k_2=1}^{r} \sum_{k_3=1}^{r} \cdots \sum_{k_{2m}=1}^{r} r^{-2m+1} P\Bigg(\left\{ X_1 \leqslant 1 - \frac{z_1}{n} \right\} \cap \\ &\bigcap_{i=2}^{m} \left\{ X_1 \leqslant r^{2i-2} + \sum_{j=1}^{i-1} r^{2j-2} k_{2j} - \sum_{j=2}^{i} r^{2j-3} k_{2j-1} - \frac{r^{2i-2} z_{2i-1}}{n} \right\} \\ &\bigcap_{i=1}^{m} \left\{ X_1 \geqslant -r^{2i-1} - \sum_{j=1}^{i-1} r^{2j-1} k_{2j+1} + \sum_{j=1}^{i} r^{2j-2} k_{2j} + \frac{r^{2i-1} z_{2i}}{n} \right\} \Bigg) \\ &= S_1 + S_2 + \cdots + S_{2m}, \end{aligned} \tag{3.7.25}$$

where we have decomposed the full $(2m-1)$-tuple sum on the first line by grouping, once again, its terms into sums S_i, $1 \leqslant i \leqslant 2m$, according to the following rules:

Sum S_{2m} consists of a single term with $k_2 = k_4 = \cdots = k_{2m} = r$ and $k_3 = k_5 = \cdots = k_{2m-1} = 1$. If we select n so large that $z_1/n + r^{j-1} z_j/n \in (0, 1)$ for all $2 \leqslant j \leqslant 2m$ (this choice of n will remain in force for the duration of the proof), then

$$\begin{aligned} S_{2m} &= r^{-2m+1} P\left\{ X_1 \leqslant 1 - \frac{z_1}{n},\, X_1 \geqslant \max_{1 \leqslant i \leqslant m} \frac{r^{2i-1} z_{2i}}{n} \right\} \\ &= r^{-2m+1} \left(1 - \frac{z_1}{n} - \frac{\max_{1 \leqslant i \leqslant m} r^{2i-1} z_{2i}}{n} \right). \end{aligned} \tag{3.7.26}$$

Sum S_{2m-1} consists of r terms with $k_2 = k_4 = \cdots = k_{2m-2} = 1$ and $k_3 = k_5 = \cdots = k_{2m-1} = r$, while k_{2m} can take any value in $\{1, 2, \ldots, r\}$. We have

$$\begin{aligned} S_{2m-1} &= \frac{r}{r^{2m-1}} P\left\{ X_1 \leqslant 1 - \frac{z_1}{n},\, X_1 \leqslant 1 - \max_{2 \leqslant i \leqslant m} \frac{r^{2i-2} z_{2i-1}}{n} \right\} \\ &= r^{-2m+2} \left(1 - \frac{\max_{1 \leqslant i \leqslant m} r^{2i-2} z_{2i-1}}{n} \right). \end{aligned} \tag{3.7.27}$$

For $1 \leqslant l \leqslant m-1$, sum S_{2l} runs over all $k_2, k_3, \ldots, k_{2m} \in \{1, 2, \ldots, r\}$ such that $k_2 = \cdots = k_{2l} = r$ and $k_3 = \cdots = k_{2l-1} = 1$, but $(k_{2l+1}, k_{2l+2}) \neq (1, r)$. Clearly, this sum has $(r^2-1)r^{2m-2l-2}$ terms and

$$
\begin{aligned}
S_{2l} &= \frac{r^2-1}{r^{2l+1}}\left(1 - \frac{z_1}{n} - \max_{1\leqslant i\leqslant l} \frac{r^{2i-1}z_{2i}}{n}\right) \\
&= \frac{r^2-1}{r^{2l+1}}\left(1 - \frac{z_1}{n}\right) - \frac{(r^2-1)\max_{1\leqslant i\leqslant l} r^{2i} z_{2i}}{r^{2l+2}n} \\
&:= A_{2l} - B_{2l}.
\end{aligned} \tag{3.7.28}
$$

For $2 \leqslant l \leqslant m-1$, sum S_{2l-1} is taken over all $k_2, k_3, \ldots, k_{2m} \in \{1, 2, \ldots, r\}$ such that $k_2 = \cdots = k_{2l-2} = 1$ and $k_3 = \cdots = k_{2l-1} = r$, but $(k_{2l}, k_{2l+1}) \neq (1, r)$. This sum has $(r^2-1)r^{2m-2l-1}$ terms and

$$
\begin{aligned}
S_{2l-1} &= \frac{r^2-1}{r^{2l}}\left(1 - \max_{1\leqslant i\leqslant l} \frac{r^{2i-2}z_{2i-1}}{n}\right) \\
&= \frac{r^2-1}{r^{2l}} - \frac{(r^2-1)\max_{1\leqslant i\leqslant l} r^{2i-1} z_{2i-1}}{r^{2l+1}n} \\
&:= A_{2l-1} - B_{2l-1}.
\end{aligned} \tag{3.7.29}
$$

Finally, sum S_1 runs over all $k_2, k_3, \ldots, k_{2m} \in \{1, 2, \ldots, r\}$ such that both $k_2 \neq r$ and $(k_2, k_3) \neq (1, r)$. This sum has $r^{2m-1} - r^{2m-2} - r^{2m-3}$ identical terms, so that

$$
S_1 = \left(1 - r^{-1} - r^{-2}\right)\left(1 - z_1/n\right). \tag{3.7.30}
$$

We proceed now in a manner similar to the steps we took to prove Theorem 3.7.1. Let $j = 1, 2, \ldots$ and consider a pair (x, y) such that $0 < xr^{-2j} \leqslant y < xr^{-(2j-2)}$. We consider $l \in \{2, \ldots, 2m-2\}$. The following cases are possible in (3.7.28) and (3.7.29).

Case 1 If for some $k \in \{0, 1, \ldots, j-1\}$, $c_{2l-2k} = 1$, while $c_{2l-2k+2} = \cdots = c_{2l} = 0$, then $\max_{1\leqslant i\leqslant l} r^{2i} z_{2i} = r^{2l-2k}x$ and hence B_{2l} is equal to

$$
\frac{(r^2-1)x}{r^{2k+2}n}. \tag{3.7.31}
$$

Similarly, if for some $k \in \{0, 1, \ldots, j-1\}$, $c_{2l-2k-1} = 1$, while $c_{2l-2k+1} = \cdots = c_{2l-1} = 0$, then $\max_{1\leqslant i\leqslant l} r^{2i-1} z_{2i-1} = r^{2l-2k-1}x$ and B_{2l-1} is still given by (3.7.31).

Case 2 If $c_{2d} = 0$ for all even integers $2d$ such that $\max\{2l-2j+2, 2\} \leqslant 2d \leqslant 2l$, then $\max_{1\leqslant i\leqslant l} r^{2i} z_{2i} = r^{2l}y$, and B_{2l} is equal to

$$\frac{(r^2-1)y}{r^2 n}. \tag{3.7.32}$$

If $c_{2d-1}=0$ for all odd integers $2d-1$ such that $\max\{2l-2k-1,1\}\leqslant 2d-1\leqslant 2l-1$, then $\max_{1\leqslant i\leqslant l} r^{2i-1}z_{2i-1}=r^{2l-1}y$, and B_{2l-1} is still given by (3.7.32).

Let us denote by $\Pi_k(10\ldots0)$ the pattern *one followed by k zeroes on consecutive positions of the same parity*, where $k\in\{0,1,\ldots,j-1\}$. Each occurence of such pattern in sequence $c_2,\ldots,c_{2m-2}$ contributes exactly

$$\frac{(r^2-1)x}{r^{2k+2}n}$$

to sum $B_2+\cdots+B_{2m-2}$. Each occurence of the pattern "j zeroes on consecutive positions of the same parity" contributes exactly

$$\frac{(r^2-1)y}{r^2 n}$$

to that sum. The number of times the pattern $\Pi_k(10\ldots0)$ occurs in sequence $c_1,\ldots,c_{2m-2}$ is equal to $\tilde f_{2m-2,k}-\tilde f_{2m-2,k+1}-\delta$, where $\delta\in\{0,1,2\}$ depends on the initial strings of zeroes on even and odd positions in sequence (c_n), with the convention $\tilde f_{2m,0}=2m$. As in the proof of Theorem 3.7.1 we conclude that

$$a_{2m}=1-\frac{(r^2-1)x}{n}\sum_{k=0}^{j-1}\frac{\tilde f_{2m-2,k}-\tilde f_{2m-2,k+1}}{r^{2k+2}}-\frac{(r^2-1)y}{r^2n}\tilde f_{2m-2,j}+O(1/n).$$

Since $m=m_n\to\infty$ as $n\to\infty$, we conclude by (3.7.8) that

$$\begin{aligned}
&\frac{n}{2m}\left(1-P\left(\tilde M_{2m}\leqslant 1-\frac{x}{n},\ M_{2m}\leqslant 1-\frac{y}{n}\right)\right)\\
&=(r^2-1)x\sum_{k=0}^{j-1}\left(\frac{\tilde f_{2m-2,k}}{2m}-\frac{\tilde f_{2m-2,k+1}}{2m}\right)r^{-(2k+2)}+(r^2-1)y\frac{\tilde f_{2m-2,j}}{2mr^2}+o(1)\\
&\to (r^2-1)x\sum_{k=0}^{j-1}\frac{\tilde f_k-\tilde f_{k+1}}{r^{2k+2}}+(r^2-1)y\frac{\tilde f_j}{r^2}
\end{aligned}$$

as $n\to\infty$, so (3.7.10) holds, and the statement of part (b) of the theorem follows from (3.7.11). □

3.7.4 Condition $D(u_n, v_n)$

In this subsection we shall prove the claim made at the beginning of Sect. 3.7.3. Specifically, if $x, y > 0$ are in the range $0 < xr^{-j} \leqslant y < xr^{-(j-1)}$ for some positive integer j, then under the assumption that the limit f_k in (3.7.6) exists for all $k \in \{1, \dots, j\}$, statement (3.7.10) for positively correlated uniform $AR(1)$ processes implies that the limit in (3.7.3) exists and is given by (3.7.11). The argument in the negatively correlated case (under the assumption that the appropriate limits in (3.7.8) exist) is similar.

We claim that positively correlated uniform $AR(1)$ processes satisfy condition $D(u_n, v_n)$ from Definition 3.3.1 with $u_n = 1 - y/n$, $v_n = 1 - x/n$, where $x > 0$, $y > 0$. Moreover, we can choose $\alpha_{n,l}$ to be independent of $l \geqslant 1$, depending instead on the cardinality of $A_1 \cup A_2$. Specifically, inequality (3.3.1) holds with the right-hand side given by

$$\alpha_{n,m} = \frac{\max(x, y)}{n} \frac{\ln(m \max(x, y))}{\ln r} + \frac{1}{n} \frac{r}{1 - 1/r}, \tag{3.7.33}$$

where $m = card(A_1 \cup A_2)$.

In order to establish the above fact, let us denote

$$\widetilde{A}_1(u_n) = \bigcap_{j \in A_1} \{X_j \leqslant u_n\}, \quad \widetilde{A}_2(v_n) = \bigcap_{j \in A_2} \{X_j \leqslant v_n\},$$

$$\widetilde{B}_1(u_n) = \bigcap_{j \in B_1} \{X_j \leqslant u_n\}, \quad \widetilde{B}_2(v_n) = \bigcap_{j \in B_2} \{X_j \leqslant v_n\},$$

$$\Delta_n(A_1, A_2, B_1, B_2) = P\left(\widetilde{A}_1(u_n)\widetilde{A}_2(v_n)\widetilde{B}_1(u_n)\widetilde{B}_2(v_n)\right) - P\left(\widetilde{A}_1(u_n)\widetilde{A}_2(v_n)\right) \cdot P\left(\widetilde{B}_1(u_n)\widetilde{B}_2(v_n)\right).$$

Since process (X_n) is obviously associated, it follows that

$$\begin{aligned} 0 \leqslant \Delta_n(A_1, A_2, B_1, B_2) &= P\left(\widetilde{A}_1(u_n)\widetilde{A}_2(v_n)\right) \times \\ &\times \left\{ P\left(\widetilde{B}_1(u_n)\widetilde{B}_2(v_n) | \widetilde{A}_1(u_n)\widetilde{A}_2(v_n)\right) - P\left(\widetilde{B}_1(u_n)\widetilde{B}_2(v_n)\right) \right\} \\ &:= P\left(\widetilde{A}_1(u_n)\widetilde{A}_2(v_n)\right)(P_1 - P_2). \end{aligned}$$

Let $i_* = \max(A_1 \cup A_2)$. To bound the difference $P_1 - P_2$ we will use a coupling argument and start by outlining its general structure. Let $X_0, \hat{X}_0, \varepsilon_i, i = 1, 2, \dots$ be random variables defined on some probability space, satisfying the following conditions. Both X_0 and $\hat{X}_0$ take values in $(0, 1)$, $\hat{X}_0$ has the standard uniform distribution, and $\hat{X}_0 - X_0 \leqslant \vartheta$ a.s. for some non-random $\vartheta \in (0, 1)$. Furthermore, $(X_0, \hat{X}_0)$ are independent of i.i.d sequence (ε_i) of random variables taking values

$\{0, 1/r, \ldots, (r-1)/r\}$ with equal probabilities. Let us define

$$X_i = r^{-i} X_0 + \sum_{j=1}^{i} r^{-(i-j)} \varepsilon_j, \; i = 1, 2, \ldots$$

$$\hat{X}_i = r^{-i} \hat{X}_0 + \sum_{j=1}^{i} r^{-(i-j)} \varepsilon_j, \; i = 1, 2, \ldots.$$

Let C and D be two disjoint finite subsets of $\{l, l+1, \ldots\}$. Then

$$\begin{aligned} & P\{X_i \leqslant u_n, i \in C, \; X_i \leqslant v_n, i \in D\} \\ & \leqslant P\{\hat{X}_i - r^{-i}\vartheta \leqslant u_n, i \in C, \; \hat{X}_i - r^{-i}\vartheta \leqslant v_n, i \in D\} \\ & \leqslant P\{\hat{X}_i \leqslant u_n, i \in C, \; \hat{X}_i \leqslant v_n, i \in D\} \\ & + \sum_{i \in C} P\{u_n < \hat{X}_i \leqslant u_n + r^{-i}\vartheta\} + \sum_{i \in D} P\{v_n < \hat{X}_i \leqslant v_n + r^{-i}\vartheta\}. \end{aligned}$$

Therefore,

$$\begin{aligned} & P\{X_i \leqslant u_n, i \in C, \; X_i \leqslant v_n, i \in D\} - P\{\hat{X}_i \leqslant u_n, i \in C, \; \hat{X}_i \leqslant v_n, i \in D\} \\ & \leqslant \sum_{i=l}^{\infty} \min\left(r^{-i}\vartheta, \; \max(x, y)/n\right) = \sum_{l \leqslant i \leqslant \ln(n\vartheta)/\ln r} \cdot + \sum_{i > \ln(n\vartheta)/\ln r} \cdot \\ & \leqslant \frac{\max(x, y)}{n} \frac{\ln(n\vartheta)}{\ln r} + \frac{1}{n} \frac{r}{1 - 1/r}. \end{aligned} \tag{3.7.34}$$

We now use (3.7.34) to estimate the difference $P_1 - P_2$. To this end we need to couple two random variables, X_{i_*} with its unconditional standard uniform distribution, and the same X_{i_*} with its conditional law given $X_i \leqslant u_n, i \in A_1$ and $X_i \leqslant v_n, i \in A_2$. In the coupling the latter random variable is X_0 in the computation leading to (3.7.34), and the former random variable is $\hat{X}_0$. We need to couple these random variables in such a way that the resulting difference ϑ is small.

Note that the conditional law of X_{i_*} given $X_i \leqslant u_n, i \in A_1$ and $X_i \leqslant v_n, i \in A_2$ is uniform on some subset of $(0, 1)$, consisting of some finite collection of intervals. Each condition of the type $X_i \leqslant u_n, i \in A_1$ removes some subintervals of $(0, 1)$ of total length y/n from the support of the conditional law of X_{i_*}, while each condition of the type $X_i \leqslant v_n, i \in A_2$ removes some subintervals of $(0, 1)$ of total length x/n from the same support. If $m = card(A_1 \cup A_2)$, then the total number of conditions is m, and the conditional law of X_{i_*} given $X_i \leqslant u_n, i \in A_1$ and $X_i \leqslant v_n, i \in A_2$ is uniform on a finite union of subintervals of $(0, 1)$ of total length at least $1 - (m/n) \max(x, y)$. According to Lemma 3.7.6 below, we can achieve the coupling of X_0 and $\hat{X}_0$ with $\vartheta = (m/n) \max(x, y)$. Therefore, from (3.7.34),

$$P_1 - P_2 \leqslant \frac{\max(x, y)}{n} \frac{\ln(m \max(x, y))}{\ln r} + \frac{1}{n} \frac{r}{1 - 1/r}$$

and, hence, inequality (3.3.1) holds with the bound in (3.7.33).

Lemma 3.7.6 *Let $0 < \gamma < 1$ and let X_0 be a random variable with uniform distribution on the finite union of disjoint subintervals of $(0, 1)$ of total length γ. Then there is a coupling of X_0 with a standard uniform random variable $\hat{X}_0$ such that $\hat{X}_0 - X_0 \leqslant 1 - \gamma$ a.s..*

Proof Let the support of X_0 be the disjoint union of $(h_i, h_i + p_i)$, $i \in \{1, \dots, k\}$, with $0 \leqslant h_1 < h_1 + p_1 < h_2 < h_2 + p_2 < \dots < h_k < h_k + p_k \leqslant 1$ and $p_1 + p_2 + \dots + p_k = \gamma$. We couple X_0 and $\hat{X}_0$ as follows: Generate X_0. If $X_0 \in (h_i, h_i + p_i)$ for some $i \in \{1, \dots, k\}$, we set

$$\hat{X}_0 = X_0/\gamma + (p_1 + \dots + p_{i-1} - h_i)/\gamma.$$

It is elementary to check that $\hat{X}_0$ has the standard uniform distribution. If $X_0 \in (h_i, h_i + p_i)$ for some $i \in \{1, \dots, k\}$, we have

$$\begin{aligned}
\hat{X}_0 - X_0 &= (1/\gamma - 1)\, X_0 + (p_1 + \dots + p_{i-1} - h_i)/\gamma \\
&\leqslant (1/\gamma - 1)\, (h_i + p_i) + (p_1 + \dots + p_{i-1} - h_i)/\gamma \\
&= -\, (h_i + p_i) + (p_1 + \dots + p_{i-1})/\gamma \\
&\leqslant -\, (p_1 + \dots + p_{i-1} + p_i) + (p_1 + \dots + p_{i-1} + p_i)/\gamma \\
&= (1/\gamma - 1)\, (p_1 + \dots + p_{i-1} + p_i) \\
&\leqslant (1/\gamma - 1)\, \gamma = 1 - \gamma,
\end{aligned}$$

as required. □

We are now ready to prove that (3.7.10) implies that limit $G(x, y)$ in (3.7.3) exists and (3.7.11) holds. Let $m = m_n \to \infty$ be a sequence of positive integers such that $m_n/n \to 0$ and (3.7.10) holds. Choose another sequence of positive integers, $l = l_n \to \infty$ such that $l_n/m_n \to 0$ as $n \to \infty$. Let $k = [n/(m + l)]$. For $0 \leqslant y \leqslant x$ we consider the difference

$$\Delta = P\left\{\widetilde{M}_n \leqslant 1 - \frac{x}{n},\ M_n \leqslant 1 - \frac{y}{n}\right\} - \left(P\left\{\widetilde{M}_m \leqslant 1 - \frac{x}{n},\ M_m \leqslant 1 - \frac{y}{n}\right\}\right)^k.$$

We will prove that $\Delta \to 0$ as $n \to \infty$. This will clearly imply that the limit in (3.7.3) exists and is given by (3.7.11). Decompose $\mathbb{N}_n = \{1, 2, \dots, n\} = (I_1 \cup J_1) \cup (I_2 \cup J_2) \cup \dots \cup (I_k \cup J_k) \cup I_{k+1}$ with

$$I_i = \{(m + l)(i - 1) + 1, \dots, (m + l)(i - 1) + m\}, \quad 1 \leqslant i \leqslant k,$$

$$J_i = \{mi + l(i - 1) + 1, \dots, mi + l(i - 1) + l\}, \quad 1 \leqslant i \leqslant k,$$

$$I_{k+1} = \{(m+l)k+1, \ldots, n\}.$$

For $1 \leqslant i \leqslant k$ we set

$$\begin{aligned}
\widetilde{M}(I_i) &= \max\{X_j |\, j \in I_i, c_j = 1\}, \\
M(I_i) &= \max\{X_j |\, j \in I_i\}, \\
A_i &= \left\{\widetilde{M}(I_i) \leqslant 1 - x/n, M(I_i) \leqslant 1 - y/n\right\}.
\end{aligned}$$

Using the notation $A_1 A_2 \ldots A_k$ for the intersection of k events, it is clear that $|\Delta|$ does not exceed

$$\begin{aligned}
&\left| P\left\{\widetilde{M}_n \leqslant 1 - \frac{x}{n},\ M_n \leqslant 1 - \frac{y}{n}\right\} \right. \\
&\quad \left. - P\left\{\widetilde{M}_{(m+l)k} \leqslant 1 - \frac{x}{n},\ M_{(m+l)k} \leqslant 1 - \frac{y}{n}\right\} \right| \\
&\quad + \left| P\left\{\widetilde{M}_{(m+l)k} \leqslant 1 - \frac{x}{n},\ M_{(m+l)k} \leqslant 1 - \frac{y}{n}\right\} - P(A_1 A_2 \ldots A_k) \right| \\
&\quad + \left| P(A_1 A_2 \ldots A_k) - \left(P(A_1)\right)^k \right| \\
&:= \Delta_1 + \Delta_2 + \Delta_3.
\end{aligned}$$

Since $1 - x/n \leqslant 1 - y/n$,

$$\begin{aligned}
&\left\{\widetilde{M}_{(m+l)k} \leqslant 1 - \frac{x}{n},\ M_{(m+l)k} \leqslant 1 - \frac{y}{n}\right\} \setminus \left\{\widetilde{M}_n \leqslant 1 - \frac{x}{n},\ M_n \leqslant 1 - \frac{y}{n}\right\} \\
&\subset \bigcup_{j=(m+l)k+1}^{n} \left\{X_j > 1 - \frac{x}{n}\right\},
\end{aligned}$$

hence,

$$\Delta_1 \leqslant (m+l) P\left\{X_1 > 1 - x/n\right\} = (m+l)x/n \to 0 \quad \text{as } n \to \infty.$$

Similarly, $\Delta_2 \leqslant lk\, P\left\{X_1 > 1 - x/n\right\} = lkx/n \to 0$ as $n \to \infty$ Furthermore, we can write

$$\begin{aligned}
\Delta_3 &\leqslant |P(A_1 A_2 \ldots A_k) - P(A_1) P(A_2) \ldots P(A_k)| \\
&\qquad + \left| P(A_1) P(A_2) \ldots P(A_k) - \left(P(A_1)\right)^k \right| \\
&:= \Delta_3' + \Delta_3''.
\end{aligned}$$

Since each set I_i contains m numbers, by a repeated application of $D(u_n, v_n)$ bound (3.3.1) with (3.7.33), we obtain

$$\Delta_3' \leqslant k\alpha_{n,m} \leqslant \frac{\max(x, y)}{m} \frac{\ln(m \max(x, y))}{\ln r} + \frac{1}{m} \frac{r}{1 - 1/r} \to 0$$

since $m \to \infty$ as $n \to \infty$, so it remains to consider Δ_3''.

For $1 \leqslant i \leqslant k$ let

$$f_{m,j}^{(i)} = \sum_{i=(m+l)(i-1)+1}^{(m+l)(i-1)+m-j+1} \mathbf{1}\{c_i = c_{i+1} = \cdots = c_{i+j-1} = 0\}$$

be the number of times the pattern $00\ldots0$ (j consecutive zeros) appears in the stretch of sequence $(c_n,\ n \in I_i)$. Since the limit in (3.7.6) is assumed to exist, we know that $f_{m,j}^{(1)}/m = f_{m,j}/m \to f_j$ as $m \to \infty$ and, similarly, for $i \geqslant 2$,

$$\begin{aligned}
\frac{f_{m,j}^{(i)}}{m} &= \frac{f_{(m+l)(i-1)+m,j}}{(m+l)(i-1)+m} \cdot \frac{(m+l)(i-1)+m}{m} \\
&\quad - \frac{f_{(m+l)(i-1),j}}{(m+l)(i-1)} \cdot \frac{(m+l)(i-1)}{m} \\
&\to f_j \cdot i - f_j \cdot (i-1) = f_j
\end{aligned}$$

as $m \to \infty$. In the proof of Theorem 3.7.1 we showed that this implies that

$$\begin{aligned}
P(A_i) &= 1 - \frac{(r-1)x}{n} \sum_{d=0}^{j-1} \frac{f_{m,d}^{(i)} - f_{m,d+1}^{(i)}}{r^{d+1}} - \frac{(r-1)y}{rn} f_{m,j}^{(i)} + O(1/n) \\
&= \exp\left\{ -\left(\frac{(r-1)x}{n} \sum_{d=0}^{j-1} \frac{f_{m,d}^{(i)} - f_{m,d+1}^{(i)}}{r^{d+1}} - \frac{(r-1)y}{rn} f_{m,j}^{(i)} \right) \right\} + O(1/n),
\end{aligned}$$

$i = 1, 2, \ldots$. Consequently,

$$\begin{aligned}
&P(A_1)P(A_2)\ldots P(A_k) \\
&= \exp\left\{ -\sum_{i=1}^{k} \left(\frac{(r-1)x}{n} \sum_{d=0}^{j-1} \frac{f_{m,d}^{(i)} - f_{m,d+1}^{(i)}}{r^{d+1}} - \frac{(r-1)y}{rn} f_{m,j}^{(i)} \right) \right\} + O\left(\tfrac{k}{n}\right) \\
&= \exp\left\{ -\left(\frac{(r-1)x}{n} \sum_{d=0}^{j-1} \frac{f_{n,d} - f_{n,d+1}}{r^{d+1}} - \frac{(r-1)y}{rn} f_{n,j} \right) + O\left(\tfrac{kl}{n}\right) \right\} + O\left(\tfrac{k}{n}\right) \\
&\to \exp\{ -H(x, y) \}
\end{aligned}$$

as $n \to \infty$. Since we also have $\left(P(A_1)\right)^k \to e^{-H(x,y)}$ as $n \to \infty$, we have proved that $\Delta_3'' \to 0$. □

3.8 Linear Processes with Heavy-Tailed Innovations

3.8.1 Preliminaries

Linear processes or infinite moving averages with heavy-tailed innovations are determined as follows: Let $(X_n)_{n \geqslant 0}$ be a random sequence given by

$$X_n = \sum_{i=0}^{\infty} \psi_i Z_{n-i}, \tag{3.8.1}$$

where Z_n, $n \in \{\dots, -2, -1, 0, 1, 2, \dots\}$, is a sequence of i.i.d. random variables with regularly varying balanced tails, i.e.,

$$P\{|Z_0| > t\} = t^{-\alpha} L(t), \quad t > 0, \tag{3.8.2}$$

for some $\alpha > 0$ and slowly varying function L, and

$$\lim_{t \to \infty} \frac{P\{Z_0 > t\}}{P\{|Z_0| > t\}} = p, \quad \lim_{t \to \infty} \frac{P\{Z_0 < -t\}}{P\{|Z_0| > t\}} = q, \tag{3.8.3}$$

for some nonnegative constants p and q such that $p + q = 1$, and $(\psi_i)_{i \geqslant 0}$ is a sequence of real numbers satisfying the condition

$$\sum_{i=0}^{\infty} |\psi_i|^{\delta} < \infty, \tag{3.8.4}$$

where δ is a constant such that $0 < \delta < \min\{\alpha, 1\}$. The series in (3.8.1) converges almost surely under the conditions imposed. Let us denote

$$a_n = \inf\left\{x : P\{|Z_0| > x\} < \frac{1}{n}\right\}, \quad n \in \mathbb{N}. \tag{3.8.5}$$

Then we have the following asymptotic relation

$$\lim_{n \to \infty} n P\{a_n^{-1} Z_0 \in \cdot\} = \mu(\cdot), \tag{3.8.6}$$

where μ is a measure on σ-algebra generated by open sets in $[-\infty, \infty] \setminus \{0\}$ that is given by

$$\mu(dx) = p\alpha x^{-\alpha-1}\mathbf{1}_{(0,\infty]}(x) + q\alpha(-x)^{-\alpha-1}\mathbf{1}_{[-\infty,0)}(x). \quad (3.8.7)$$

Let $(c_i)_{i\geqslant 1}$ be a deterministic $0-1$ sequence and let us denote

$$M_n = \max_{1\leqslant i\leqslant n} X_i, \quad \widetilde{M}_n = \max_{1\leqslant i\leqslant n:\, c_i=1} X_i, \quad \widehat{M}_n = \max_{1\leqslant i\leqslant n:\, c_i=0} X_i. \quad (3.8.8)$$

Example 3.8.1 (Autoregressive Model) Let $(X_n)_{n\geqslant 0}$ be a random sequence given by

$$X_n = \phi X_{n-1} + Z_n, \quad (3.8.9)$$

where $|\phi| < 1$ and $(Z_n)_{n\geqslant 0}$ is a sequence of i.i.d. random variables satisfying (3.8.2)–(3.8.3). Then, there is a strictly stationary solution of the recurrence Eq. (3.8.9) of the form (3.8.1), where $\psi_i = \phi^i$, $i = 0, 1, 2, \ldots$. Note that if $0 < \phi < 1$, then $(\psi_i)_{i\geqslant 0}$ is a strictly decreasing sequence, i.e., $\psi_0 > \psi_1 > \psi_2 > \cdots > 0$. If $-1 < \phi < 0$, then the terms of sequence $(\psi_i)_{i\geqslant 0}$ are alternately positive and negative: $\psi_{2i} > 0$, $\psi_{2i+1} < 0$, for each $i \in \mathbb{N}_0 = \{0, 1, 2, \ldots\}$. △

Example 3.8.2 (General) ARMA(k, l) *processes* are defined by

$$X_n + \sum_{i=1}^{k} \alpha_i X_{n-i} = \sum_{i=0}^{l} \beta_i Z_{n-i}, \quad \beta_0 = 1. \quad (3.8.10)$$

Under the assumption that the roots of $z^n + \sum_{i=1}^{k} \alpha_i z^{n-i} = 0$ lie inside the unit circle, stochastic difference equation (3.8.10) has a stationary solution of the form (3.8.1). △

3.8.2 Point Process Theory Approach

Extreme value theory for infinite moving averages with heavy-tailed innovations has been considered by several authors, *see* for example Rootzén [196], Leadbetter, Lidbgren and Rootzén [143], Davis and Resnick [51], and Resnick [192]. Many results were obtained using the point process theory approach. Some of them are listed in Section 4.5 of Resnick [192]. *See* also Exercises and Supplements at the end of this chapter.

We shall formulate here a convergence result for point processes based on $(X_n)_{n\geqslant 1}$. If we put $\psi_i = 0$ for $i < 0$, Proposition 2.7 of Resnick [192] takes the following form.

Theorem 3.8.3 (Davis and Resnick [51]) *Let* $(X_n)_{n\geqslant 0}$ *be a random sequence such that conditions* (3.8.1)–(3.8.4) *are satisfied. Then, we have*

$$\sum_{m=1}^{\infty} \varepsilon_{(m/n, X_m/a_n)} \Longrightarrow \sum_{m} \sum_{i=0}^{+\infty} \varepsilon_{(t_m, \psi_i j_m)}, \tag{3.8.11}$$

in $M_p([0,\infty) \times ([-\infty,\infty] \setminus \{0\}))$, *where* $\sum_{i=0}^{+\infty} \varepsilon_{(t_m, j_m)}$ *is PRM with mean measure* $dt\, \mu(dx)$, *and* μ *is given by* (3.8.5)–(3.8.7).

We shall use Theorem 3.8.3 and some of its consequences in what follows.

In the next two subsections we shall consider linear processes (3.8.1) with assumptions and notation (3.8.2)–(3.8.7). We shall study relations between maxima in complete and incomplete samples. As mentioned, we shall consider the case when the sequence of coefficients $(\psi_i)_{i\geqslant 0}$ is strictly decreasing and tends to 0, and the case when the terms of this sequence alternately change the sign.

3.8.3 Linear Processes with Positive Coefficients

Suppose that $1 = \psi_0 > \psi_1 > \psi_2 > \cdots$, and $p > 0$, where parameter p is given by (3.8.3), and let us consider a deterministic 0–1 sequence $(c_i)_{i\geqslant 1}$.

A finite subsequence of the form $100\ldots01$ (or $00\ldots01$ at the beginning of sequence $c_1c_2c_3\ldots$), where the number of 0's is equal to $k \in \mathbb{N}$, will be called a k-block of 0's. We similarly define k-blocks of 1's. Our goal is to determine the limiting distribution of properly normalized random vector $(\widetilde{M}_n, M_n)$ as a function of $(c_i)_{i\geqslant 1}$. Certain additional conditions will be imposed on sequence $(c_i)_{i\geqslant 1}$, and the blocks of 0's will play a crucial role in reaching this goal. Note once again that here we consider non-overlapping k-blocks of 0's and 1's.

Let $f_{n,0}^{(k)}$ be the number of k-blocks of 0's among the first n terms of sequence $(c_i)_{i\geqslant 1}$. Let us denote

$$p_0^{(k)} = \lim_{n\to\infty} \frac{f_{n,0}^{(k)}}{n}, \quad k \geqslant 1, \tag{3.8.12}$$

if such limits exist. Additionally, we suppose that there exists the limit

$$p_1 = \lim_{n\to\infty} \frac{1}{n} \sum_{i=1}^{n} c_i, \quad p_0 = 1 - p_1. \tag{3.8.13}$$

For every set $A \subset \mathbb{N}$ we also define the density of set A in set $\mathbb{N}$ to be equal to

$$\lim_{n\to\infty} \frac{|A \cap \{1, 2, \ldots, n\}|}{n}, \tag{3.8.14}$$

if such limit exists, where $|A|$ is the notation for the number of elements of set A. Note that the following equality holds:

$$p_0 = \sum_{k=1}^{\infty} k p_0^{(k)}. \tag{3.8.15}$$

It is worth mentioning that the following two statements also hold:

- $\sum_{k=r}^{\infty} p_0^{(k)}$ is the asymptotic relative frequency of blocks of the form $100\ldots0$, where 1 is followed by at least r zeroes.
- $\sum_{k=r}^{\infty} (k-r+1) p_0^{(k)}$ is the asymptotic relative frequency of blocks of zeroes of length r, where such blocks may partially overlap.

Now we are ready to formulate the main result related to the case of linear processes with positive coefficients.

Theorem 3.8.4 (Glavaš and Mladenović [78]) *Let $(X_n)_{n\geqslant 1}$ be a random sequence given by (3.8.1)–(3.8.4), where $1 = \psi_0 > \psi_1 > \cdots$, a_n and μ be given by (3.8.5) and (3.8.7), and $p > 0$. Let $(c_i)_{i\geqslant 1}$ be a deterministic 0–1 sequence such that there exist limit values (3.8.12) and (3.8.13), M_n and $\widetilde{M}_n$ be defined by (3.8.8), and suppose that $0 < y \leqslant x$. Then there exists the limit*

$$\lim_{n\to\infty} P\{a_n^{-1}\widetilde{M}_n \leqslant y, a_n^{-1} M_n \leqslant x\} = G(x, y), \tag{3.8.16}$$

where function $G(x, y)$ is determined as follows: If $0 < \psi_r x < y \leqslant \psi_{r-1} x$, where $r \geqslant 1$, then

$$G(x, y) = \exp\left\{-p\,(a_r x^{-\alpha} + b_r y^{-\alpha})\right\}, \tag{3.8.17}$$

where

$$a_r = \sum_{k=r}^{\infty} (k-r+1) p_0^{(k)}, \tag{3.8.18}$$

$$b_r = 1 - p_0 + \sum_{l=1}^{r-1} \psi_l^{\alpha} \sum_{k=r}^{\infty} p_0^{(k)} + \sum_{k=1}^{r-1} p_0^{(k)} \sum_{l=1}^{k} \psi_l^{\alpha}. \tag{3.8.19}$$

Proof Let us define a partition of the set of positive integers that will have a crucial role in what follows. The partition will be defined in three steps. The first step,

$$\mathbb{N} = S_0 \cup S_1, \tag{3.8.20}$$

where $S_0 = \{i \in \mathbb{N} \mid c_i = 0\}$, $S_1 = \{i \in \mathbb{N} \mid c_i = 1\}$. The second step,

$$S_0 = \bigcup_{k=1}^{\infty} S_0^{(k)}, \quad S_1 = \bigcup_{k=1}^{\infty} S_1^{(k)}. \tag{3.8.21}$$

where, for every $k \geqslant 1$,

$$S_0^{(k)} = \{i, i+1, \ldots, i+k-1 \mid A[i] \text{ and } c_i = \cdots = c_{i+k-1} = 0, c_{i+k} = 1\},$$

$$S_1^{(k)} = \{i, i+1, \ldots, i+k-1 \mid B[i] \text{ and } c_i = \cdots = c_{i+k-1} = 1, c_{i+k} = 0\},$$

$$A[i] \text{ is notation for the condition } i = 1 \text{ or } c_{i-1} = 1,$$

$$B[i] \text{ is notation for the condition } i = 1 \text{ or } c_{i-1} = 0.$$

In other words, $S_j^{(k)}$ consists of those positive integers that determine the positions of digit $j \in \{0, 1\}$ from all k-blocks of j's. The third step,

$$S_0^{(k)} = S_0^{(k,1)} \cup \cdots \cup S_0^{(k,k)}, \quad S_1^{(k)} = S_1^{(k,1)} \cup \cdots \cup S_1^{(k,k)}, \tag{3.8.22}$$

where, for every $l \in \{1, 2, \ldots, k\}$,

$$S_0^{(k,l)} = \{i + l - 1 \mid A[i] \text{ and } c_i = \cdots = c_{i+k-1} = 0, c_{i+k} = 1\}, \tag{3.8.23}$$

$$S_1^{(k,l)} = \{i + l - 1 \mid B[i] \text{ and } c_i = \cdots = c_{i+k-1} = 1, c_{i+k} = 0\}. \tag{3.8.24}$$

Hence, $S_0^{(k,l)}$ consists of those positive integers that determine the positions of l-th digit 0 in all k-blocks of 0's. Note that set $S_1^{(k,l)}$ has a similar meaning for 1's. Therefore, we finally have the following partition of the set of positive integers $\mathbb{N}$ into mutually disjoint subsets:

$$\mathbb{N} = \bigcup_{k=1}^{\infty} \bigcup_{l=1}^{k} S_0^{(k,l)} \cup \bigcup_{k=1}^{\infty} \bigcup_{l=1}^{k} S_1^{(k,l)}. \tag{3.8.25}$$

If the limit values given by (3.8.12) exist, then each of the sets $S_0^{(k,1)}, \ldots, S_0^{(k,k)}$ has a density equal to $p_0^{(k)}$. Set $S_1 = \bigcup_{k=1}^{\infty} \bigcup_{l=1}^{k} S_1^{(k,l)}$ has density p_1.

Under conditions (3.8.1)–(3.8.4) we have

$$\sum_{m \in S_0} \varepsilon_{(m/n, X_m/a_n)} \Longrightarrow \sum_{m \in S_0} \sum_{i=0}^{\infty} \varepsilon_{(t_m^{(i)}, \psi_i j_{m-i})}, \tag{3.8.26}$$

$$\sum_{m \in S_1} \varepsilon_{(m/n, X_m/a_n)} \Longrightarrow \sum_{m \in S_1} \sum_{i=0}^{\infty} \varepsilon_{(t_m^{(i)}, \psi_i j_{m-i})}, \tag{3.8.27}$$

in $M_p([0,\infty) \times ([-\infty,\infty] \setminus \{0\}))$. Random points j_{m-i}, where $m \in S_j$, $j\in\{0,1\}$, $i\in\mathbb{N}_0$, appear in the limiting point processes in (3.8.26) and (3.8.27) as a "contribution" of random variables Z_{m-i} (from the representation $X_m = \sum_{i=0}^{\infty} \psi_i Z_{m-i}$) to these limiting point processes. We shall classify points j_{m-i} according to the values of $m+i$ in different subsets of $\mathbb{N}$. Constants $t_m^{(i)}$ will not influence the form of the limiting distribution. Relations (3.8.26) and (3.8.27) can be proved in a similar way as Proposition 4.27 in Resnick [192], where the case

$$X_m = \sum_{i=-\infty}^{\infty} \psi_i Z_{m-i}$$

and $S_1 = \mathbb{N}$, was considered. *See* also Theorem 1 in Scotto, Turkman, and Anderson [205], where the case $X_m = \sum_{i=0}^{\infty} \psi_i Z_{m-i}$ and $S_1 = \{kM \mid k \in \mathbb{N}\}$ for some $M \geqslant 2$ was investigated.

The limiting point processes in (3.8.26) and (3.8.27) are independent Poisson random measures (PRM). Note also that

$$\sum_{i\in S_0^{(k,l)}} \varepsilon_{j_i}, \quad l \in \{1,2,\ldots,k\}, \quad k \geqslant 1, \quad \text{and} \quad \sum_{i\in S_1} \varepsilon_{j_i} \tag{3.8.28}$$

are independent PRM's, such that $p_0^{(k)}\mu(dx)$ is the mean measure for each of the point processes $\sum_{i\in S_0^{(k,l)}} \varepsilon_{j_i}$, $l \in \{1,2,\ldots,k\}$, and $p_1\mu(dx)$ is the mean measure for the point process $\sum_{i\in S_1} \varepsilon_{j_i}$, where measure μ is determined by (3.8.7).

The limiting point process from (3.8.26) can be represented in the domain $([0,1],\cdot)$ as the sum of point processes

$$N_{00}^{(k,l)}(\cdot) := \sum_{m\in S_0} \sum_{i\geqslant 0: m+i\in S_0^{(k,l)}} \varepsilon_{(t_m^{(i)},\psi_i j_{m-i})}([0,1]\times\cdot), \tag{3.8.29}$$

$$N_{01}^{(k,l)}(\cdot) := \sum_{m\in S_0} \sum_{i\geqslant 0: m+i\in S_1^{(k,l)}} \varepsilon_{(t_m^{(i)},\psi_i j_{m-i})}([0,1]\times\cdot), \tag{3.8.30}$$

over all $k \geqslant 1$ and $l \in \{1,2,\ldots,k\}$. Similarly, the limiting point process from (3.8.27) can be represented in the domain $([0,1],\cdot)$ as the sum of point processes

$$N_{10}^{(k,l)}(\cdot) := \sum_{m\in S_1} \sum_{i\geqslant 0: m+i\in S_0^{(k,l)}} \varepsilon_{(t_m^{(i)},\psi_i j_{m-i})}([0,1]\times\cdot), \tag{3.8.31}$$

$$N_{11}^{(k,l)}(\cdot) := \sum_{m\in S_1} \sum_{i\geqslant 0: m+i\in S_1^{(k,l)}} \varepsilon_{(t_m^{(i)},\psi_i j_{m-i})}([0,1]\times\cdot), \tag{3.8.32}$$

over all $k \geqslant 1$ and $l \in \{1, 2, \ldots, k\}$. For every $l \in \{1, \ldots, k\}$, and $k \geqslant 1$, let us define

$$\widetilde{N}_0^{(k,l)}(\cdot) := \sum_{i \in S_0^{(k,l)}} \varepsilon_{j_i}(\cdot), \quad \widetilde{N}_1^{(k,l)}(\cdot) := \sum_{i \in S_1^{(k,l)}} \varepsilon_{j_i}(\cdot). \tag{3.8.33}$$

It is necessary to consider also point process $\widetilde{N}_1$ that is given by

$$\widetilde{N}_1 := \sum_{k=1}^{\infty} \sum_{l=1}^{k} \widetilde{N}_1^{(k,l)} = \sum_{k=1}^{\infty} \sum_{l=1}^{k} \sum_{i \in S_1^{(k,l)}} \varepsilon_{j_i} = \sum_{i \in S_1} \varepsilon_{j_i}. \tag{3.8.34}$$

Note that $\widetilde{N}_0^{(k,l)}$, $l \in \{1, \ldots, k\}$, $k \geqslant 1$, and $\widetilde{N}_1$, are *independent* PRM's with mean measures $p_0^{(k)} \mu(dx)$ and $p_1 \mu(dx)$, respectively. This statement follows from the fact that $S_0^{(k,l)}$, $l \in \{1, \ldots, k\}$, $k \geqslant 1$, and S_1 are mutually disjoint sets.

Now, for $0 < y \leqslant x$ we obtain that

$$\begin{aligned}
&P\{a_n^{-1}\widetilde{M}_n \leqslant y, a_n^{-1} M_n \leqslant x\} \\
&\to P\Bigg(\bigcap_{k=1}^{\infty}\bigcap_{l=1}^{k}\Big\{N_{11}^{(k,l)}(y,\infty] = 0, N_{10}^{(k,l)}(y,\infty] = 0, \\
&\qquad\qquad N_{01}^{(k,l)}(x,\infty] = 0, N_{00}^{(k,l)}(x,\infty] = 0\Big\}\Bigg) \\
&= P\Bigg(\bigcap_{k=1}^{\infty}\bigcap_{l=1}^{k}\Big\{\widetilde{N}_1^{(k,l)}(y,\infty] = 0, \widetilde{N}_0^{(k,l)}\Big(\frac{y}{\psi_l},\infty\Big] = 0, \\
&\qquad\qquad \widetilde{N}_1^{(k,l)}\Big(\frac{x}{\psi_l},\infty\Big] = 0, \widetilde{N}_0^{(k,l)}(x,\infty] = 0\Big\}\Bigg).
\end{aligned} \tag{3.8.35}$$

Let $r \in \mathbb{N}$, and $0 < \psi_r x < y \leqslant \psi_{r-1} x$, i.e., $0 < \dfrac{y}{\psi_{r-1}} \leqslant x < \dfrac{y}{\psi_r}$. Note also that $y = \dfrac{y}{\psi_0} < \dfrac{y}{\psi_1} < \dfrac{y}{\psi_2} < \cdots$ The following statements hold:

(a) Since $0 < y \leqslant x$, and $1 = \psi_0 > \psi_1 > \psi_2 > \ldots$, it follows that for every $k \geqslant 1$, and $l \in \{1, 2, \ldots, k\}$,

$$\Big\{\widetilde{N}_1^{(k,l)}(y,\infty] = 0, \widetilde{N}_1^{(k,l)}\Big(\frac{x}{\psi_l},\infty\Big] = 0\Big\} = \Big\{\widetilde{N}_1^{(k,l)}(y,\infty] = 0\Big\}. \tag{3.8.36}$$

Relation (3.8.36) is the reason why we assume only the existence of probability p_1 in the formulation of Theorem 3.8.4 instead of the existence of $p_1^{(k)}$ for all $k \geqslant 1$. But we require the existence of $p_0^{(k)}$ for all $k \geqslant 1$.

(b1) For $l \geqslant r$, we have $x < \dfrac{y}{\psi_r} \leqslant \dfrac{y}{\psi_l}$, and hence

$$\left\{\widetilde{N}_0^{(k,l)}\left(\frac{y}{\psi_l}, \infty\right] = 0, \widetilde{N}_0^{(k,l)}(x, \infty] = 0\right\} = \left\{\widetilde{N}_0^{(k,l)}(x, \infty] = 0\right\}. \tag{3.8.37}$$

(b2) For $1 \leqslant l < r$, we have $\dfrac{y}{\psi_l} \leqslant \dfrac{y}{\psi_{r-1}} \leqslant x$, and hence

$$\left\{\widetilde{N}_0^{(k,l)}\left(\frac{y}{\psi_l}, \infty\right] = 0, \widetilde{N}_0^{(k,l)}(x, \infty] = 0\right\} = \left\{\widetilde{N}_0^{(k,l)}\left(\frac{y}{\psi_l}, \infty\right] = 0\right\}. \tag{3.8.38}$$

If $k \geqslant r > 1$, then both cases (b1) and (b2) appear. If $k < r$, then only case (b2) takes place. Obviously, if $r = 1$, then only case (b1) takes place.

Note also that for every $a > 0$, $\mu(a, \infty] = pa^{-\alpha}$, and

$$P\left\{\widetilde{N}_0^{(k,l)}(a, \infty] = 0\right\} = e^{-p_0^{(k)} pa^{-\alpha}}, \quad l \in \{1, \dots, k\} \ k \geqslant 1, \tag{3.8.39}$$

$$P\left\{\widetilde{N}_1(a, \infty] = 0\right\} = e^{-p_1 pa^{-\alpha}}. \tag{3.8.40}$$

Case $r = 1$, i.e., $0 < \psi_1 x < y \leqslant \psi_0 x = x$. Using (3.8.35), (3.8.36), and (3.8.37), it is easy to obtain that $G(x, y) = \Pi_1 \cdot \Pi_2$, where

$$\Pi_1 = P\left(\bigcap_{k=1}^{\infty}\bigcap_{l=1}^{k}\left\{\widetilde{N}_1^{(k,l)}(y, \infty] = 0\right\}\right) = P\left\{\widetilde{N}_1(y, \infty] = 0\right\},$$

$$\Pi_2 = \prod_{k=1}^{\infty}\prod_{l=1}^{k} P\left\{\widetilde{N}_0^{(k,l)}(x, \infty] = 0\right\}.$$

Case $r \geqslant 2$, and $0 < \psi_r x < y \leqslant \psi_{r-1} x$. In this case, using (3.8.35), (3.8.36), (3.8.37), and (3.8.38), we obtain that $G(x, y) = \Pi_1 \cdot \Pi_2 \cdot \Pi_3$, where

$$\Pi_1 = P\left(\bigcap_{k=1}^{\infty}\bigcap_{l=1}^{k}\left\{\widetilde{N}_1^{(k,l)}(y, \infty] = 0\right\}\right) = P\left\{\widetilde{N}_1(y, \infty] = 0\right\},$$

$$\Pi_2 = \prod_{k=r}^{\infty}\left(\prod_{l=r}^{k} P\left\{\widetilde{N}_0^{(k,l)}(x, \infty] = 0\right\} \prod_{l=1}^{r-1} P\left\{\widetilde{N}_0^{(k,l)}\left(\frac{y}{\psi_l}, \infty\right] = 0\right\}\right),$$

$$\Pi_3 = \prod_{k=1}^{r-1}\prod_{l=1}^{k} P\left\{\widetilde{N}_0^{(k,l)}\left(\frac{y}{\psi_l}, \infty\right] = 0\right\}.$$

Now, using (3.8.39), and (3.8.40), we easily obtain (3.8.17) in both cases, where $\sum_{l=1}^{r-1}$ should be considered 0 if $r = 1$. □

Remark 3.8.5 Let sequence $(c_n)_{n\geqslant 1}$ be given by $c_{4i-3} = c_{4i-2} = 1$ and $c_{4i-1} = c_{4i} = 0$ for every $i \in \mathbb{N}$. Then function $G(x, y)$ given by (3.8.17) reduces to the form

$$G(x, y) = \begin{cases} \exp\left\{-\frac{1}{2}p(x^{-\alpha} + y^{-\alpha})\right\} & \text{if } 0 < \psi_1 x < y \leqslant x, \\ \exp\left\{-\frac{1}{4}p(x^{-\alpha} + (\psi_1^\alpha + 2)y^{-\alpha}\right\} & \text{if } 0 < \psi_2 x < y \leqslant \psi_1 x, \\ \exp\left\{-\frac{1}{4}p(\psi_1^\alpha + \psi_2^\alpha + 2)y^{-\alpha}\right\} & \text{if } 0 < y \leqslant \psi_2 x. \end{cases}$$

More generally, let $(c_n)_{n\geq 1}$ be a periodic 0–1 sequence with the maximal length of blocks of 0's equal to $m \in \mathbb{N}$. Then, $p_0^{(k)} = 0$ for all $k > m$, and the area $\{(x, y) | 0 < y \leqslant x\}$ is split into $m + 1$ areas

$$A_r = \{(x, y) | 0 < \psi_r x < y \leqslant \psi_{r-1} x\}, \quad r \in \{1, 2, \ldots, m\},$$

$$A_{m+1} = \{(x, y) | 0 < y \leqslant \psi_m x\} = \bigcup_{r=m+1}^{\infty} \{(x, y) | 0 < \psi_r x < y \leqslant \psi_{r-1} x\},$$

in which function $G(x, y)$ has different forms. Note that for $(x, y) \in A_r$, where $r \in \{1, 2, \ldots, m\}$, events $\{a_n^{-1}\widetilde{M}_n \leqslant y\}$ and $\{a_n^{-1}M_n \leqslant x\}$ are asymptotically independent. Indeed, random vector $(a_n^{-1}M_n, a_n^{-1}\widetilde{M}_n)$ converges in distribution to some random vector (X, Y) whose distribution function $G(x, y)$ is determined by Theorem 3.8.4. It is easy to check that $G(x, y)$ is continuous on the half-lines $y = \psi_i x$, $x \geqslant 0$, and hence it is continuous everywhere. Consequently, marginal distribution functions that correspond to X and Y are also continuous. Projections $(x, y) \mapsto x$ and $(x, y) \mapsto y$ are continuous functions, and hence the limiting distributions of random variables $a_n^{-1}M_n$ and $a_n^{-1}\widetilde{M}_n$ coincide with the distributions of X and Y, respectively. For $(x, y) \in A_r$, where $r \in \{1, 2, \ldots, m\}$ we have

$$G(x, y) = \exp\{-pa_r x^{-\alpha}\} \cdot \exp\{-pb_r y^{-\alpha}\} = P\{X \leqslant x\}P\{Y \leqslant y\},$$

and the conclusion follows. On the other hand, for $(x, y) \in A_{m+1}$, events $\{a_n^{-1}\widetilde{M}_n \leqslant y\}$ and $\{a_n^{-1}M_n \leqslant x\}$ are asymptotically perfectly dependent in the sense that $G(x, y)$ depends only on y. Note also that sequence $a_n^{-1}(M_n - \widetilde{M}_n)$, $n = 1, 2, \ldots,$ has its own limiting distribution with support $[0, +\infty)$ that coincides with the distribution of $X - Y$.

The form of the limiting distribution function (for example in the case of periodic sequences $(c_i)_{i\geqslant 1}$) implies that both conditions $D(u_n)$ and $D'(u_n)$ cannot be satisfied for sequence $(X_n)_{n\geqslant 1}$. The unusual way of clustering of extremes leads to the different asymptotic behavior of probability $P\{a_n^{-1}\widetilde{M}_n \leqslant y, a_n^{-1}M_n \leqslant x\}$ for different pairs (x, y) as stated above.

Remark 3.8.6 Any non-random 0–1 sequence $(c_i)_{i\geqslant 1}$ can be considered the binary expansion of a real number from segment [0, 1]. Since almost all real numbers are normal, it follows that almost all sequences $(c_i)_{i\geqslant 1}$ give the same limiting distribution function, say $G_0(x, y)$, of random vector $(a_n^{-1}\widetilde{M}_n, a_n^{-1}M_n)$. Note that $G_0(x, y)$ is determined by asymptotic frequencies $p_0^{(k)} = 2^{-k-2}$, for all $k \geqslant 1$. It is easy to see that for $0 < \psi_r x < y \leqslant \psi_{r-1}x$, where $r \geqslant 1$, $G_0(x, y)$ is given by

$$G_0(x, y) = \exp\left\{ -\frac{1}{2^r}px^{-\alpha} - \left(\frac{1}{2} + \frac{1}{2^{r+1}}\sum_{l=1}^{r-1}\psi_l^\alpha + \frac{2^r-1}{2^{r+1}}\sum_{l=1}^{k}\psi_l^\alpha\right)py^{-\alpha}\right\}.$$

The remaining part of the sequences, say $\mathcal{S}$, give infinitely many additional possible limiting distributions of random vector $(a_n^{-1}\widetilde{M}_n, a_n^{-1}M_n)$. Of course, some sequences from $\mathcal{S}$ do not give the limiting distribution. All periodic sequences, as well as some non-periodic sequences for which a limiting distribution exists, belong to $\mathcal{S}$. Let $(c_i)_{i\geqslant 1}$ be a periodic sequence in $\mathcal{S}$. If we change the position of blocks of 0's inside periods without merging any two such blocks (the blocks of 1's may change lengths), then starting and resulting sequences both give the same limiting distribution of random vector $(a_n^{-1}\widetilde{M}_n, a_n^{-1}M_n)$.

Remark 3.8.7 Let us suppose that a sequence $(c_n)_{n\geqslant 1}$ is determined randomly by a sequence of independent Bernoulli trials that give 1 with probability $\widetilde{p}$, and 0 with probability $\widetilde{q}$. Then for almost every realization of sequence $(c_n)_{n\geqslant 1}$ we have

$$p_0^{(k)} = \widetilde{q}^k\widetilde{p}^2, \quad k \in \mathbb{N}, \quad \text{and} \quad 1 - p_0 = \widetilde{p}.$$

The form of the corresponding almost sure limiting distribution follows from (3.8.17), (3.8.18), and (3.8.19).

Remark 3.8.8 Since $M_n \geqslant \widetilde{M}_n$ for any n, it is easy to conclude that

$$P\{a_n^{-1}\widetilde{M}_n \leqslant y, a_n^{-1}M_n \leqslant x\} = P\{a_n^{-1}\widetilde{M}_n \leqslant y, a_n^{-1}\widehat{M}_n \leqslant x\}$$

for every (x, y) such that $0 < y \leqslant x$, and the limiting distribution of random vector $(a_n^{-1}\widetilde{M}_n, a_n^{-1}\widehat{M}_n)$ is determined by Theorem 3.8.4 as well.

3.8.4 *Linear Processes with Coefficients Alternately Changing Sign*

Suppose that $1 = |\psi_0| > |\psi_1| > |\psi_2| > \cdots$, and $\psi_{2i-1} < 0$, $\psi_{2i} > 0$ for every $i \in \mathbb{N}$. This assumption covers the case of negatively correlated autoregressive processes. Let $(c_i)_{i\geqslant 1}$ be a deterministic 0–1 sequence. In order to formulate and prove results related to the asymptotic behavior of random vector $(a_n^{-1}\widetilde{M}_n, a_n^{-1}M_n)$ in this case, we need some preliminary considerations and notation.

Let $S_0 = \{i \in \mathbb{N} \,|\, c_i = 0\}$ and $S_1 = \{i \in \mathbb{N} \,|\, c_i = 1\}$. Let us also define the sets $S_{00}, S_{01}, S_{10}, S_{11}$ as follows:

$$S_{00} = \{i \in \mathbb{N} \,|\, c_i = 0,\ i \text{ is even}\}, \tag{3.8.41}$$

$$S_{01} = \{i \in \mathbb{N} \,|\, c_i = 0,\ i \text{ is odd}\}, \tag{3.8.42}$$

$$S_{10} = \{i \in \mathbb{N} \,|\, c_i = 1,\ i \text{ is even}\}, \tag{3.8.43}$$

$$S_{11} = \{i \in \mathbb{N} \,|\, c_i = 1,\ i \text{ is odd}\}. \tag{3.8.44}$$

For every positive integer i that is considered a position of the term c_i, we define the characteristics d, e, u, v and w as follows:

$$d = d(i) = c_i \in \{0, 1\}, \tag{3.8.45}$$

$$e = e(i) = \begin{cases} 0, & \text{if } i \text{ is even}, \\ 1, & \text{if } i \text{ is odd}. \end{cases} \tag{3.8.46}$$

In order to define the characteristics $u = u(i)$, $v = v(i)$ and $w = w(i)$, we first introduce the following sets:

$$U_i = \{l \,|\, l < i,\ l \text{ and } i \text{ have different parity and } c_l = c_i\}, \tag{3.8.47}$$

$$V_i = \{l \,|\, l < i,\ l \text{ and } i \text{ have the same parity and } c_l \neq c_i\}, \tag{3.8.48}$$

$$W_i = \{l \,|\, l < i,\ l \text{ and } i \text{ have different parity and } c_l \neq c_i\}. \tag{3.8.49}$$

Now let us define

$$u = i - \max U_i, \qquad v = i - \max V_i, \qquad w = i - \max W_i. \tag{3.8.50}$$

If for some $i \in \mathbb{N}$, some of the sets U_i, V_i or W_i are empty, then the corresponding characteristics $u = u(i)$, $v = v(i)$ or $w = w(i)$ are not defined. Let S^* be the set of those positive integers for which not all three characteristics u, v and w are defined.

For every fixed $d \in \{0, 1\}$, $e \in \{0, 1\}$, $u, w \in \{1, 3, 5, \ldots\}$ and $v \in \{2, 4, 6, \ldots\}$ let us denote

$$S_{de}^{(u,v,w)} = \{i \mid c_i = d, e(i) = e, i - \max U_i = u, i - \max V_i = v, i - \max W_i = w\}. \tag{3.8.51}$$

Let $p_{de}^{(u,v,w)}$ be the density of set $S_{de}^{(u,v,w)}$ in the set of all positive integers $\mathbb{N}$, if such density exists. If for some $i \in \mathbb{N}$ both characteristics u and w are defined, then $u = 1$, $w > 1$ or $u > 1$, $w = 1$. Hence, we will be interested in the sets

$$S_{de}^{(1,2k,2l+1)} \text{ and } S_{de}^{(2l+1,2k,1)}, \quad \text{where } d, e \in \{0, 1\}, \quad k, l \in \mathbb{N}, \tag{3.8.52}$$

and their densities

$$p_{de}^{(1,2k,2l+1)} \text{ and } p_{de}^{(2l+1,2k,1)}, \quad \text{where } d, e \in \{0, 1\}, \quad k, l \in \mathbb{N}. \tag{3.8.53}$$

Note that the following statements hold: Sets $S_{00}^{(1,2k,2l+1)}$ and $S_{01}^{(1,2k,2l+1)}$, where $k, l \in \mathbb{N}$, consist of even and odd positive integers i, respectively, such that

$$\left.\begin{array}{ll} c_i = c_{i-2} = \cdots = c_{i-2k+2} = 0, & c_{i-2k} = 1, \\ c_{i-1} = c_{i-3} = \cdots = c_{i-2l+1} = 0, & c_{i-2l-1} = 1. \end{array}\right\} \tag{3.8.54}$$

Sets $S_{00}^{(2l+1,2k,1)}$ and $S_{01}^{(2l+1,2k,1)}$, where $k, l \in \mathbb{N}$, consist of even and odd positive integers i, respectively, such that

$$\left.\begin{array}{ll} c_i = c_{i-2} = \cdots = c_{i-2k+2} = 0, & c_{i-2k} = 1, \\ c_{i-1} = c_{i-3} = \cdots = c_{i-2l+1} = 1, & c_{i-2l-1} = 0. \end{array}\right\} \tag{3.8.55}$$

Sets $S_{10}^{(1,2k,2l+1)}$ and $S_{11}^{(1,2k,2l+1)}$, where $k, l \in \mathbb{N}$, consist of even and odd positive integers i, respectively, such that

$$\left.\begin{array}{ll} c_i = c_{i-2} = \cdots = c_{i-2k+2} = 1, & c_{i-2k} = 0, \\ c_{i-1} = c_{i-3} = \cdots = c_{i-2l+1} = 1, & c_{i-2l-1} = 0. \end{array}\right\} \tag{3.8.56}$$

Sets $S_{10}^{(2l+1,2k,1)}$ and $S_{11}^{(2l+1,2k,1)}$, where $k, l \in \mathbb{N}$, consist of even and odd positive integers i, respectively, such that

$$\left.\begin{array}{ll} c_i = c_{i-2} = \cdots = c_{i-2k+2} = 1, & c_{i-2k} = 0, \\ c_{i-1} = c_{i-3} = \cdots = c_{i-2l+1} = 0, & c_{i-2l-1} = 1. \end{array}\right\} \tag{3.8.57}$$

We shall consider the following partition of $\mathbb{N}$ into mutually disjoint sets:

$$\mathbb{N} = \bigcup_{k=1}^{\infty} \bigcup_{l=1}^{\infty} \bigcup_{d=0}^{1} \bigcup_{e=0}^{1} \left(S_{de}^{(1,2k,2l+1)} \cup S_{de}^{(2l+1,2k,1)} \right) \cup S^*. \tag{3.8.58}$$

Remark 3.8.9 For almost all deterministic 0–1 sequences $(c_i)_{i\geqslant 1}$, we have that for every $d, e \in \{0, 1\}$, and every $k, l \in \mathbb{N}$,

$$p_{de}^{(1,2k,2l+1)} = p_{de}^{(2l+1,2k,1)} = 2^{-(k+l+1)}. \tag{3.8.59}$$

If densities $p_{de}^{(1,2k,2l+1)}$ and $p_{de}^{(2l+1,2k,1)}$ are given by (3.8.59), then

$$\sum_{k=1}^{\infty}\sum_{l=1}^{\infty}\sum_{d=0}^{1}\sum_{e=0}^{1}\left(p_{de}^{(1,2k,2l+1)} + p_{de}^{(2l+1,2k,1)}\right) = 1, \tag{3.8.60}$$

and the density of set S^* is equal to 0. We shall formulate the general result under the assumption that all densities (3.8.53) exist and (3.8.60) holds, and without the assumption that (3.8.59) necessarily holds. Hence, the result will cover all deterministic 0–1 sequences $(c_i)_{i\geqslant 1}$ for which condition (3.8.59) holds, and some other deterministic sequences, for example all periodic sequences, as well.

If the limiting distribution function

$$\lim_{n\to\infty} P\{a_n^{-1}\widetilde{M}_n \leqslant y, a_n^{-1}M_n \leqslant x\} = G(x, y) \tag{3.8.61}$$

exists, it can be represented in the form $G(x, y) = \Pi^+\Pi^-$, where

$$\Pi^+ = \exp\left\{-p\left(a^+x^{-\alpha} + b^+y^{-\alpha}\right)\right\}, \tag{3.8.62}$$

$$\Pi^- = \exp\left\{-q\left(a^-x^{-\alpha} + b^-y^{-\alpha}\right)\right\}. \tag{3.8.63}$$

Here Π^+ is determined as the contribution of the part of measure $\mu(dx)$ that is concentrated on the positive real half-line, and Π^- is determined by the part of measure $\mu(dx)$ that is concentrated on the negative real half-line.

Moreover, constants a^+ and b^+ may take different values, say a_r^+ and b_r^+, for $(x, y) \in A_r, r = 1, 2, \ldots$, where

$$A_r = \{(x, y) \,|\, 0 < \psi_{2r}x < y \leqslant \psi_{2r-2}x\}. \tag{3.8.64}$$

Similarly, constants a^- and b^- may take different values, say a_r^- and b_r^-, for $(x, y) \in B_r, r = 1, 2, \ldots$, where

$$B_r = \left\{(x, y) \,\middle|\, 0 < \frac{|\psi_{2r+1}|}{|\psi_1|}x < y \leqslant \frac{|\psi_{2r-1}|}{|\psi_1|}x\right\}. \tag{3.8.65}$$

Under some wide conditions we shall give explicit expressions for a_r^+, b_r^+, a_r^-, and b_r^-, where $r = 1, 2, \ldots$

Theorem 3.8.10 *Let $(X_n)_{n\geqslant 1}$ be a random sequence given by (3.8.1)–(3.8.4), where $1 = |\psi_0| > |\psi_1| > |\psi_2| > \cdots$, and $\psi_{2i-1} < 0$, $\psi_{2i} > 0$ for every $i \in \mathbb{N}$, and suppose that a_n and μ are given by (3.8.5) and (3.8.7).*

(a) Suppose that all densities (3.8.53) of the sets (3.8.52) exist, and that equality (3.8.60) holds. Then, random vector $(a_n^{-1}\widetilde{M}_n, a_n^{-1}M_n)$ converges in distribution as $n \to \infty$, and the form of the limiting distribution function $G(x, y) = \Pi^+\Pi^-$ is given by (3.8.62)–(3.8.63).

(b) For every $r \in \mathbb{N}$, and $(x, y) \in A_r$, where A_r is given by (3.8.64), the coefficients a_r^+ and b_r^+ are determined by

$$a_r^+ = \sum_{k=r}^{\infty} \widetilde{p}_0^{(2k)}, \quad b_r^+ = 1 - p_0 + \sum_{k=1}^{r-1} \widetilde{p}_0^{(2k)} \psi_{2k}^{\alpha}, \tag{3.8.66}$$

where $1 - p_0 = p_1$, p_1 is given by (3.8.13), and $\widetilde{p}_0^{(2k)}$ is the density of set

$$S_0^{(2k)} = \{i \mid c_i = c_{i-2} = \cdots = c_{i-2k+2} = 0, c_{i-2k} = 1\}. \tag{3.8.67}$$

Also, a_r^+ is the density of set $\{i \mid c_i = c_{i-2} = \cdots = c_{i-2r+2} = 0\}$.

(c) For every $r \in \mathbb{N}$, and $(x, y) \in B_r$, where B_r is given by (3.8.65), coefficients a_r^- and b_r^- are given by

$$a_r^- = |\psi_1|^{\alpha}\sigma_r, \quad b_r^- = |\psi_1|^{\alpha}\sigma_0 + \sum_{l=1}^{r-1} |\psi_{2l+1}|^{\alpha}\sigma_l^*, \tag{3.8.68}$$

where σ_r, σ_0, and σ_l^* are respectively the densities of sets

$$\{i \mid c_{i-1} = c_{i-3} = \cdots = c_{i-2r+1} = 0\}, \quad \{i \mid c_{i-1} = 1\}, \quad \text{and}$$
$$\{i \mid c_{i-1} = c_{i-3} = \cdots = c_{i-2l+1} = 0, c_{i-2l-1} = 1\}.$$

Proof Again we shall consider the limiting point processes in (3.8.26) and (3.8.27). For every $d_1, d_2 \in \{0, 1\}$, and every

$$(u, v, w) \in \{(1, 2k, 2l + 1) \mid k, l \in \mathbb{N}\} \cup \{(2l + 1, 2k, 1) \mid k, l \in \mathbb{N}\}, \tag{3.8.69}$$

let us also introduce the following point processes:

$$N_{d_1 0 d_2 0}^{(u,v,w)}(\cdot) := \sum_{m \in S_{d_1 0}} \sum_{i \geqslant 0:\, m+2i \in S_{d_2 0}^{(u,v,w)}} \varepsilon_{(t_m^{(2i)}, \psi_{2i} j_{m-2i})}([0, 1] \times \cdot),$$

$$N_{d_1 1 d_2 1}^{(u,v,w)}(\cdot) := \sum_{m \in S_{d_1 1}} \sum_{i \geqslant 0: m+2i \in S_{d_2 1}^{(u,v,w)}} \varepsilon_{(t_m^{(2i)}, \psi_{2i} j_{m-2i})}([0,1] \times \cdot),$$

and

$$N_{d_1 0 d_2 1}^{(u,v,w)}(\cdot) := \sum_{m \in S_{d_1 0}} \sum_{i \geqslant 1: m+2i-1 \in S_{d_2 1}^{(u,v,w)}} \varepsilon_{(t_m^{(2i-1)}, \psi_{2i-1} j_{m-2i+1})}([0,1] \times \cdot),$$

$$N_{d_1 1 d_2 0}^{(u,v,w)}(\cdot) := \sum_{m \in S_{d_1 1}} \sum_{i \geqslant 1: m+2i-1 \in S_{d_2 0}^{(u,v,w)}} \varepsilon_{(t_m^{(2i-1)}, \psi_{2i-1} j_{m-2i+1})}([0,1] \times \cdot).$$

Note that $m + 2i$ may take the same value $a \in \mathbb{N}$ for different $m \in \mathbb{N}$ and $i \in \mathbb{N}_0$. Hence, the same positive $j_a = j_{m-2i}$ may appear in the expression that determines the point process $N_{d_1 0 d_2 0}^{(1,2k,2l+1)}(\cdot)$, multiplied with different positive coefficients ψ_{2i}. It is important to know the maximal value of these coefficients, say ψ_{2i}^{max}, because inequality $\psi_{2i}^{max} j_{m-2i} \leqslant x$ implies that $\psi_{2i} j_{m-2i} \leqslant x$ for any other of them. The same conclusion holds for point processes

$$N_{d_1 0 d_2 0}^{(2l+1,2k,1)}(\cdot), \quad N_{d_1 1 d_2 1}^{(1,2k,2l+1)}(\cdot), \quad \text{and} \quad N_{d_1 1 d_2 1}^{(2l+1,2k,1)}(\cdot).$$

Similarly, the same negative $j_a = j_{m-2i+1}$ may appear in the expression that determines point process $N_{d_1 0 d_2 1}^{(1,2k,2l+1)}(\cdot)$, multiplied with different negative coefficients ψ_{2i-1}. Here it is important to know the minimal value of these coefficients (maximal in absolute value), say ψ_{2i-1}^{min}, because inequality $\psi_{2i-1}^{min} j_{m-2i+1} \geqslant -x$ implies that $\psi_{2i-1} j_{m-2i+1} \geqslant -x$ for any other of them. The same conclusion holds for point processes

$$N_{d_1 0 d_2 1}^{(2l+1,2k,1)}(\cdot), \quad N_{d_1 1 d_2 0}^{(1,2k,2l+1)}(\cdot), \quad \text{and} \quad N_{d_1 1 d_2 0}^{(2l+1,2k,1)}(\cdot).$$

The limiting point processes on the right-hand side of (3.8.26) and (3.8.27) can be represented in the domain $([0, 1] \times \cdot)$ in the following form, respectively:

$$\sum_{k=1}^{\infty} \sum_{l=1}^{\infty} \sum_{e_1=0}^{1} \sum_{d=0}^{1} \sum_{e_2=0}^{1} \left(N_{0e_1 d e_2}^{(1,2k,2l+1)}(\cdot) + N_{0e_1 d e_2}^{(2l+1,2k,1)}(\cdot) \right), \tag{3.8.70}$$

$$\sum_{k=1}^{\infty} \sum_{l=1}^{\infty} \sum_{e_1=0}^{1} \sum_{d=0}^{1} \sum_{e_2=0}^{1} \left(N_{1e_1 d e_2}^{(1,2k,2l+1)}(\cdot) + N_{1e_1 d e_2}^{(2l+1,2k,1)}(\cdot) \right). \tag{3.8.71}$$

For $0 < y \leqslant x$ we obtain that the limit value of $P\{a_n^{-1} \widetilde{M}_n \leqslant y, M_n \leqslant x\}$ is equal to

$$P\Big(\bigcap_{k=1}^{\infty}\bigcap_{l=1}^{\infty}\bigcap_{e_1=0}^{1}\bigcap_{d=0}^{1}\bigcap_{e_2=0}^{1}\left\{N_{0e_1de_2}^{(1,2k,2l+1)}(x,\infty]=0,\,N_{0e_1de_2}^{(2l+1,2k,1)}(x,\infty]=0\right\}\cap$$

$$\bigcap_{k=1}^{\infty}\bigcap_{l=1}^{\infty}\bigcap_{e_1=0}^{1}\bigcap_{d=0}^{1}\bigcap_{e_2=0}^{1}\left\{N_{1e_1de_2}^{(1,2k,2l+1)}(y,\infty]=0,\,N_{1e_1de_2}^{(2l+1,2k,1)}(y,\infty]=0\right\}\Big). \tag{3.8.72}$$

For every $d, e \in \{0, 1\}$ and $k, l \in \mathbb{N}$, let us define

$$\widetilde{N}_{de}^{(1,2k,2l+1)}(\cdot) := \sum_{i\in S_{de}^{(1,2k,2l+1)}} \varepsilon_{j_i}(\cdot), \tag{3.8.73}$$

$$\widetilde{N}_{de}^{(2l+1,2k,1)}(\cdot) := \sum_{i\in S_{de}^{(2l+1,2k,1)}} \varepsilon_{j_i}(\cdot). \tag{3.8.74}$$

Note that point processes $\widetilde{N}_{de}^{(1,2k,2l+1)}$ and $\widetilde{N}_{de}^{(2l+1,2k,1)}$, where $d, e \in \{0, 1\}$, $k, l \in \mathbb{N}$, are independent PRM's with mean measures $p_{de}^{(1,2k,2l+1)}\mu(dx)$ and $p_{de}^{(2l+1,2k,1)}\mu(dx)$, $d, e \in \{0, 1\}$, $k, l \in \mathbb{N}$, respectively.
For every $x > 0$, we have $\mu(x, \infty] = px^{-\alpha}$, and $\mu[-\infty, -x) = qx^{-\alpha}$, and therefore for every

$$(u, v, w) \in \{(1, 2k, 2l+1) \mid k, l \in \mathbb{N}\} \cup \{(2l+1, 2k, 1) \mid k, l \in \mathbb{N}\},$$

and $d, e \in \{0, 1\}$, we have

$$P\left\{\widetilde{N}_{de}^{(u,v,w)}(x,\infty]=0\right\} = \exp\left\{-p_{de}^{(u,v,w)}px^{-\alpha}\right\}, \tag{3.8.75}$$

$$P\left\{\widetilde{N}_{de}^{(u,v,w)}[-\infty,-x)=0\right\} = \exp\left\{-p_{de}^{(u,v,w)}qx^{-\alpha}\right\}. \tag{3.8.76}$$

Contribution of the Positive Half-line Let $r \in \mathbb{N}$, and $(x, y) \in A_r$, i.e.,

$$0 < \frac{y}{\psi_{2r-2}} \leqslant x < \frac{y}{\psi_{2r}}. \tag{3.8.77}$$

Note that $y = \frac{y}{\psi_0} < \frac{y}{\psi_2} < \frac{y}{\psi_4} < \dots$

1. Let us consider point processes $N_{0000}^{(1,2k,2l+1)}$, $N_{1000}^{(1,2k,2l+1)}$, $N_{0000}^{(2l+1,2k,1)}$, and $N_{1000}^{(2l+1,2k,1)}$. The contribution of these processes in (3.8.72) will be expressed in terms of processes $\widetilde{N}_{00}^{(1,2k,2l+1)}$ and $\widetilde{N}_{00}^{(2l+1,2k,1)}$. Namely, for both $u = 1$, $w = 2l + 1$, and $u = 2l + 1$, $w = 1$, we obtain that

$$\begin{aligned}
&\left\{N_{0000}^{(u,2k,w)}(x,\infty]=0,\, N_{1000}^{(u,2k,w)}(y,\infty]=0\right\}\\
&=\left\{\widetilde{N}_{00}^{(u,2k,w)}(x,\infty]=0,\, \widetilde{N}_{00}^{(u,2k,w)}\left(\frac{y}{\psi_{2k}},\infty\right]=0\right\}\\
&=\begin{cases}\left\{\widetilde{N}_{00}^{(u,2k,w)}\left(\frac{y}{\psi_{2k}},\infty\right]=0\right\}, & \text{if } k\leqslant r-1,\\ \left\{\widetilde{N}_{00}^{(u,2k,w)}(x,\infty]=0\right\}, & \text{if } k\geqslant r.\end{cases}
\end{aligned}\tag{3.8.78}$$

Note that for $r=1$, only the case $k\geqslant r$ appears in (3.8.78).

2. Let us consider point processes $N_{0010}^{(1,2k,2l+1)}$, $N_{1010}^{(1,2k,2l+1)}$, $N_{0010}^{(2l+1,2k,1)}$, and $N_{1010}^{(2l+1,2k,1)}$. For both

$$(u,v,w)=(1,2k,2l+1),\quad\text{and}\quad (u,v,w)=(2l+1,2k,1),$$

we obtain that

$$\begin{aligned}
&\left\{N_{0010}^{(u,v,w)}(x,\infty]=0,\, N_{1010}^{(u,v,w)}(y,\infty]=0\right\}\\
&=\left\{\widetilde{N}_{10}^{(u,v,w)}\left(\frac{x}{\psi_{2k}},\infty\right],\, \widetilde{N}_{10}^{(u,v,w)}(y,\infty]=0\right\}\\
&=\left\{\widetilde{N}_{10}^{(u,v,w)}(y,\infty]=0\right\}.
\end{aligned}\tag{3.8.79}$$

3. Now we consider point processes $N_{0101}^{(1,2k,2l+1)}$, $N_{1101}^{(1,2k,2l+1)}$, $N_{0101}^{(2l+1,2k,1)}$, and $N_{1101}^{(2l+1,2k,1)}$. For both $u=1$, $w=2l+1$, and $u=2l+1$, $w=1$, we obtain that

$$\begin{aligned}
&\left\{N_{0101}^{(u,2k,w)}(x,\infty]=0,\, N_{1101}^{(u,2k,w)}(y,\infty]=0\right\}\\
&=\left\{\widetilde{N}_{01}^{(u,2k,w)}(x,\infty]=0,\, \widetilde{N}_{01}^{(u,2k,w)}\left(\frac{y}{\psi_{2k}},\infty\right]=0\right\}\\
&=\begin{cases}\left\{\widetilde{N}_{01}^{(u,2k,w)}\left(\frac{y}{\psi_{2k}},\infty\right]=0\right\}, & \text{if } k\leqslant r-1,\\ \left\{\widetilde{N}_{01}^{(u,2k,w)}(x,\infty]=0\right\}, & \text{if } k\geqslant r.\end{cases}
\end{aligned}\tag{3.8.80}$$

4. Finally, we consider point processes $N_{0111}^{(1,2k,2l+1)}$, $N_{1111}^{(1,2k,2l+1)}$, $N_{0111}^{(2l+1,2k,1)}$, and $N_{1111}^{(2l+1,2k,1)}$. For both

$$(u,v,w)=(1,2k,2l+1),\quad\text{and}\quad (u,v,w)=(2l+1,2k,1),$$

we obtain that

$$\begin{aligned}
&\left\{N_{0111}^{(u,v,w)}(x,\infty]=0,\,N_{1111}^{(u,v,w)}(y,\infty]=0\right\}\\
&=\left\{\widetilde{N}_{11}^{(u,v,w)}\left(\frac{x}{\psi_{2k}},\infty\right],\,\widetilde{N}_{11}^{(u,v,w)}(y,\infty]=0\right\}\\
&=\left\{\widetilde{N}_{11}^{(u,v,w)}(y,\infty]=0\right\}.
\end{aligned}\tag{3.8.81}$$

Contribution of the Negative Half-line Let $r\in\mathbb{N}$, and $(x,y)\in B_r$, i.e.,

$$0<\frac{y}{|\psi_{2r-1}|}\leqslant\frac{x}{|\psi_1|}<\frac{y}{|\psi_{2r+1}|}.\tag{3.8.82}$$

Note that $y=\dfrac{y}{|\psi_1|}<\dfrac{y}{|\psi_3|}<\cdots<\dfrac{y}{|\psi_{2r-1}|}\leqslant\dfrac{x}{|\psi_1|}<\dfrac{y}{|\psi_{2r+1}|}<\dfrac{y}{|\psi_{2r+3}|}\cdots$

5. Let us consider point processes $N_{0001}^{(1,2k,2l+1)}$, $N_{1001}^{(1,2k,2l+1)}$, $N_{0001}^{(2l+1,2k,1)}$, and $N_{1001}^{(2l+1,2k,1)}$. The contribution of these processes in (3.8.72) will be expressed in terms of processes $\widetilde{N}_{01}^{(1,2k,2l+1)}$ and $\widetilde{N}_{01}^{(2l+1,2k,1)}$. Namely, we obtain that

$$\begin{aligned}
&\left\{N_{0001}^{(1,2k,2l+1)}(x,\infty]=0,\,N_{1001}^{(1,2k,2l+1)}(y,\infty]=0\right\}\\
&=\left\{\widetilde{N}_{01}^{(1,2k,2l+1)}\left[-\infty,-\frac{x}{|\psi_1|}\right)=0,\,\widetilde{N}_{01}^{(1,2k,2l+1)}\left[-\infty,-\frac{y}{|\psi_{2l+1}|}\right)=0\right\}\\
&=\begin{cases}\left\{\widetilde{N}_{01}^{(1,2k,2l+1)}\left[-\infty,-\dfrac{y}{|\psi_{2l+1}|}\right)=0\right\}, & \text{if } l\leqslant r-1,\\[2ex] \left\{\widetilde{N}_{01}^{(1,2k,2l+1)}\left[-\infty,-\dfrac{x}{|\psi_1|}\right)=0\right\}, & \text{if } l\geqslant r,\end{cases}
\end{aligned}\tag{3.8.83}$$

and

$$\begin{aligned}
&\left\{N_{0001}^{(2l+1,2k,1)}(x,\infty]=0,\,N_{1001}^{(2l+1,2k,1)}(y,\infty]=0\right\}\\
&=\left\{\widetilde{N}_{01}^{(2l+1,2k,1)}\left[-\infty,-\frac{x}{|\psi_{2l+1}|}\right)=0,\,\widetilde{N}_{01}^{(2l+1,2k,1)}\left[-\infty,-\frac{y}{|\psi_1|}\right)=0\right\}\\
&=\left\{\widetilde{N}_{01}^{(2l+1,2k,1)}\left[-\infty,-\frac{y}{|\psi_1|}\right)=0\right\},\quad\text{for every } l\in\mathbb{N}.
\end{aligned}\tag{3.8.84}$$

6. Let us consider point processes $N_{0011}^{(1,2k,2l+1)}$, $N_{1011}^{(1,2k,2l+1)}$, $N_{0011}^{(2l+1,2k,1)}$, and $N_{1011}^{(2l+1,2k,1)}$. For these processes we obtain that

$$\left\{N_{0011}^{(1,2k,2l+1)}(x,\infty]=0,\,N_{1011}^{(1,2k,2l+1)}(y,\infty]=0\right\}$$

$$= \left\{ \widetilde{N}_{11}^{(1,2k,2l+1)} \left[-\infty, -\frac{x}{|\psi_{2l+1}|} \right) = 0, \widetilde{N}_{11}^{(1,2k,2l+1)} \left[-\infty, -\frac{y}{|\psi_1|} \right) = 0 \right\}$$

$$= \left\{ \widetilde{N}_{11}^{(1,2k,2l+1)} \left[-\infty, -\frac{y}{|\psi_1|} \right) = 0 \right\}, \quad \text{for every } l \in \mathbb{N}, \tag{3.8.85}$$

and

$$\left\{ N_{0011}^{(2l+1,2k,1)}(x,\infty] = 0, N_{1011}^{(2l+1,2k,1)}(y,\infty] = 0 \right\}$$

$$= \left\{ \widetilde{N}_{11}^{(2l+1,2k,1)} \left[-\infty, -\frac{x}{|\psi_1|} \right) = 0, \widetilde{N}_{11}^{(2l+1,2k,1)} \left[-\infty, -\frac{y}{|\psi_{2l+1}|} \right) = 0 \right\}$$

$$= \begin{cases} \left\{ \widetilde{N}_{11}^{(2l+1,2k,1)} \left[-\infty, -\frac{y}{|\psi_{2l+1}|} \right) = 0 \right\}, \text{ if } l \leqslant r-1, \\ \left\{ \widetilde{N}_{11}^{(2l+1,2k,1)} \left[-\infty, -\frac{x}{|\psi_1|} \right) = 0 \right\}, \quad \text{if } l \geqslant r. \end{cases} \tag{3.8.86}$$

7. Let us consider point processes $N_{0100}^{(1,2k,2l+1)}$, $N_{1100}^{(1,2k,2l+1)}$, $N_{0100}^{(2l+1,2k,1)}$, and $N_{1100}^{(2l+1,2k,1)}$ For these processes we obtain the following relations:

$$\left\{ N_{0100}^{(1,2k,2l+1)}(x,\infty] = 0, N_{1100}^{(1,2k,2l+1)}(y,\infty] = 0 \right\}$$

$$= \left\{ \widetilde{N}_{00}^{(1,2k,2l+1)} \left[-\infty, -\frac{x}{|\psi_1|} \right) = 0, \widetilde{N}_{00}^{(1,2k,2l+1)} \left[-\infty, -\frac{y}{|\psi_{2l+1}|} \right) = 0 \right\}$$

$$= \begin{cases} \left\{ \widetilde{N}_{00}^{(1,2k,2l+1)} \left[-\infty, -\frac{y}{|\psi_{2l+1}|} \right) = 0 \right\}, \text{ if } l \leqslant r-1, \\ \left\{ \widetilde{N}_{00}^{(1,2k,2l+1)} \left[-\infty, -\frac{x}{|\psi_1|} \right) = 0 \right\}, \quad \text{if } l \geqslant r, \end{cases} \tag{3.8.87}$$

and

$$\left\{ N_{0100}^{(2l+1,2k,1)}(x,\infty] = 0, N_{1100}^{(2l+1,2k,1)}(y,\infty] = 0 \right\}$$

$$= \left\{ \widetilde{N}_{00}^{(2l+1,2k,1)} \left[-\infty, -\frac{x}{|\psi_{2l+1}|} \right) = 0, \widetilde{N}_{00}^{(2l+1,2k,1)} \left[-\infty, -\frac{y}{|\psi_1|} \right) = 0 \right\}$$

$$= \left\{ \widetilde{N}_{00}^{(2l+1,2k,1)} \left[-\infty, -\frac{y}{|\psi_1|} \right) = 0 \right\}, \quad \text{for every } l \in \mathbb{N}. \tag{3.8.88}$$

8. Finally, for point processes $N_{0110}^{(1,2k,2l+1)}$, $N_{1110}^{(1,2k,2l+1)}$, $N_{0110}^{(2l+1,2k,1)}$, and $N_{1110}^{(2l+1,2k,1)}$ we obtain that

$$
\begin{aligned}
&\left\{N_{0110}^{(1,2k,2l+1)}(x,\infty]=0,\, N_{1110}^{(1,2k,2l+1)}(y,\infty]=0\right\}\\
&=\left\{\widetilde{N}_{10}^{(1,2k,2l+1)}\left[-\infty,-\frac{x}{|\psi_{2l+1}|}\right)=0,\, \widetilde{N}_{10}^{(1,2k,2l+1)}\left[-\infty,-\frac{y}{|\psi_{1}|}\right)=0\right\}\\
&=\left\{\widetilde{N}_{10}^{(1,2k,2l+1)}\left[-\infty,-\frac{y}{|\psi_{1}|}\right)=0\right\}, \quad \text{for every } l\in\mathbb{N},
\end{aligned}
\tag{3.8.89}
$$

and

$$
\begin{aligned}
&\left\{N_{0110}^{(2l+1,2k,1)}(x,\infty]=0,\, N_{1110}^{(2l+1,2k,1)}(y,\infty]=0\right\}\\
&=\left\{\widetilde{N}_{10}^{(2l+1,2k,1)}\left[-\infty,-\frac{x}{|\psi_{1}|}\right)=0,\, \widetilde{N}_{10}^{(2l+1,2k,1)}\left[-\infty,-\frac{y}{|\psi_{2l+1}|}\right)=0\right\}\\
&=\begin{cases}\left\{\widetilde{N}_{10}^{(2l+1,2k,1)}\left[-\infty,-\frac{y}{|\psi_{2l+1}|}\right)=0\right\}, & \text{if } l\leqslant r-1,\\[2mm] \left\{\widetilde{N}_{10}^{(2l+1,2k,1)}\left[-\infty,-\frac{x}{|\psi_{1}|}\right)=0\right\}, & \text{if } l\geqslant r.\end{cases}
\end{aligned}
\tag{3.8.90}
$$

(b) Relations (3.8.72) and (3.8.78)–(3.8.81) imply that

$$
\Pi^{+}=\Pi_{00}^{+}\cdot\Pi_{10}^{+}\cdot\Pi_{01}^{+}\cdot\Pi_{11}^{+},
\tag{3.8.91}
$$

where

$$
\begin{aligned}
\Pi_{00}^{+}=\prod_{l=1}^{\infty}\Bigg(&\prod_{k=1}^{r-1}\exp\left\{-py^{-\alpha}\psi_{2k}^{\alpha}\left(p_{00}^{(1,2k,2l+1)}+p_{00}^{(2l+1,2k,1)}\right)\right\}\times\\
&\times\prod_{k=r}^{\infty}\exp\left\{-px^{-\alpha}\left(p_{00}^{(1,2k,2l+1)}+p_{00}^{(2l+1,2k,1)}\right)\right\}\Bigg),
\end{aligned}
\tag{3.8.92}
$$

$$
\Pi_{10}^{+}=\prod_{l=1}^{\infty}\prod_{k=1}^{\infty}\exp\left\{-py^{-\alpha}\left(p_{10}^{(1,2k,2l+1)}+p_{10}^{(2l+1,2k,1)}\right)\right\},
\tag{3.8.93}
$$

$$
\begin{aligned}
\Pi_{01}^{+}=\prod_{l=1}^{\infty}\Bigg(&\prod_{k=1}^{r-1}\exp\left\{-py^{-\alpha}\psi_{2k}^{\alpha}\left(p_{01}^{(1,2k,2l+1)}+p_{01}^{(2l+1,2k,1)}\right)\right\}\times\\
&\times\prod_{k=r}^{\infty}\exp\left\{-px^{-\alpha}\left(p_{01}^{(1,2k,2l+1)}+p_{01}^{(2l+1,2k,1)}\right)\right\}\Bigg),
\end{aligned}
\tag{3.8.94}
$$

$$\Pi_{11}^{+} = \prod_{l=1}^{\infty}\prod_{k=1}^{\infty} \exp\left\{-py^{-\alpha}\left(p_{11}^{(1,2k,2l+1)} + p_{11}^{(2l+1,2k,1)}\right)\right\}. \tag{3.8.95}$$

Using property (3.8.54) of sets $S_{00}^{(1,2k,2l+1)}$ and $S_{01}^{(1,2k,2l+1)}$, and property (3.8.55) of sets $S_{10}^{(2l+1,2k,1)}$ and $S_{11}^{(2l+1,2k,1)}$, we obtain that

$$\widetilde{p}_0^{(2k)} := \sum_{l=1}^{\infty}\left(p_{00}^{(1,2k,2l+1)} + p_{00}^{(2l+1,2k,1)} + p_{01}^{(1,2k,2l+1)} + p_{01}^{(2l+1,2k,1)}\right) \tag{3.8.96}$$

is the density of set $S_0^{(2k)}$.

Similarly, using property (3.8.56) of sets $S_{10}^{(1,2k,2l+1)}$ and $S_{11}^{(1,2k,2l+1)}$, and property (3.8.57) of sets $S_{10}^{(2l+1,2k,1)}$ and $S_{11}^{(2l+1,2k,1)}$, we obtain that

$$\sum_{k=1}^{\infty}\sum_{l=1}^{\infty}\left(p_{10}^{(1,2k,2l+1)} + p_{10}^{(2l+1,2k,1)} + p_{11}^{(1,2k,2l+1)} + p_{11}^{(2l+1,2k,1)}\right) = p_1.$$

Now it is easy to obtain that equalities (3.8.66) follow from (3.8.91)–(3.8.95).

(c) It follows from (3.8.72) and (3.8.83)–(3.8.90) that Π^- can be represented in the form

$$\Pi^- = \exp\left\{-qx^{-\alpha}|\psi_1|^\alpha\sigma_r - qy^{-\alpha}\left(|\psi_1|^\alpha\sigma_0 + \sum_{l=1}^{r-1}|\psi_{2l+1}|^\alpha\sigma_l^*\right)\right\}, \tag{3.8.97}$$

where

$$\sigma_r = \sum_{k=1}^{\infty}\sum_{l=r}^{\infty}\left(p_{00}^{(1,2k,2l+1)} + p_{01}^{(1,2k,2l+1)} + p_{10}^{(2l+1,2k,1)} + p_{11}^{(2l+1,2k,1)}\right), \tag{3.8.98}$$

$$\sigma_0 = \sum_{k=1}^{\infty}\sum_{l=1}^{\infty}\left(p_{00}^{(2l+1,2k,1)} + p_{01}^{(2l+1,2k,1)} + p_{10}^{(1,2k,2l+1)} + p_{11}^{(1,2k,2l+1)}\right), \tag{3.8.99}$$

$$\sigma_l^* = \sum_{k=1}^{\infty}\left(p_{00}^{(1,2k,2l+1)} + p_{01}^{(1,2k,2l+1)} + p_{10}^{(2l+1,2k,1)} + p_{11}^{(2l+1,2k,1)}\right). \tag{3.8.100}$$

Using property (3.8.55) of sets $S_{00}^{(2l+1,2k,1)}$ and $S_{01}^{(2l+1,2k,1)}$, and property (3.8.56) of sets $S_{10}^{(1,2k,2l+1)}$ and $S_{11}^{(1,2k,2l+1)}$, we obtain that the sum on the right-hand side of (3.8.99) is equal to the density of set $\{i \mid c_{i-1} = 1\}$, i.e., σ_0 is the density of set $\{i \mid c_{i-1} = 1\}$.

Similarly, using property (3.8.54) of sets $S_{00}^{(1,2k,2l+1)}$ and $S_{01}^{(1,2k,2l+1)}$, and property (3.8.57) of sets $S_{10}^{(2l+1,2k,1)}$ and $S_{11}^{(2l+1,2k,1)}$, we obtain that σ_r and σ_l^* are respectively the densities of sets

$$\{i \mid c_{i-1} = c_{i-3} = \cdots = c_{i-2r+1} = 0\},$$
$$\{i \mid c_{i-1} = c_{i-3} = \cdots = c_{i-2l+1} = 0, c_{i-2l-1} = 1\}.$$

□

Remark 3.8.11 Note that we considered case $0 < y \leqslant x$ in Theorems 3.8.4 and 3.8.10. Cases $y \leqslant 0 \leqslant x$ and $y \leqslant x \leqslant 0$ can be considered in a similar way with more technical details.

Remark 3.8.12 Comments and remarks similar to those given after Theorem 3.8.4 can also be formulated here. Let us note only the following: Densities $p_{de}^{(1,2k,2l+1)}$ and $p_{de}^{(2l+1,2k,1)}$ of sets $S_{de}^{(1,2k,2l+1)}$ and $S_{de}^{(2l+1,2k,1)}$ can also be interpreted as relative asymptotic frequencies of the patterns given by (3.8.54)–(3.8.57). As in Remark 3.8.7 let us suppose that $(c_n)_{n\geqslant 1}$ is determined randomly by a sequence of independent Bernoulli trials that give 1 with probability $\widetilde{p}$, and 0 with probability $\widetilde{q}$, $\widetilde{p} + \widetilde{q} = 1$. Then, for almost every realization of sequence $(c_n)_{n\geqslant 1}$ we have

$$p_{00}^{(1,2k,2l+1)} = p_{01}^{(1,2k,2l+1)} = \frac{1}{2}\,\widetilde{p}^{\,2}\widetilde{q}^{\,k+l},$$

$$p_{00}^{(2l+1,2k,1)} = p_{01}^{(2l+1,2k,1)} = \frac{1}{2}\,\widetilde{p}^{\,l+1}\widetilde{q}^{\,k+1},$$

$$p_{00}^{(1,2k,2l+1)} = p_{01}^{(1,2k,2l+1)} = \frac{1}{2}\,\widetilde{p}^{\,k+l}\widetilde{q}^{\,2},$$

$$p_{00}^{(2l+1,2k,1)} = p_{01}^{(2l+1,2k,1)} = \frac{1}{2}\,\widetilde{p}^{\,k+1}\widetilde{q}^{\,l+1},$$

and $1 - p_0 = \widetilde{p}$, $p_0^{(2k)} = \widetilde{q}^{\,k}\,\widetilde{p}$, $a_r^+ = \widetilde{q}^{\,k}$, $\sigma_r = \widetilde{q}^{\,r}$, $\sigma_0 = \widetilde{p}$, and $\sigma_l^* = \widetilde{q}^{\,l}\,\widetilde{p}$. The form of the almost sure limiting distribution easily follows from the general result.

Remark 3.8.13 Theorems 3.7.1, 3.7.5, 3.8.4, and 3.8.10 lead us to the following conclusion: There are some patterns of digits 0 and 1 in sequence $(c_n)_{\geqslant 1}$ with stable asymptotic frequencies, and/or some subsets of the positions of digits in this sequence for which the density in $\mathbb{N}$ exist that play a crucial role in the asymptotic behavior of properly normalized random vector $(\widetilde{M}_n, M_n)$. These patterns and subsets depend on the process considered.

Remark 3.8.14 Extremes in incomplete samples from moving averages of random variables from the domain of attraction of the Gumbel distribution were considered by Glavaš and Mladenović [79].

3.8.5 On Extremes in Non-Stationary Sequences

If $(X_n)_{n\geqslant 1}$ is a stationary random sequence, then a sequence of observed random variables is also stationary only in very special cases. Therefore results that provide the asymptotic behavior of the maxima in incomplete samples from stationary sequences are also related to extreme value theory in non-stationary sequences.

Extreme values in non-stationary sequences of identically distributed random variables subject to suitable long-range dependence restrictions were investigated by Auld and Papastathopoulos [14]. Extremes in non-stationary sequences were also considered by Hüsler [112], Eastoe and Tawn [57], and in Chapter 6 of Coles [47].

Exercises and Supplements

3.1 Scotto, Turkman, and Anderson [205]: Let us suppose that sequence $(X_n)_{n\geqslant 0}$ is defined by (3.8.1), such that conditions (3.8.2)–(3.8.7) are satisfied, and consider the sub-sampled moving average process

$$Y_k = X_{kM} = \sum_{i=0}^{\infty} \psi_i Z_{kM-i},$$

where $M > 1$ is a systematic sampling interval.

(a) The asymptotic distribution of $M_n^{(Y)} = \max\{Y_k \mid 1 \leqslant k \leqslant [n/M]\}$ is given by the following equality

$$\lim_{n\to\infty} P\{M_n^{(Y)} \leqslant a_n x\} = \exp\left\{ -\sum_{i=0}^{M-1} \frac{1}{M} \left(p(\psi_i^{(+)})^{\alpha} + q(\psi_i^{(-)})^{\alpha} \right) x^{-\alpha} \right\},$$

where constants $\psi_i^{(+)}$ and $\psi_i^{(-)}$ are given by

$$\psi_i^{(+)} = \max_{0\leqslant j<\infty} \max\{\psi_{jM+i}, 0\}, \quad \psi_i^{(-)} = \max_{0\leqslant j<\infty} \max\{-\psi_{jM+i}, 0\}.$$

(b) The extremal index ϑ_M of the sub-sampled series is given by

$$\vartheta_M = \frac{\sum_{i=0}^{M-1} \left\{ p(\psi_i^{(+)})^{\alpha} + q(\psi_i^{(-)})^{\alpha} \right\}}{\sum_{i=0}^{M-1} \left\{ p\left(\sum_{j=0}^{\infty} (\psi_{jM+i}^{+})^{\alpha}\right) + q\left(\sum_{j=0}^{\infty} (\psi_{jM+i}^{-})^{\alpha}\right) \right\}}.$$

3.2 Kudrov and Piterbarg [138]: Let X_n, $n \in \mathbb{Z}$, be a Gaussian sequence with zero mean and unit variance, and let $(m_n)_{n\geqslant 1}$ be a non-random sequence such that

$m_n = o(\log n)$ as $n \to \infty$. Let us define

$$M_n(X) = \max_{1 \leqslant i \leqslant n} X_i + m_i d_n, \quad M_{n,k}(X) = \max_{i \in G_n} X_i + m_i d_n,$$

where $G_n \subset \mathbb{N}_n = \{1, \ldots, n\}$, $|G_n| = [n/k]$, $k \geqslant 1$. Let $Q_n \subset \mathbb{Z}$, $n \in \mathbb{N}$, be a sequence of finite sets such that $|Q_n| \sim an$, as $n \to \infty$, for some $a > 0$. By definition, sequence $(c_k)_{k \geqslant 1}$ is called pseudo-stationary with respect to $(Q_n)_{n \geqslant 1}$, if for every $x \in \mathbb{R}$, there exists the limit

$$C(x) = \lim_{n \to \infty} \frac{|\{i \mid c_i \leqslant x, 1 \leqslant i \leqslant n\}|}{|Q_n|}.$$

We call $C(x)$ the distribution function of $(c_n)_{n \geqslant 1}$. Note that $C(x)$ is not necessarily a probability distribution function. We assume that $(m_i)_{i \geqslant 1}$ is pseudo-stationary with respect to both sequences of sets $(G_n)_{n \geqslant 1}$ and $(\mathbb{N}_n)_{n \geqslant 1}$. and denote by $G(x)$ and $N(x)$ the corresponding distribution functions. The Laplace transform of distribution function M is defined by $\widetilde{M}(c) = \int_{-\infty}^{\infty} e^{cx}\, dM(x)$. Let us denote

$$F_\lambda^{N,G}(x, y) = \frac{\widetilde{G}(1)}{A(k, x)} E \exp\left\{-A(k, x) e^{-y-\lambda+\xi\sqrt{2\lambda}}\right\},$$

where $A(k, x) := ke^{-x}\widetilde{N}(1) + (1 - e^{-x})\widetilde{G}(1)$, and ξ is a standard normal random variable. We also take $d_n = a_n = (2 \ln n)^{-1/2}$.

If the covariance function $r_n = E(X_1 X_{1+n})$, $n \in \mathbb{N}$, satisfies the condition $r_n \ln n \to \lambda$ for some $\lambda \geqslant 0$, and

$$\int_{-\infty}^{\infty} e^{3|x|}\, d(S(x) + G(x)) < \infty,$$

then for any $x \geqslant 0$, and $y \in \mathbb{R}$, the following equality holds

$$\lim_{n \to \infty} P\{M_n(X) - M_{n,k}(X) \leqslant a_n x, M_{n,k}(X) \leqslant b_{n,k} + a_n y\} = F_\lambda^{N,G}(x, y),$$

where $b_{n,k} = \sqrt{2 \ln(n/k)} - \dfrac{1}{2\sqrt{2 \ln n}}(\ln 4\pi + \ln \ln n)$.

3.3 Tan and Wang [217]: Suppose that the following conditions are satisfied:

(a) $(X_n)_{n \geqslant 1}$ is a sequence of i.i.d. random variables with marginal distribution function $F(x) = P\{X_1 \leqslant x\}$, and $F \in DA(G)$ for some extreme value distribution function G, i.e., for some constants $a_n > 0$ and $b_n \in \mathbb{R}$, $n \in \mathbb{N}$, and every real x, $F^n(a_n x + b_n) \to G(x)$ as $n \to \infty$.
(b) $(\varepsilon_k)_{k \geqslant 1}$ is a sequence of indicators independent of $(X_n)_{n \geqslant 1}$, where ε_k is the indicator of the event that X_k is observed. Let us denote $S_n = \varepsilon_1 + \cdots + \varepsilon_n$, and suppose that

$$S_n/n \xrightarrow{P} p \in [0, 1], \quad \text{as } n \to \infty.$$

Let us denote $M_n = \max\{X_1, \ldots, X_n\}$, $\widetilde{M}_n = \max\{X_i \mid 1 \leqslant i \leqslant n, \varepsilon_i = 1\}$ if $S_n \geqslant 1$, and $\widetilde{M}_n = \inf\{t \mid F(t) > 0\}$ if $S_n = 0$. Then for any $x \geqslant 0$ and $y \in \mathbb{R}$, the following equalities hold:

$$\lim_{n\to\infty} P\{M_n - \widetilde{M}_n \leqslant a_n x, \widetilde{M}_n \leqslant a_n y + b_n\} = \int_{-\infty}^{y} G^{1-p}(x+z)\, dG^p(z),$$

$$\lim_{n\to\infty} P\{M_n - \widetilde{M}_n \leqslant a_n x\} = G^{1-p} * G^p(x),$$

where $H_1 * H_2$ denotes the convolution of H_1 and H_2.

3.4 Tan and Wang [217]: Suppose that the following conditions are satisfied:

(a) $(X_n)_{n\geqslant 1}$ is a stationary Gaussian sequence with zero mean, unit variance, and covariance function $r_n = E(X_1 X_{n+1})$, $n \geqslant 0$.
(b) $(\varepsilon_k)_{k\geqslant 1}$ is a sequence of indicators independent of $(X_n)_{n\geqslant 1}$, where ε_k is the indicator of the event that X_k is observed. Let us denote $S_n = \varepsilon_1 + \cdots + \varepsilon_n$, and suppose that

$$\frac{S_n}{n} \xrightarrow{P} p \in [0, 1], \quad \text{as } n \to \infty.$$

Let us denote $M_n = \max\{X_1, \ldots, X_n\}$, $\widetilde{M}_n = \max\{X_i \mid 1 \leqslant i \leqslant n, \varepsilon_i = 1\}$ if $S_n \geqslant 1$, and $\widetilde{M}_n = \inf\{t \mid F(t) > 0\}$ if $S_n = 0$. Let us assume that constants a_n and b_n are given by (2.3.78)–(2.3.79). Then the following two statements hold:

If $r_n \ln n \to 0$ as $n \to \infty$, and $G_0(x)$ is the standard Gumbel distribution function, then for any $x \geqslant 0$ and $y \in \mathbb{R}$,

$$\begin{aligned}
&\lim_{n\to\infty} P\{M_n - \widetilde{M}_n \leqslant a_n x, \widetilde{M}_n \leqslant a_n y + b_n\} \\
&= \int_{-\infty}^{y} G_0^{1-p}(x+z)\, dG_0^p(z) \\
&= \frac{p}{(1-p)e^{-x} + p} \exp\left\{-e^{-y}\left((1-p)e^{-x} + p\right)\right\}.
\end{aligned}$$

If $r_n \ln n \to \lambda \in (0, \infty)$ as $n \to \infty$, and ϕ is notation for the standard normal density, then for any $x \geqslant 0$ and $y \in \mathbb{R}$,

$$\begin{aligned}
&\lim_{n\to\infty} P\{M_n - \widetilde{M}_n \leqslant a_n x, \widetilde{M}_n \leqslant a_n y + b_n\} \\
&= \int_{-\infty}^{\infty} \int_{-\infty}^{y} p \exp\left\{-\left((1-p)e^{-x} + p\right)e^{-t-\lambda+z\sqrt{2\lambda}}\right\} e^{-t-\lambda+z\sqrt{2\lambda}} \phi(z)\, dt\, dz.
\end{aligned}$$

3.5 Peng, Weng and Nadarajah [179]: Let $(X_n)_{n\geqslant 1}$ be a stationary Gaussian sequence with zero mean, unit variance and correlation sequence $r_n = E(X_1X_{n+1})$, $n \geqslant 0$, such that $r_n \log n \to 0$ as $n \to \infty$.

$(\varepsilon_k)_{k\geqslant 1}$ is a sequence of indicators independent of $(X_n)_{n\geqslant 1}$, where ε_k is the indicator of the event that X_k is observed. Let us denote $S_n = \varepsilon_1 + \cdots + \varepsilon_n$, and suppose that

$$\frac{S_n}{n} \xrightarrow{P} p \in (0, 1], \qquad \text{as } n \to \infty.$$

Let us also introduce the following notation:

$$\begin{aligned}
M_n &= \max\{X_1, \ldots, X_n\}, \\
m_n &= \min\{X_1, \ldots, X_n\}, \\
\widetilde{M}_n &= \max\{X_i \mid 1 \leqslant i \leqslant n, \varepsilon_i = 1\}, \quad \text{if } \; S_n \geqslant 1, \\
\widetilde{m}_n &= \min\{X_i \mid 1 \leqslant i \leqslant n, \varepsilon_i = 1\}, \quad \text{if } \; S_n \geqslant 1,
\end{aligned}$$

and $\widetilde{M}_n = \widetilde{m}_n = \inf\{t \mid F(t) > 0\}$ if $S_n = 0$. Let us assume that constants a_n and b_n are given by (2.3.78)–(2.3.79), and denote $u_n(x) = a_n x + b_n$ for every $x \in \mathbb{R}$.

(a) For all $x_1, x_2, y_1, y_2 \in \mathbb{R}$, such that $x_2 < x_1$ and $y_2 < y_1$, the following equality holds:

$$\begin{aligned}
&\lim_{n\to\infty} P\left\{-u_n(y_2) \leqslant \widetilde{m}_n \leqslant \widetilde{M}_n \leqslant u_n(x_2), -u_n(y_1) \leqslant m_n \leqslant M_n \leqslant u_n(x_1)\right\} \\
&= \exp\left(-pe^{-(x_2+y_2)}\right) \exp\left(-(1-p)e^{-(x_1+y_1)}\right).
\end{aligned}$$

(b) For all $x, y \in \mathbb{R}$, we have

$$\lim_{n\to\infty} P\left\{-u_n(y) \leqslant \widetilde{m}_n \leqslant \widetilde{M}_n \leqslant u_n(x)\right\} = exp\left(-pe^{-(x+y)}\right).$$

Especially, for all $x \in \mathbb{R}$,

$$\lim_{n\to\infty} P\left\{\max_{1\leqslant k\leqslant n,\, \varepsilon_k=1} |X_k| \leqslant u_n(x + \ln 2)\right\} = \exp\left(-pe^{-x}\right).$$

Chapter 4
Combinatorial Problems and Extreme Values

4.1 Introduction

In this chapter we shall consider some extremal properties of several random combinatorial configurations. In all cases considered we shall introduce the uniform probability distribution on the set Ω of all possible combinatorial configurations of interest. Then every numerical characteristic of the elements of Ω becomes a random variable. The main focus will be on certain extremal characteristics of random elements from set Ω. We will be interested in the limiting behavior of such random variables as set Ω grows in a specified sense. Probabilistic methods have been used in combinatorics starting probably with the papers of Erdös, *see*, for example, Erdös and Szekeres [62], Erdös and Renyi [60], Erdös and Spencer [61], and the paper by Goncharov [232].

In Sect. 4.2 we shall consider the set of random permutations of a set consisting of n elements. In Sect. 4.3 we shall provide results related to the coupon collector's problem and its various extensions. The polynomial scheme will be considered in Sect. 4.4. Results related to random trees and random forests will be provided in Sect. 4.5. Results of random partitions of finite sets will be given in Sect. 4.6. Finally, the geometric properties of samples of random vectors will be the subject of Sect. 4.7.

4.2 Random Permutations

Let $\mathbb{N}_n = \{1, 2, \ldots, n\}$, and let Ω_n be the set of all permutations of set $\mathbb{N}_n$. A permutation of set $\mathbb{N}_n$ may have different representations. For example, $\omega = 43251867$ or $\omega = (4, 3, 2, 5, 1, 8, 6, 7)$ is the so-called *one line representation* of a permutation of set $\mathbb{N}_8$. This representation means that $\omega(1) = 5$, $\omega(2) = 3$,

P. Mladenović, *Extreme Values In Random Sequences*,
Springer Series in Operations Research and Financial Engineering,
https://doi.org/10.1007/978-3-031-57412-2_4

$\omega(3) = 2$, etc. The same permutation, considered as function $\mathbb{N}_8 \to \mathbb{N}_8$, can be represented as follows:

$$\omega : \begin{pmatrix} 1\,2\,3\,4\,5\,6\,7\,8 \\ 4\,3\,2\,5\,1\,8\,6\,7 \end{pmatrix}. \tag{4.2.1}$$

Every permutation ω can be represented as a product of cycles. This representation is not unique. Permutation ω that is given by (4.2.1) can be written in the form $\omega = (145)(23)(687)$. The same permutation is obtained if we arrange the entries of any cycle in another cyclic order and then rearrange the order of the cycles. For example, $\omega = (768)(23)(451)$. The uniqueness of the representation of a permutation as a product of cycles can be achieved in different ways. We provide here two possibilities.

The First Canonical Representation The first cycle starts with element 1, the second cycle starts with the smallest element that is not included in the first cycle, the third cycle starts with the smallest element that is not included in the previous two cycles, etc. For permutation ω given by (4.2.1) we obtain $\omega = (145)(23)(687)$.

The Second Canonical Representation Every cycle starts with the greatest entry of this cycle, and then the cycles are arranged such that their first elements appear in increasing order. For permutation ω given by (4.2.1) we obtain $\omega = (32)(514)(876)$.

Each of these canonical representations can be better suited to some specific problems.

Throughout this section we shall assume that every permutation $\omega \in \Omega_n$ has probability $1/n!$.

4.2.1 The Maximal Step

For every $j \in \mathbb{N}_n$, and $\omega = (a_1, \ldots, a_n) \in \Omega_n$ let us define

$$X_{nj}(\omega) = |a_j - a_{j+1}|, \tag{4.2.2}$$

where $a_{n+1} = a_1$. Then, $X_{n1}, \ldots, X_{nn}$ is a sequence of *dependent* random variables that satisfy the condition of strict stationarity. For every $j \in \mathbb{N}_n$ the marginal distribution of random variable X_{nj} is given by

$$P\{X_{nj} = k\} = \frac{2(n-k)}{n(n-1)}, \quad k \in \{1, 2, \ldots, n\}.$$

Let us denote $M_n' = \max\{X_{n1}, \ldots, X_{n,n-1}\}$ and $M_n = \max\{X_{n1}, \ldots, X_{nn}\}$. Note that M_n includes a_n and a_1 as adjacent terms of permutation ω. It is easy to prove that random variables M_n' and M_n have the same limiting distribution as $n \to \infty$, i.e., the Weibull distribution $G_{2,2}(x)$.

Theorem 4.2.1 (Mladenović [157]) *The limit distribution of random variable M_n is given by*

$$\lim_{n\to\infty} P\{M_n \leqslant x\sqrt{n}+n\} = \begin{cases} e^{-x^2}, & \text{if } x < 0, \\ 1, & \text{if } x \geqslant 0. \end{cases} \tag{4.2.3}$$

Proof Let $X^*_{n1}, \ldots, X^*_{nn}$ be a sequence of n independent random variables having the same distribution as random variables $X_{n1}, \ldots, X_{nn}$, and $F_n(x) = P\{X_{nj} \leqslant x\} = P\{X^*_{nj} \leqslant x\}$, $j \in \mathbb{N}$. Let us denote $M^*_n = \max\{X^*_{n1}, \ldots, X^*_{nn}\}$, and $A_{nj} = \{X_{nj} > u_n\}$. □

Lemma 4.2.2 *The limit distribution of random variable M^*_n is given by*

$$\lim_{n\to\infty} P\{M^*_n \leqslant x\sqrt{n}+n\} = \begin{cases} e^{-x^2}, & \text{if } x < 0, \\ 1, & \text{if } x \geqslant 0, \end{cases} \tag{4.2.4}$$

Proof For all positive integers $m \in \{1, 2, \ldots, n-1\}$ the following equalities hold:

$$F_n(m) = \frac{2}{n(n-1)}\left\{mn - \frac{m(m+1)}{2}\right\},$$

$$1 - F_n(m) = 1 - \frac{2m}{n-1} + \frac{m(m+1)}{n(n-1)}.$$

Let us denote $u_n = u_n(x) = x\sqrt{n}+n$. For $x < 0$ we obtain $n(1-F_n(u_n)) \to x^2$ as $n \to \infty$, and for every $x \geqslant 0$ and every positive integer n we get $n(1-F_n(u_n)) = 0$. Consequently, (4.2.4) follows by Theorem 2.1.4. □

Lemma 4.2.3 *For $x < 0$ and $u_n = x\sqrt{n}+n$ the following asymptotic relations hold as $n \to \infty$:*

$$P(A_{nj}) \sim \frac{x^2}{n},$$

$$P(A_{n1}A_{n2}) \sim \frac{2(-x)^3}{3n^{3/2}},$$

$$P(A_{n1}A_{nj}) \sim \frac{x^4}{n^2}, \quad j \in \{3, \ldots, n-1\},$$

$$P(A_{n1}A_{n2}A_{n3}) = O\left(\frac{1}{n^2}\right).$$

. . . .

Proof Straightforward exercise. □

Lemma 4.2.4 *For $x < 0$ and $u_n = x\sqrt{n}+n$ there exists a positive constant $C_1(x)$, such that for every positive integer $k \leqslant n$, and all $1 \leqslant j_1 < j_2 < \cdots < j_k \leqslant n$, the following inequality holds:*

$$\left| P\left(\bigcap_{r=1}^{k} \overline{A}_{nj_r}\right) - \prod_{r=1}^{k} P(\overline{A}_{nj_r}) \right| \leqslant \frac{C_1(x)}{\sqrt{n}}. \tag{4.2.5}$$

Proof The following equalities hold:

$$P\left(\bigcap_{r=1}^{k} \overline{A}_{nj_r}\right) - \prod_{r=1}^{k} P(\overline{A}_{nj_r}) = 1 - P\left(\bigcup_{r=1}^{k} A_{nj_r}\right) - \prod_{r=1}^{k}(1 - P(A_{nj_r}))$$

$$= 1 - \sum_{r=1}^{k} P(A_{nj_r}) + \sum_{1\leqslant r<s\leqslant k} P(A_{nj_r} A_{nj_s}) - \sum_{1\leqslant r<s<t\leqslant k} P(A_{nj_r} A_{nj_s} A_{nj_t}) + \cdots$$

$$-1 + \sum_{r=1}^{k} P(A_{nj_r}) - \sum_{1\leqslant r<s\leqslant k} P(A_{nj_r})P(A_{nj_s})$$

$$+ \sum_{1\leqslant r<s<t\leqslant k} P(A_{nj_r})P(A_{nj_s})P(A_{nj_t}) + \cdots$$

$$= \sum_{1\leqslant r<s\leqslant k} \{P(A_{nj_r} A_{nj_s}) - P(A_{nj_r})P(A_{nj_s})\} -$$

$$- \sum_{1\leqslant r<s<t\leqslant k} \{P(A_{nj_r} A_{nj_s} A_{nj_t}) - P(A_{nj_r})P(A_{nj_s})P(A_{nj_t})\} + \cdots$$

Using the definition of random variables X_{nj} and events A_{nj}, $1 \leqslant j \leqslant n$, and the equality $u_n = x\sqrt{n} + n$, where $x < 0$, we obtain that

$$\sum_{j=1}^{n} I_{A_{nj}} \leqslant C_0(x) \cdot \sqrt{n},$$

where $I_{A_{nj}}$ is the indicator of event A_{nj}, i.e., $I_{A_{nj}} = 1$ if $X_{nj} > u_n$, and $I_{A_{nj}} = 0$ otherwise. In other words, the number of exceedances of X_{nj} over threshold u_n is at most $O(\sqrt{n})$. Inequality (4.2.5) now follows from Lemma 4.2.3. □

Lemma 4.2.5 *Let us suppose that $x < 0$ and $u_n = x\sqrt{n} + n$. Then there exists a positive constant $C_2(x)$, such that for all positive integers k and l, where $k + l \leqslant n$, and all $1 \leqslant j_1 < j_2 < \cdots < j_k < j_{k+1} < \cdots < j_{k+l} \leqslant n$, the following inequality holds:*

$$\left| P\left(\bigcap_{r=1}^{k+l} \overline{A}_{nj_r} \right) - P\left(\bigcap_{r=1}^{k} \overline{A}_{nj_r} \right) P\left(\bigcap_{r=k+1}^{k+l} \overline{A}_{nj_r} \right) \right| \leqslant \frac{C_2(x)}{\sqrt{n}}, \tag{4.2.6}$$

i.e., Leadbetter's condition $D(u_n)$ is satisfied.

Proof Lemma 4.2.5 is a consequence of Lemma 4.2.4 and the following inequality:

$$\begin{aligned} &\left| P\left(\bigcap_{r=1}^{k+l} \overline{A}_{nj_r} \right) - P\left(\bigcap_{r=1}^{k} \overline{A}_{nj_r} \right) P\left(\bigcap_{r=k+1}^{k+l} \overline{A}_{nj_r} \right) \right| \\ &\leqslant \left| P\left(\bigcap_{r=1}^{k+l} \overline{A}_{nj_r} \right) - \prod_{r=1}^{k+l} P(\overline{A}_{nj_r}) \right| \\ &\quad + \prod_{r=1}^{k} P(\overline{A}_{nj_r}) \cdot \left| \prod_{r=k+1}^{k+l} P(\overline{A}_{nj_r}) - P\left(\bigcap_{r=k+1}^{k+l} \overline{A}_{nj_r} \right) \right| \\ &\quad + \left| P\left(\bigcap_{r=1}^{k} \overline{A}_{nj_r} \right) - \prod_{r=1}^{k} P(\overline{A}_{nj_r}) \right| \cdot P\left(\bigcap_{r=k+1}^{k+l} \overline{A}_{nj_r} \right). \end{aligned}$$

□

Lemma 4.2.6 *If $u_n = x\sqrt{n} + n$ and $x < 0$, then condition $D'(u_n)$ is satisfied, i.e.,*

$$\lim_{k\to\infty} \limsup_{n\to\infty} n \cdot \sum_{j=2}^{[n/k]} P(A_{n1}A_{nj}) = 0.$$

Proof Using Lemma 4.2.3 we obtain that for every positive integer k,

$$n \cdot \sum_{j=2}^{[n/k]} P(A_{n1}A_{nj}) \sim n \cdot \frac{2(-x)^3}{3n^{3/2}} + n \cdot \frac{n}{k} \cdot \frac{x^4}{n^2} \sim -\frac{2x^3}{3\sqrt{n}} + \frac{x^4}{k},$$

as $n \to \infty$, and condition $D'(u_n)$ follows immediately. □

Proof of Theorem 4.2.1 This theorem follows from Lemmas 4.2.2, 4.2.5, 4.2.6, and Theorem 2.7.13. □

4.2.2 Maximum of the Sum of Consecutive Terms

Let n and k be positive integers such that $k < n$. For every $j \in \mathbb{N}_n$, and $\omega = (a_1, \ldots, a_n) \in \Omega_n$ let us define

$$X_{nj}(\omega) = a_j + a_{j+1} + \cdots + a_{j+k-1}, \tag{4.2.7}$$

where $a_{n+j} = a_j$ for $j \in \{1, 2, \ldots, k\}$. Then $X_{n1}, \ldots, X_{nn}$ is a sequence of *dependent* random variables that satisfy the condition of strict stationarity. If we denote $M_n' = \max\{X_{n1}, \ldots, X_{n,n-k+1}\}$ and $M_n = \max\{X_{n1}, \ldots, X_{nn}\}$, then it is easy to see that for a fixed k the maxima M_n' and M_n have the same limiting distribution as $n \to \infty$. Let us formulate the result related to maximum M_n.

Theorem 4.2.7 (Glavaš, Jocković and Mladenović [76]) *The limit distribution of random variable M_n is given by*

$$\lim_{n\to\infty} P\{M_n \leqslant (k!\,n^{k-1})^{1/k}x + kn\} = \begin{cases} e^{-(-x)^k}, & \text{if } x < 0, \\ 1, & \text{if } x \geqslant 0. \end{cases} \tag{4.2.8}$$

The limiting distribution function on the right-hand side of (4.2.8) is the Weibull distribution function $G_{2,k}(x)$. Theorem 4.2.7 can be proved in a similar way as Theorem 4.2.1 using arguments from extreme value theory. Let $X_{n1}^*, \ldots, X_{nn}^*$ be a sequence of n independent and identically distributed random variables, such that $F_n(x) := P\{X_{nj} \leqslant x\} = P\{X_{nj}^* \leqslant x\}$, and $M_n^* = \max\{X_{n1}^*, \ldots, X_{nn}^*\}$.

The proof of Theorem 4.2.7 is based on the following lemmas provided here without proof.

Lemma 4.2.8 *For all positive integers $m \in \left\{\frac{k(k+1)}{2}, \ldots, kn - \frac{k(k-1)}{2}\right\}$ the following equality holds*

$$F_n(m) = P\{X_{nj} \leqslant m\} = \frac{k! \cdot (H_k(z))^{(m)}_{|z=0}}{m! \cdot n(n-1)\ldots(n-k+1)}, \tag{4.2.9}$$

where function $H_k(z)$ is given by

$$H_k(z) = \frac{z^{k(k+1)/2}}{(1-z)^2(1-z^2)(1-z^3)\ldots(1-z^k)}. \tag{4.2.10}$$

Note here the following facts:

(a) $H_k(z))^{(m)}_{|k=0}/m!$ is the number of partitions of integers no greater than m into exactly k distinct positive integers.
(b) The number of partitions of m into k distinct positive integers is equal to the number of partitions of $m - \frac{k(k+1)}{2}$ into positive integers no greater than k, with ordinary generating function

$$L_k(z) = \frac{1}{(1-z)(1-z^2)(1-z^3)\ldots(1-z^k)}. \tag{4.2.11}$$

(c) $F_n(m) = P\{X_{nj} \leqslant m\}$ is equal to the number of partitions considered in (b) multiplied by the number of permutations of these k numbers and divided by the number of k-arrangements of n elements.

Lemma 4.2.9 *The limit distribution of random variable M_n^* is given by*

$$\lim_{n\to\infty} P\{M_n^* \leqslant (k!\,n^{k-1})^{1/k}x + kn\} = \begin{cases} e^{-(-x)^k}, & \text{if } x < 0, \\ 1, & \text{if } x \geqslant 0. \end{cases}$$

Lemma 4.2.10 *Let us denote $u_n = u_n(x) = (k!\,n^{k-1})^{1/k}x + kn$, and $A_{nj} = \{X_{nj} > u_n\}$, where $j \in \{1, 2, \ldots, n\}$.*

(a) *There exists a constant $C(x, k)$ such that for all positive integers d and l, where $d + l \leqslant n$, and all $1 \leqslant j_1 < j_2 < \cdots < j_d < j_{d+1} < \cdots < j_{d+l} \leqslant n$, the following inequality holds:*

$$\left| P\Big(\bigcap_{r=1}^{d+l} \overline{A}_{nj_r}\Big) - P\Big(\bigcap_{r=1}^{d} \overline{A}_{nj_r}\Big) P\Big(\bigcap_{r=d+1}^{d+l} \overline{A}_{nj_r}\Big) \right| \leqslant \frac{C(x,k)}{n^{2/k}},$$

i.e., Leadbetter's condition $D(u_n)$ is satisfied.

(b) *Condition $D'(u_n)$ is also satisfied, i.e.,*

$$\lim_{k\to\infty} \limsup_{n\to\infty} n \cdot \sum_{j=2}^{[n/k]} P(A_{n1}A_{nj}) = 0.$$

Theorem 4.2.7 is an immediate consequence of Lemmas 4.2.9 and 4.2.10 and Theorem 2.7.13. □

4.2.3 Cycles of Random Permutations

In this subsection we provide some general results related to the representation of random permutations as a product of cycles.

Let us suppose that every permutation $\omega \in \Omega_n$ is given in the second canonical representation. For every $j \in \mathbb{N}_n$ let us define random variable X_{nj} as follows: $X_{nj}(\omega) = 1$ if position j is the end of a cycle of permutation ω, and $X_{nj}(\omega) = 0$ otherwise. Obviously, $X_{n1}(\omega)+\cdots+X_{nn}(\omega)$ is the number of cycles of permutation ω. It is easy to prove that the distribution of random variable X_{nj} is given by

$$P\{X_{nj} = 1\} = \frac{1}{n-j+1}, \quad P\{X_{nj} = 0\} = \frac{n-j}{n-j+1},$$

and that $X_{n1}, \ldots, X_{nn}$ are independent random variables.

Theorem 4.2.11 *Let us denote $S_n = X_{n1} + \cdots + X_{nn}$. Then for every $x \in \mathbb{R}$,*

$$\lim_{n\to\infty} P\left\{\frac{S_n - \ln n}{\sqrt{\ln n}} \leqslant x\right\} = \frac{1}{\sqrt{2\pi}} \int_{-\infty}^{x} e^{-t^2/2}\, dt.$$

Theorem 4.2.12 *Let $Y_{nl}(\omega)$ be the number of cycles of length l of a permutation $\omega \in \Omega_n$. For every fixed $l \in \{1, 2, \ldots\}$ the limiting distribution of random variable Y_{nl} as $n \to \infty$ is the Poisson distribution with parameter $\lambda = 1/l$, i.e., for every $k \in \mathbb{N}_0 = \{0, 1, 2, \ldots\}$,*

$$\lim_{n\to\infty} P\{Y_{nl} = k\} = e^{-1/l} \frac{1}{l^k k!}.$$

Theorem 4.2.13 *Let us suppose that $l_1, l_2, \ldots, l_n$ are natural numbers such that $l_1 + 2l_2 + \cdots + nl_n = n$. Then the following equality holds:*

$$P\{Y_{nj} = l_j,\ j = 1, 2, \ldots, n\} = \prod_{j=1}^{n} \frac{1}{j^{l_j} l_j!}.$$

Theorem 4.2.14 *Let us suppose that $j_1, j_2, \ldots, j_s$ are distinct natural numbers, and $l_1, l_2, \ldots, l_s$ are natural numbers. As $n \to \infty$ the following asymptotic equality holds:*

$$P\{Y_{nj_1} = l_1,\ Y_{nj_2} = l_2,\ \ldots,\ Y_{nj_s} = l_s\} =$$
$$= \frac{j_1^{-l_1} j_2^{-l_2} \cdots j_s^{-l_s}}{l_1! l_2! \ldots l_s!} \exp\left\{-\frac{1}{j_1} - \frac{1}{j_2} - \cdots - \frac{1}{j_s}\right\} + o(1).$$

Let $L_1 \leqslant L_2 \leqslant \cdots \leqslant L_{S_n}$ be the lengths of the cycles of a random permutation of set $\mathbb{N}_n$ ordered in nondecreasing order, where S_n is the number of cycles.

Theorem 4.2.15 *Let m be a fixed positive integer. For every positive integer r*

$$P\{L_m \leqslant r\} = \sum_{k=m}^{\infty} e^{-\lambda_r} \frac{\lambda_r^k}{k!} + o(1), \quad n \to \infty,$$

where $\lambda_r = 1 + \frac{1}{2} + \cdots + \frac{1}{r}$.

4.2.4 Poisson-Dirichlet Distribution and the Longest Cycles

The *Poisson-Dirichlet distribution* appears in various applications in combinatorics, analytic number theory, population genetics and ecology, etc. It is defined on the simplex $\Delta \subset \mathbb{R}^\infty$, where

$$\Delta := \left\{ (x_1, x_2, \ldots) \,|\, x_i \geqslant 0 \;\text{ for all }\; i \geqslant 1, \; \sum_{n=1}^{\infty} x_i = 1 \right\}. \tag{4.2.12}$$

Let us give the definition and provide results related to the asymptotic distribution of the longest cycles in a random permutation of an n-set. For more details about the properties and possible applications of the Poisson-Dirichlet distribution, *see* Kingman [130, 131], Billibgsley [27], Holst [110], Griffits [85], Pitman [185], Pitman and Yor [186], and the references therein.

Let us consider a Poisson process $\Pi = \{(T_i, X_i) \,|\, i \geqslant 1\}$ with state space $[0, \infty) \times (0, \infty)$ and intensity measure $dt(e^{-x}/x)dx$. Let us define

$$S_\vartheta = \sum_{i=1}^{\infty} X_i I_{\{T_i < \vartheta\}}.$$

The number of points of Π in the set $A_{\vartheta,z} = \{(t, x) \,|\, 0 \leqslant t \leqslant \vartheta, z < x\}$ is a Poisson random variable $\mathcal{P}(\vartheta E_1(z))$, where $E_1(z)$ is given by

$$E_1(z) = \int_z^\infty \frac{e^{-x}}{x}\, dx = \int_1^\infty \frac{e^{-zx}}{x}\, dx.$$

With probability one $\Pi \cap A_{\vartheta,0} = \{(T_i, X_{(i)}) \,|\, i \geqslant 1\}$, where $T_1, T_2, T_3, \ldots$ are i.i.d. random variables with uniform distribution on $[0, \vartheta]$, and $X_{(1)} > X_{(2)} > X_{(3)} \cdots > 0$ are points of a non-homogenous Poisson process on $(0, \infty)$ with intensity measure $\vartheta e^{-x}/x\, dx$. Additionally, sequences $(T_i)_{i \geqslant 1}$ and $(X_{(i)})_{i \geqslant 1}$ are independent. The density of random variable $X_{(i)}$ is

$$f_{X_{(i)}}(x) = \frac{\vartheta e^{-x}}{x} \frac{(\vartheta E_1(x))^{i-1}}{(i-1)!} e^{-\vartheta E_1(x)}, \quad x > 0,$$

and $S_\vartheta = \sum_{i=1}^{\infty} X_{(i)}$ is a random variable with the Laplace transform

$$E\left(e^{-tS_\vartheta}\right) = \exp\left\{ -\int_0^\infty (1 - e^{-tx}) \frac{\vartheta e^{-x}}{x}\, dx \right\} = \frac{1}{(1+t)^\vartheta}, \qquad |t| < 1,$$

and consequently random variable S_ϑ has $\Gamma(\vartheta)$ distribution with density $x^{\vartheta-1}e^{-x}/\Gamma(\vartheta)$ if $x > 0$, and 0 otherwise. Random process $\{S_\vartheta \,|\, \vartheta > 0\}$ is called *Moran's subordinator*. Let us denote $L_{(i)} := X_{(i)}/S_\vartheta$ for $i \geqslant 1$. Then we have

$$L_{(1)} > L_{(2)} > \cdots > 0, \quad \sum_{i=1}^{\infty} L_{(i)} = 1.$$

Definition 4.2.16 The *Poisson–Dirichlet distribution* with parameter $\vartheta > 0$, notation $PD(\vartheta)$, is the distribution of random vector $(L_{(1)}, L_{(2)}, \ldots)$ on simplex Δ determined by 4.2.12.

Increments of Moran's subordinator $\{S_\vartheta \mid \vartheta > 0\}$ are independent stationary and gamma distributed. For a fixed $\vartheta > 0$, and $\alpha = \vartheta/n$ let us define $Y_j := S_{j\alpha} - S_{(j-1)\alpha}$, $j \in \{1, \ldots, n\}$. Then, $Y_1, \ldots, Y_n$ are independent random variables with $\Gamma(\alpha)$ distribution. Let us define $S_\vartheta^{(n)} := Y_1 + \cdots + Y_n$, and $Z_i := Y_i/S_\vartheta^{(n)}$ for $i \in \{1, 2, \ldots, n\}$. Then $S_\vartheta^{(n)}$ and $(Z_1, \ldots, Z_n)$ are independent and random vector $(Z_1, \ldots, Z_n)$ has a symmetric Dirichlet distribution $D(\alpha, \ldots, \alpha)$ on the simplex $\{(x_1, \ldots, x_n) \mid x_i \geqslant 0, \sum_{i=1}^n x_i = 1\}$ with density $x_1^{\alpha-1} \ldots x_n^{\alpha-1}\Gamma(\vartheta)/(\Gamma(\alpha))^n$. Hence, random variable Z_i is beta distributed for every $i \in \{1, \ldots, n\}$.

It can be proved that for order statistics $Z_{(1)} > Z_{(2)} > \cdots > Z_{(n)}$ the following relation holds as $n \to \infty$:

$$(Z_{(1)}, Z_{(2)}, \ldots, Z_{(n)}, 0, 0, \ldots) \Rightarrow (L_{(1)}, L_{(2)}, \ldots).$$

The following two theorems give the distributions of random variable $L_{(1)}$ and random vector $(L_{(1)}, \ldots, L_{(r)})$ for fixed $r \in \mathbb{N}$.

Theorem 4.2.17 (Griffits [85]) *For $x > 0$ the distribution function of random variable $L_{(1)}$ is given by*

$$P\{L_{(1)} \leqslant x\} = 1 + \sum_{k=1}^{[1/x]} \frac{(-\vartheta)^k}{k!} \int_x^1 \cdots \int_x^1 \frac{(1 - y_1 - \cdots - y_k)_+^{\vartheta-1}}{y_1 \ldots y_k} \, dy_1 \ldots dy_k.$$

Theorem 4.2.18 (Watterson [223]) *For every fixed $r \in \mathbb{N}$ the probability density function $f_{L_{(1)},\ldots,L_{(r)}}(z_1, \ldots, z_r)$ of random vector $(L_{(1)}, \ldots, L_{(r)})$ is equal to*

$$\frac{\vartheta^r (1 - z_1 - \cdots - z_r)^{\vartheta-1}}{z_1 \ldots z_r} P\left\{L_{(1)} \leqslant \frac{z_r}{1 - z_1 - \cdots - z_r}\right\},$$

for $z_1 > \cdots > z_r > 0$, $z_1 + \cdots + z_r < 1$, and $f_{L_{(1)},\ldots,L_{(r)}} = 0$ otherwise.

The *Chinese restaurant process* is a generator of random permutations. We consider a box containing a black ball of weight ϑ, a pile containing n white balls of weight 1 each, and a list with written numbers on it that is empty initially. First we choose the black ball from the box, put it back into the box, put the white ball numbered 1 into the box, and write 1 on the list. At step j, where $2 \leqslant j \leqslant n$, we choose a ball from the box with a probability that is proportional to its weight (the only black ball with probability $\vartheta/(\vartheta+j-1)$, and every white ball with probability $1/(\vartheta+j-1)$), put it back into the box, put the white ball numbered j into the box, and write the number j on the list according to the following rule:

- If the black ball is drawn at step j, then number j is written to the left of the list.
- If the white ball numbered $r \in \{1, \ldots, j-1\}$ is drawn at step j, then j is written just to the right of number r on the list.

After n steps a permutation of numbers $1, \ldots, n$ is written on the list.

Example 4.2.19 Let $n = 6$. Then in the 1st, 2nd, 3rd, 4th, 5th, and 6th steps we choose the elements respectively from sets:

$$\{\bullet\}, \ \{1, \bullet\}, \ \{1, 2, \bullet\}, \ \{1, 2, 3, \bullet\}, \ \{1, 2, 3, 4, \bullet\}, \ \{1, 2, 3, 4, 5, \bullet\}.$$

Let us suppose that the sequence of chosen elements is the following: $\bullet$, 1, 1, $\bullet$, 4, and 3. The list of written numbers grows as follows: 1, 12, 132, 4132, 45,132, 451,362. The final content of the box is $\{1, 2, 3, 4, 5, 6, \bullet\}$. The permutation obtained is 451,362. The black ball is chosen in the first and the fourth steps. If 1 and 4 are considered the starting elements of the cycles, then $(45)(1362) = (1362)(45)$ is the representation of the obtained permutation as a product of the cycles. $\triangle$

Remark 4.2.20 The *Chinese restaurant process* can be alternatively introduced the following way: Let us consider a restaurant with infinitely many tables that are labeled $1, 2, \ldots$, and infinitely many seats around each table. Customers enter the restaurant. The first customer chooses the first table. Let $\vartheta > 0$ be a parameter. For every $j \geqslant 2$, the j-th customer chooses the first unoccupied table with probability $\vartheta/(\vartheta + j - 1)$, and an occupied table with probability $i/(\vartheta + j - 1)$, where i is the number of people already sitting at this table. Additionally, a unique way of choosing the seats around the tables is determined. Let us suppose that exactly $k \leqslant n$ tables are occupied after n customers entered the restaurant. Thus, we have obtained a permutation of set $\mathbb{N}_n$ with exactly k cycles.

Let us now suppose that the black ball is drawn at steps $j_1, j_2, \ldots, j_k$, where $1 = j_1 < j_2 < \cdots < j_k \leqslant n$. The probability of the obtained permutation is

$$\frac{\vartheta^k}{\vartheta(\vartheta+1)(\vartheta+2)\ldots(\vartheta+n-1)}.$$

Let us define that the cycles start at numbers $j_1, j_2, \ldots, j_k$, i.e., the number of cycles of the realized permutation is equal to k. The number of cycles of a random permutation is random variable X_n with the distribution

$$P\{X_n = k\} = \begin{bmatrix} n \\ k \end{bmatrix} \frac{\vartheta^k}{\vartheta(\vartheta+1)(\vartheta+2)\ldots(\vartheta+n-1)}, \qquad k \in \{1, 2, \ldots, n\},$$

where $\left[{n \atop k}\right]$ is the number of permutations of an n-set with k cycles. These numbers are called *Stirling numbers of the first kind*.

Remark 4.2.21 Note that the probability of a random permutation obtained in a Chinese restaurant process depends on n, ϑ and the number of cycles k. The particular choice of elements that entered the cycles and the arrangements of elements inside the cycles are irrelevant for this probability. This property is called *exchangeability*.

Theorem 4.2.22 *Let $L_1^{(n)}, L_2^{(n)}, \dots, L_{X_n}^{(n)}$ be cycle-lengths in a random permutation of set $\mathbb{N}_n = \{1, 2, \dots, n\}$, and let us arrange them in decreasing order $L_{(1)}^{(n)} \geqslant L_{(2)}^{(n)} \geqslant \dots \geqslant L_{(X_n)}^{(n)}$. Then*

$$\left(\frac{L_{(1)}^{(n)}}{n}, \frac{L_{(2)}^{(n)}}{n}, \dots, \frac{L_{(X_n)}^{(n)}}{n}, 0, 0, \dots\right) \Rightarrow (L_{(1)}, L_{(2)}, \dots),$$

where $(L_{(1)}, L_{(2)}, \dots)$ is a random element with $PD(\vartheta)$ distribution on Δ.

Note that $\vartheta = 1$ corresponds to the case when all permutations of set $\mathbb{N}_n$ have probability $1/n!$. The limit distribution of the longest cycle is given by Theorem 4.2.17. The joint limit distribution of r longest cycles is given by Theorem 4.2.18. The proof of Theorem 4.2.22 is given, for example, in Billingsley [27] and Holst [110].

4.2.5 The Longest Increasing Subsequence

Let us consider a permutation $\omega \in \Omega_n$. We say that $\omega(i_1), \omega(i_2), \dots, \omega(i_k)$ is an increasing subsequence in permutation ω if $i_1 < i_2 < \dots < i_k$ and $\omega(i_1) < \omega(i_2) < \dots < \omega(i_k)$, and that k is the length of this subsequence. Let $l_n(\omega)$ be the length of the longest increasing subsequence. For example, for $\omega = 43251867$, the longest subsequences are 2567, 3567, 4567, and hence $l_8(\omega) = 4$.

The properties of random variable l_n for fixed n, and the asymptotic properties as $n \to \infty$ have been studied by several authors. Using the results of Monte Carlo simulations Ulam [220] conjectured that there exists the limit value

$$c := \lim_{n\to\infty} \frac{E(l_n)}{\sqrt{n}}. \tag{4.2.13}$$

Using the arguments of Erdös and Szekeres [62] we can show that $E(l_n) \geqslant \sqrt{n - 1}/2$. The existence of limit value (4.2.13) was established in Hammersley [100], while Vershik and Kerov [221] proved that $c = 2$. On the basis of their Monte Carlo simulations Oldyzko and Rains [175] suggested that

$$\lim_{n\to\infty} \frac{Var\,(l_n)}{n^{1/3}} = c_0 \approx 0.819, \tag{4.2.14}$$

$$\lim_{n\to\infty} \frac{E(l_n) - 2\sqrt{n}}{n^{1/6}} = c_1 \approx -1.758. \tag{4.2.15}$$

In order to formulate a result that gives the asymptotic distribution of random variable l_n after appropriate centering and scaling, we shall first introduce auxiliary function $u(x)$ that is the solution of the Painlevé II equation,

$$u_{xx} = 2u^3 + xu, \quad u \sim -A(x) \text{ as } x \to \infty, \tag{4.2.16}$$

where $A(x)$ is the Airy function satisfying $A(x) \sim \dfrac{e^{-(2/3)x^{3/2}}}{2\sqrt{\pi}x^{1/4}}$ as $x \to \infty$. The existence and uniqueness of the solution of Eq. (4.2.16) was proved in Hastings and McLeod [104]. The asymptotics of this solution are given by

$$u(x) = -A(x) + O\left(\frac{e^{-(4/3)x^{3/2}}}{x^{1/4}}\right) \quad \text{as } x \to +\infty, \tag{4.2.17}$$

$$u(x) = -\sqrt{\frac{-x}{2}}\left(1 + O\left(\frac{1}{x^2}\right)\right) \quad \text{as } x \to -\infty. \tag{4.2.18}$$

The *Tracy-Widom distribution* is defined via function $u(x)$ as follows:

$$F(x) = \exp\left\{-\int_x^\infty (t-x)u^2(t)\,dt\right\}. \tag{4.2.19}$$

Now we are ready to formulate the result that gives the asymptotic distribution of the longest increasing subsequence in random permutations.

Theorem 4.2.23 (Baik, Deift, and Johansson [16]) *Let X be a random variable with distribution function F given by* (4.2.19). *Then the random variable $n^{-1/6}(l_n - 2\sqrt{n})$ converges in distribution to X as $n \to \infty$, i.e., for every $x \in \mathbb{R}$, the following equality holds:*

$$\lim_{n\to\infty} P\left\{\frac{l_n - 2\sqrt{n}}{n^{1/6}} \leqslant x\right\} = F(x). \tag{4.2.20}$$

Theorem 4.2.24 (Baik, Deift, and Johansson [16]) *Let X be a random variable with distribution function F given by* (4.2.19). *Then the moments of random variable $n^{-1/6}(l_n - 2\sqrt{n})$ converge to those of random variable X as $n \to \infty$, and particularly the following equalities hold:*

$$\lim_{n\to\infty} \frac{Var(l_n)}{n^{1/3}} = \int_{-\infty}^{\infty} t^2\,dF(t) - \left(\int_{-\infty}^{\infty} t\,dF(t)\right)^2, \tag{4.2.21}$$

$$\lim_{n\to\infty} \frac{E(l_n) - 2\sqrt{n}}{n^{1/6}} = \int_{-\infty}^{\infty} t\,dF(t). \tag{4.2.22}$$

Numerical computations show that the constants on the right-hand sides of equalities (4.2.21) and (4.2.22) are approximately 0.8132 and -1.7711. The proof of Theorems 4.2.23 and 4.2.24 can be found in the 60-page long paper by Baik, Deift, and Johansson [16]. More information about the history of this problem and about the references, methods used and relations with other problems can be found in Aldous and Diaconis [5, 6], and Baik, Deift, and Johansson [16].

4.3 The Coupon Collector's Problem

4.3.1 Introduction

In this section we consider an old combinatorial problem known as the coupon collector's problem and several of its extensions. The formulation of the problem is the following: Let us suppose that we sample with replacement from set $\mathbb{N}_n := \{1, 2, \ldots, n\}$ under the assumption that each element of $\mathbb{N}_n$ has probability $1/n$ of being drawn in each trial, and we are interested in the waiting time until all elements of $\mathbb{N}_n$ are collected. Let us denote by M_n this waiting time, i.e., M_n is the number of trials until all elements of $\mathbb{N}_n$ are collected. The natural questions that arise here are the following:

1. What is the distribution of random variable M_n?
2. What is the expected value $E(M_n)$?
3. What is the limiting distribution of M_n as $n \to \infty$?

First, let us give the answers to questions 1 and 2.

Theorem 4.3.1 *The distribution of random variable M_n is given by*

$$P\{M_n = k\} = \sum_{i=0}^{n-2} \binom{n-1}{i} \left(1 - \frac{i+1}{n}\right)^{k-1}, \quad k \in \{n, n+1, \ldots\}. \tag{4.3.1}$$

Proof Let $k \geqslant n$ be a fixed natural number. The outcome in the first k trials is of the form $\omega = c_1 c_2 \ldots c_k$, where $c_1, c_2, \ldots, c_k \in \mathbb{N}_n$. The probability of every outcome $\omega = c_1 c_2 \ldots c_k$ is $p(\omega) = 1/n^k$. We are interested in outcomes ω such that exactly $n-1$ distinct elements of $\mathbb{N}_n$ appear among $c_1, c_2, \ldots, c_{k-1}$, and c_k is the missing n-th element. Let A_n be the event that each of the natural numbers $1, 2, \ldots, n-1$ appears in the first $k-1$ trials, and the number n does not appear.

Let S be the set of $(k-1)$-arrangements of elements $1, \ldots, n-1$, and S_k be the set of $(k-1)$-arrangements of elements of set $\{1, \ldots, n-1\} \setminus \{k\}$. Then we have $A_n = S \setminus (S_1 \cup S_2 \cup \cdots \cup S_{n-1})$. Set S consists of $(n-1)^{k-1}$ elements. Each of sets $S_1, S_2, \ldots, S_{n-1}$ consists of $(n-2)^{k-1}$ elements, the intersection of every two of them consists of $(n-3)^{k-1}$ elements, etc.

By the inclusion-exclusion principle it follows that

$$|A_n| = |S \setminus (S_1 \cup S_2 \cup \cdots \cup S_{n-1})| = \sum_{i=0}^{n-2} \binom{n-1}{i} (n-i-1)^{k-1}.$$

The probability of event A_n is

$$P(A_n) = \sum_{i=0}^{n-2} \binom{n-1}{i} \left(1 - \frac{i+1}{n}\right)^{k-1}.$$

The probability that exactly $n-1$ distinct numbers appear in the first $k-1$ trials is $nP(A_n)$. The probability that the missed number appears in the k-th trial is $1/n$. Hence, $P\{M_n = k\} = \frac{1}{n} \cdot n \cdot P(A_n) = P(A_n)$. □

Theorem 4.3.2 *The expected value $E(M_n)$ is given by*

$$E(M_n) = n\left(1 + \frac{1}{2} + \cdots + \frac{1}{n}\right). \tag{4.3.2}$$

Proof Note that random variable M_n can be represented in the form $M_n = \sum_{i=1}^{n} X_i$, where $X_1 = 1$, and, for every $i \in \{2, \ldots, n\}$, X_i is a random variable with geometric $\Gamma\left(\frac{n-i+1}{n}\right)$ distribution and mathematical expectation $E(X_i) = \frac{n}{n-i+1}$. Hence, the result follows. □

The answer to the third question, first published by Erdös and Rényi in their paper [60], is given by:

$$\lim_{n\to\infty} P\{M_n \leqslant n(x + \ln n)\} = e^{-e^{-x}}. \tag{4.3.3}$$

Let us denote by X_{nj} the number of trials necessary to obtain element $j \in \mathbb{N}_n$. Then, $M_n = \max\{X_{n1}, \ldots, X_{nn}\}$. This representation of random variable M_n gives the first explanation of the fact that one of the extreme value distributions appears in the asymptotic relation (4.3.3). Note that random variables $X_{n1}, \ldots, X_{nn}$ that appear in the above representation are dependent. Erdös and Rényi proved equality (4.3.3) using moment generating functions. Several different proofs of this asymptotic relation can be found in the literature. In the next subsection we shall provide the proof based on the extreme value theory.

Remark 4.3.3 The coupon collector's problem can be formulated in terms of the distribution of *indistinguishable* particles into n boxes that are labeled $1, 2, \ldots, n$. Once a time unit we put a particle into one of these boxes, supposing that each box has probability $1/n$ to be chosen to have a particle put into it. Let us consider the question: How many particles should be distributed in order to reach the position that each box contains at least one particle for the first time? Obviously, this number of particles is a random variable whose probability distribution coincides with those of random variable M_n that represents the waiting time for all elements to appear once in the above formulation of the coupon collector's problem.

The following formulations are also popular.

(a) *The coupon collector's problem.* Suppose that a person is collecting coupons for their picture book. At each purchase every coupon comes with probability $1/n$. Let M_n be the number of purchases until all coupons are collected. And we are interested in the questions related to random variable M_n that are formulated above.
(b) *The collector's brotherhood problem.* Let us suppose that a few brothers are collecting coupons for their picture books, and that n is the total number of coupons. At each purchase every coupon comes with probability $1/n$. If the oldest brother does not have the arriving item, he sticks it into his book. If the arriving coupon is not new for the oldest brother he gives it to his younger brother. The younger brother uses the same procedure: if the arriving item is new for him, he sticks it into his own picture book; otherwise, he gives the item to the next younger brother, etc. The oldest brother is labeled 0, the younger brother 1, the next younger brother 2, etc. Let T_k, where $k \geqslant 0$, be the waiting time (i.e., the number of purchases) until the collector numbered k completes their picture book. The questions that arise here are the following: What is the distribution of random variable T_k? What is the expected value $E(T_k)$? What is the limiting distribution of T_k as $n \to \infty$? For results related to the collector's brotherhood problem *see* Foata and Zeilberger [69] and the references therein.

However, we shall always use the formulation as it is given at the beginning of this subsection.

Many generalizations and extensions of the coupon collector's problem have been studied using different methods. The methods of characteristic and moment generating functions in the theory of summing random variables were used, for example, in the papers [22, 24, 60, 67, 107, 199]. An approach by embedding in Poisson processes was used in [108, 109], and [19]. The method based on extreme value theory and the conditions of weak dependence $D(u_n)$ and $D'(u_n)$ was used in [157, 159], and [162]. The point process theory approach was used in [77].

In the next two subsections we shall consider some generalizations and extensions in more details, and the proof of the Erdös–Rényi result (4.3.3) will also be given. A review of further existing results and references will be provided at the end of this section.

4.3.2 Waiting Time for All Elements to Appear r Times

Let us consider the waiting time for every element of $\mathbb{N}_n$ to appear at least r times, where r is a natural number.

More formally, let Z_1, Z_2, Z_3, ..., be a sequence of independent random variables with a common uniform distribution over the set $\mathbb{N}_n = \{1, 2, \ldots, n\}$. For $1 \leqslant j \leqslant n$ and a fixed positive integer r, let us denote by X_{nj} the waiting time until an event of the form $\{Z_k = j\}$ occurs exactly r times as a run in the process Z_1, Z_2, $Z_3, \ldots$ Then $X_{n1}, \ldots, X_{nn}$ are *dependent* random variables such that for every $j \in \{1, 2, \ldots, n\}$ the probability distribution of X_{nj} is given by

$$P\{X_{nj} = k\} = \binom{k-1}{r-1}\left(1 - \frac{1}{n}\right)^{k-r}\frac{1}{n^r}, \quad k = r, r+1, \ldots, \tag{4.3.4}$$

i.e., X_{nj} has the negative binomial distribution $NB(r, 1/n)$ with parameters r and $1/n$. For $r = 1$ this distribution becomes a geometric distribution with parameter $1/n$.

Note that sequence $X_{n1}, \ldots, X_{nn}$ satisfies the condition of strict stationarity. The probability distribution of random vector $(X_{nj_1}, \ldots, X_{nj_k})$ depends only on k and not on the particular choice of integers $j_1, \ldots, j_k$, where $1 \leqslant j_1 < \cdots < j_k \leqslant n$. Let us denote

$$M_n = M_n(r) = \max\{X_{n1}, \ldots, X_{nn}\}. \tag{4.3.5}$$

Note also that M_n is the waiting time until all events of the form $\{Z_k = j\}$, $j \in \mathbb{N}_n$, occur at least r times as a run in the process $Z_1, Z_2, Z_3, \ldots$

For $r = 1$ the expected value of $M_n = M_n(1)$ is given by Theorem 4.3.2. The general formula for the expected value of $M_n = M_n(r)$ is obtained in Newman and Shepp [172].

Theorem 4.3.4 *The expected value of $M_n = M_n(r)$ is given by*

$$E(M_n(r)) = n \int_0^\infty \left\{1 - (1 - S_r(t)e^{-t})^n\right\} dt,$$

where $S_r(t) = 1 + \dfrac{t}{1!} + \dfrac{t^2}{2!} + \cdots + \dfrac{t^{r-1}}{(r-1)!}$.

Proof *See* Newman and Shepp [172] □

Theorem 4.3.5 *Let us denote*

$$u_n = n\{x + \ln n + (r-1)\ln\ln n - \ln(r-1)!\}. \tag{4.3.6}$$

Then the following statements hold:

(a) *The limiting distribution of random variable M_n is given by:*

$$\lim_{n\to\infty} P\{M_n \leqslant u_n\} = \exp(-e^{-x}). \tag{4.3.7}$$

(b) *For $r = 1$, we have*

$$|P\{M_n \leqslant u_n\} - \exp(-e^{-x})| = O\left(\frac{\ln n}{n}\right), \quad \text{as } n \to \infty. \tag{4.3.8}$$

(c) *For $r \geqslant 2$, the following asymptotic relation holds as $n \to \infty$:*

$$P\{M_n \leqslant u_n\} - \exp(-e^{-x}) \sim -(r-1)^2 e^{-x} \exp(-e^{-x}) \cdot \frac{\ln\ln n}{\ln n}. \tag{4.3.9}$$

Remark 4.3.6 For $r = 1$ equality (4.3.7) becomes (4.3.3) which was proved firstly in Erdös and Rényi [60] by the method of moment generating functions. For $r \geqslant 2$ equality (4.3.7) was proved in Holst [109] by the method of embedding in Poisson processes. The rates of convergence (4.3.8) and (4.3.9) were obtained by Mladenović [155, 159]. We shall provide here the proof of Theorem 4.3.5 that is based on Mladenović [155, 159].

First, we shall formulate the result in i.i.d. settings analogous to Theorem 4.3.5. Let $X^*_{n1}, \ldots, X^*_{nn}$ be a sequence of *independent* random variables that are defined on the same probability space and have the same negative binomial distribution with parameters $r \in \mathbf{N}$ and $p = 1/n$, i.e., for every $j \in \{1, 2, \ldots, n\}$,

$$P\{X^*_{nj} = k\} = \binom{k-1}{r-1}\left(1 - \frac{1}{n}\right)^{k-r} \frac{1}{n^r}, \quad k = r, r+1, \ldots.$$

Theorem 4.3.7 *For $M^*_n = \max\{X^*_{n1}, \ldots, X^*_{nn}\}$ and $u_n(x)$ given by (4.3.6) the following statements hold:*

(a) *For any real number x the following equality holds:*

$$\lim_{n\to\infty} P\{M^*_n \leqslant u_n\} = \exp(-e^{-x}). \tag{4.3.10}$$

(b) *If $r = 1$, then for any real x the following asymptotic relation holds as $n \to \infty$:*

$$P\{M^*_n \leqslant u_n\} - \exp(-e^{-x}) \sim \frac{e^{-x}\exp(-e^{-x})}{2} \cdot \frac{\ln n}{n}. \tag{4.3.11}$$

(c) *If $r \geqslant 2$, then the rate of convergence of M^*_n is given by the following asymptotic relation as $n \to \infty$:*

$$P\{M^*_n \leqslant u_n\} - \exp(-e^{-x}) \sim -(r-1)^2 e^{-x} \exp(-e^{-x}) \cdot \frac{\ln\ln n}{\ln n}. \tag{4.3.12}$$

Proof of Theorem 4.3.7

(a) Let F_n be the common distribution function of random variables X_{nj}, and X^*_{nj}, $1 \leqslant j \leqslant n$, and m a positive integer. Note that

$$\begin{aligned}1 - F_n(m) = P\{X_{nj} > m\} &= \sum_{k=m+1}^{\infty} \binom{k-1}{r-1}\left(1-\frac{1}{n}\right)^{k-r}\frac{1}{n^r} \\ &= \frac{1}{n^r(r-1)!}\sum_{k=m+1}^{\infty}(k-1)(k-2)\dots(k-r+1)q^{k-r} \\ &= \frac{1}{n^r(r-1)!}\cdot\frac{d^{r-1}}{dq^{r-1}}\frac{q^m}{1-q}\bigg|_{q=1-1/n}.\end{aligned}$$

First, we shall determine the asymptotic behavior of tail $1 - F_n(m)$ as $n \to \infty$ and $m/n \to \infty$. Using Leibniz's formula we get

$$\begin{aligned}1 - F_n(m) &= \frac{1}{n^r(r-1)!}\sum_{k=0}^{r-1}\binom{r-1}{k}\frac{d^{r-1-k}q^m}{dq^{r-1-k}}\cdot\frac{d^k}{dq^k}\frac{1}{1-q}\bigg|_{q=1-1/n} \\ &= \frac{1}{n^r(r-1)!}\bigg\{m(m-1)\dots(m-r+2)\frac{q^{m-r+1}}{1-q} + q^m\frac{(r-1)!}{(1-q)^r} + \\ &+ \sum_{k=1}^{r-2}\binom{r-1}{k}m(m-1)\dots(m-r+k+2)\frac{q^{m-r+k+1}k!}{(1-q)^{k+1}}\bigg\}\bigg|_{q=1-\frac{1}{n}} \\ &= \frac{m^{r-1}}{n^{r-1}(r-1)!}\left(1-\frac{1}{n}\right)^{m-r+1}(1+R(m,n)),\end{aligned}$$

where $R(m, n) \to 0$ as $n \to \infty$ and $m/n \to \infty$. A positive integer m will be determined from the condition

$$\frac{m^{r-1}}{n^{r-1}(r-1)!}\left(1-\frac{1}{n}\right)^{m-r+1}(1+R(m,n)) \sim \frac{1}{n}e^{-x}, \quad n \to \infty, \tag{4.3.13}$$

which is equivalent to

$$\begin{aligned}(r-1)\ln m - (r-1)\ln n - \ln(r-1)! + (m-r+1)\ln\left(1-\frac{1}{n}\right) \\ + \ln(1+R(m,n)) \sim -x - \ln n, \quad \text{as } n \to \infty.\end{aligned} \tag{4.3.14}$$

The asymptotic relation (4.3.14) holds if

$$(r-1)\ln m - (r-1)\ln n - \ln(r-1)! - \frac{m}{n} + x + \ln n \to 0, \quad n \to \infty.$$

We shall determine a positive integer m to be equal to $[u_n]$, where sequence $u_n = u_n(x)$ satisfies the relation

$$\frac{u_n}{n} - x + \ln(r-1)! + (r-2)\ln n - (r-1)\ln u_n \to 0, \quad n \to \infty. \tag{4.3.15}$$

It is easy to verify that for u_n given by (4.3.6) the asymptotic relation (4.3.15) holds. It follows from Theorem 2.1.4 that the limiting distribution of M_n^* is given by equality $\lim_{n\to\infty} P\{M_n^* \leqslant u_n\} = \exp(-e^{-x})$.

□

Let $m \leqslant n$ be a positive integer. For $s_1, \ldots, s_m \in \{0, 1, \ldots, r-1\}$, let us introduce the following notation:

$$p_{nm}(s_1, \ldots, s_m) := P\Bigg(\bigcap_{j=1}^{m} \Bigg\{\sum_{i=1}^{[u_n]} I\{Z_i = j\} = s_j\Bigg\}\Bigg), \tag{4.3.16}$$

$$\begin{aligned} p_{nm} &:= P\{X_{n1} > u_n, \ldots, X_{nm} > u_n\} \\ &= \sum_{s_1=0}^{r-1} \ldots \sum_{s_m=0}^{r-1} p_{nm}(s_1, \ldots, s_m). \end{aligned} \tag{4.3.17}$$

Lemma 4.3.8 *Let m be a fixed positive integer and $s = s_1 + \cdots + s_m$. Then the following asymptotic relation holds as $n \to \infty$:*

$$\begin{aligned} p_{nm}(s_1, \ldots, s_m) = \frac{e^{-mx}((r-1)!)^m}{s_1! \ldots s_m!}\left(\frac{u_n}{n}\right)^s \times \\ \times \frac{1}{n^m (\ln n)^{m(r-1)}} \cdot \left\{1 - \frac{m^2 \ln n}{2n} + o\left(\frac{\ln n}{n}\right)\right\}. \end{aligned} \tag{4.3.18}$$

Proof Let us denote $r_n = u_n - [u_n]$, $s = s_1 + \cdots + s_m$. Then

$$\begin{aligned} &p_{nm}(s_1, \ldots, s_m) = \\ &= \binom{[u_n]}{s_1}\binom{[u_n]-s_1}{s_2}\ldots\binom{[u_n]-s_1-\cdots-s_{m-1}}{s_m}\frac{1}{n^s}\left(1-\frac{m}{n}\right)^{[u_n]-s} \\ &= \frac{(u_n - r_n)\ldots(u_n - r_n - s + 1)}{s_1! \ldots s_m!}\frac{1}{n^s} \cdot \exp\left\{(u_n - r_n - s)\ln\left(1 - \frac{m}{n}\right)\right\} \end{aligned}$$

and using (4.3.6) we obtain (4.3.18). □

Lemma 4.3.9 *Let m be a fixed positive integer. Then, we have as $n \to \infty$:*

$$p_{nm} - p_{n,m-1}p_{n1} = -\frac{e^{-mx}}{n^{m+1}} \cdot \{(m-1)\ln n + o(\ln n)\}, \tag{4.3.19}$$

$$p_{nm} = \frac{e^{-mx}}{n^m}(1 + o(1)). \tag{4.3.20}$$

Proof Equality (4.3.19) is a straightforward consequence of (4.3.17) and (4.3.18). Note that for a fixed m and all sufficiently large n the probability $p_{nm}(s_1, \dots, s_m)$ takes the greatest possible value if $s_1 = \dots = s_m = r - 1$. Consequently, equality (4.3.20) follows also from (4.3.17) and (4.3.18). □

Proof of Theorem 4.3.7 (b)–(c) Let us denote $\tau_n = n(1 - F_n(u_n))$ and $\tau = e^{-x}$. Then we have

$$P\{M_n^* \leqslant u_n\} - e^{-\tau} = P\{M_n^* \leqslant u_n\} - e^{-\tau_n} + e^{-\tau_n} - e^{-\tau}. \tag{4.3.21}$$

Using Theorem 2.4.2 from Leadbetter, Lindgren and Rootzén [143], *see* also Exercise 2.5, we obtain that

$$P\{M_n^* \leqslant u_n\} - e^{-\tau_n} \sim \frac{e^{-2x}\exp(-e^{-x})}{2} \cdot \frac{1}{n}, \quad n \to \infty. \tag{4.3.22}$$

Tail $1 - F_n(u_n)$ is given by

$$1 - F_n(u_n) = P\{X_{nj}^* > u_n\} = P\{X_{nj} > u_n\} = \sum_{s=0}^{r-1} p_{n1}(s),$$

where

$$p_{n1} = \frac{e^{-x}(r-1)!}{s! \cdot n \cdot (\ln n)^{r-1}} \left(\frac{u_n}{n}\right)^s \left\{1 - \frac{\ln n}{2n} + o\left(\frac{\ln n}{n}\right)\right\}.$$

Let us first consider the case $r = 1$. The constant (4.3.6) has the form $u_n = n(x + \ln n)$. Let us denote $r_n = n(x + \ln n) - [n(x + \ln n)]$. Then we have the following equalities

$$\begin{aligned}
\tau_n &= nP\{X_{n1}^* > n(x + \ln n)\} = n(1 - 1/n)^{[n(x+\ln n)]} \\
&= \exp\left\{\ln n + \big(n(x + \ln n) - r_n\big) \cdot \left(-\frac{1}{n} - \frac{1}{2n^2} + o\left(\frac{1}{n^2}\right)\right)\right\} \\
&= \exp\left\{-x - \frac{x + \ln n - 2r_n}{2n}(1 + o(1))\right\},
\end{aligned}$$

and consequently we obtain

$$\tau - \tau_n \sim e^{-x} \cdot \frac{x + \ln n - 2r_n}{2n}(1 + o(1)). \tag{4.3.23}$$

Using (4.3.21)–(4.3.23) we obtain the asymptotic relation (4.3.11).

Now let us consider the case $r \geqslant 2$. Since $u_n/n = \ln n + (r-1)\ln\ln n + o(\ln\ln n)$, we obtain the following asymptotic relations as $n \to \infty$:

$$\left(\frac{u_n}{n}\right)^s = (\ln n)^s \cdot \left\{1 + \frac{s(r-1)\ln\ln n}{\ln n} + o\left(\frac{\ln\ln n}{\ln n}\right)\right\},$$

$$p_{n1}(s) = \frac{e^{-x}(r-1)!}{s! \cdot n \cdot (\ln n)^{r-1-s}} \cdot \left\{1 + \frac{s(r-1)\ln\ln n}{\ln n} + o\left(\frac{\ln\ln n}{\ln n}\right)\right\}, \tag{4.3.24}$$

$$\tau_n = n\sum_{s=0}^{r-1} p_{n1}(s) = e^{-x} \cdot \left\{1 + \frac{(r-1)^2\ln\ln n}{\ln n} + o\left(\frac{\ln\ln n}{\ln n}\right)\right\}, \tag{4.3.25}$$

and finally

$$e^{-\tau_n} - e^{-\tau} = -(r-1)^2 e^{-x}\exp(-e^{-x})\frac{\ln\ln n}{\ln n} + o\left(\frac{\ln\ln n}{\ln n}\right). \tag{4.3.26}$$

Now the asymptotic relation (4.3.12) follows from relations (4.3.21), (4.3.22), (4.3.24), (4.3.25), and (4.3.26). □

Lemma 4.3.10 *Let x be a real number and k and l positive integers such that $k + l \leqslant n$. Then Leadbetter's $D(u_n)$ mixing condition is satisfied, i.e., there exists a constant $C(x)$ such that the following inequalities hold:*

$$\left|P\left(\bigcap_{j=1}^{k+l}\{X_{nj} \leqslant u_n\}\right) - P\left(\bigcap_{j=1}^{k}\{X_{nj} \leqslant u_n\}\right)P\left(\bigcap_{j=k+1}^{k+l}\{X_{nj} \leqslant u_n\}\right)\right|$$
$$\leqslant C(x)\cdot\min\{k,l\}\frac{\ln n}{n^2} \leqslant C(x)\frac{\ln n}{n}. \tag{4.3.27}$$

Proof Let us denote

$$\Delta_n(k,1) = P\left(\bigcap_{j=1}^{k+1}\{X_{nj} \leqslant u_n\}\right) - P\left(\bigcap_{j=1}^{k}\{X_{nj} \leqslant u_n\}\right)\cdot P\{X_{n,k+1} \leqslant u_n\},$$
$$D_j = \{X_{nj} > u_n\}, \quad \text{for } j = 1, 2, \ldots, k; \qquad A = \{X_{n,k+1} > u_n\}.$$

Then we have

$$\begin{aligned}
\Delta_n(k,1) &= P(\overline{D}_1 \ldots \overline{D}_k\overline{A}) - P(\overline{D}_1 \ldots \overline{D}_k)P(\overline{A}) \\
&= 1 - P(D_1 \cup \cdots \cup D_k \cup A) - (1 - P(D_1 \cup \cdots \cup D_k))(1 - P(A)) \\
&= P((D_1 \cup \cdots \cup D_k) \cap A) - P(D_1 \cup \cdots \cup D_k)P(A) \\
&= P(D_1A \cup \cdots \cup D_kA) - P(D_1 \cup \cdots \cup D_k)P(A).
\end{aligned}$$

Using the inclusion-exclusion principle we obtain that

$$\Delta_n(k,1) = \sum_{m=2}^{k+1}(-1)^m\binom{k}{m-1}(p_{nm} - p_{n,m-1}p_{n1}). \tag{4.3.28}$$

Using (4.3.19) and (4.3.28) we get inequality (4.3.27), first for integers k and $l = 1$, and then for arbitrary k and l, where $k + l \leqslant n$. □

Lemma 4.3.11 *If u_n is given by* (4.3.6), *then condition $D'(u_n)$ is satisfied, i.e., the following equality holds:*

$$\lim_{k\to\infty}\limsup_{n\to\infty} n \cdot \sum_{j=2}^{[n/k]} P\{X_{n1} > u_n, X_{nj} > u_n\} = 0. \tag{4.3.29}$$

Proof Using equality (4.3.20) we obtain the following relation

$$n \cdot \sum_{j=2}^{[n/k]} P\{X_{n1} > u_n, X_{nj} > u_n\} = n\left(\left[\frac{n}{k}\right] - 1\right)\frac{e^{-2x}}{n^2}(1 + o(1))$$

$$\sim \frac{e^{-2x}}{k} \quad \text{as } n \to \infty,$$

from which equality (4.3.29) follows immediately as $k \to \infty$. □

Proof of Theorem 4.3.5
(a) It follows from Theorem 4.3.7 (a), Lemmas 4.3.10, 4.3.11, and Theorem 2.7.13 that the maximum M_n has the same limiting distribution as $n \to \infty$ (with the same normalizing constants) as the maximum M_n^*, i.e., (4.3.7) holds.
(b)–(c) Let us estimate the difference $|P\{M_n \leqslant u_n\} - P\{M_n^* \leqslant u_n\}|$. Using the inclusion-exclusion principle we obtain that

$$P\{M_n \leqslant u_n\} = \sum_{m=0}^{n}(-1)^m\binom{n}{m}p_{nm}, \tag{4.3.30}$$

where $p_{n0} = 1$. We also have the following equality

$$P\{M_n^* \leqslant u_n\} = \{1-(1-F_n(u_n))\}^n = \sum_{m=0}^{n}(-1)^m\binom{n}{m}(1-F_n(u_n))^m, \tag{4.3.31}$$

where $1 - F_n(u_n) = \sum_{s_1=0}^{r-1} p_{n1}(s_1) = p_{n1}$. Using (4.3.30) and (4.3.31) we obtain

$$P\{M_n \leqslant u_n\} - P\{M_n^* \leqslant u_n\} = \sum_{m=2}^{n}(-1)^m\binom{n}{m}\left\{p_{nm} - \left(\sum_{s_1=0}^{r-1} p_{n1}(s_1)\right)^m\right\}. \tag{4.3.32}$$

It is easy to see that the following equality holds:

$$p_{nm} - \left(\sum_{s_1=0}^{r-1} p_{n1}(s_1)\right)^m$$
$$= \sum_{s_1=0}^{r-1} \dots \sum_{s_m=0}^{r-1} \{p_{nm}(s_1, \dots, s_m) - p_{n1}(s_1) \dots p_{n1}(s_m)\}. \qquad (4.3.33)$$

Now, using relations (4.3.18) and (4.3.24), we get that for each fixed m the following asymptotic relation holds as $n \to \infty$:

$$p_{nm}(s_1, \dots, s_m) - p_{n1}(s_1) \dots p_{n1}(s_m)$$
$$= -\frac{e^{-mx}((r-1)!)^m}{s_1! \dots s_m!} \frac{(\ln n)^s}{n^{m+1}(\ln n)^{m(r-1)}} \left\{\frac{m(m-1)}{2} \ln n + o(\ln n)\right\}, \quad (4.3.34)$$

where $s = s_1 + \dots + s_m$. If n is sufficiently large, then we get from (4.3.34) that the value of $|p_{nm}(s_1, \dots, s_m) - p_{n1}(s_1) \dots p_{n1}(s_m)|$ is the greatest possible if $s_1 = \dots = s_m = r - 1$. Consequently we obtain from (4.3.33) that for fixed m the following relation holds as $n \to \infty$:

$$p_{nm} - \left(\sum_{s_1=0}^{r-1} p_{n1}(s_1)\right)^m \sim -\frac{e^{-mx}}{n^{m+1}} \left\{\frac{m(m-1)}{2} \ln n + o(\ln n)\right\}. \qquad (4.3.35)$$

It follows from (4.3.32) and (4.3.35) that

$$|P\{M_n \leqslant u_n\} - P\{M_n^* \leqslant u_n\}| = O\left(\frac{\ln n}{n}\right), \quad \text{as } n \to \infty. \qquad (4.3.36)$$

Finally, (4.3.8) follows from (4.3.11) and (4.3.36), and (4.3.9) follows from (4.3.12) and (4.3.36). □

Remark 4.3.12 Theorem 4.3.7 can be extended to the case when parameter r is not necessarily a natural number.

Let $X_{n1}^*, X_{n2}^*, \dots, X_{nn}^*$ be independent random variables with a common negative binomial distribution with parameters $r > 0$ and $p \in (0, 1)$, where r is not necessarily an integer, i.e.,

$$P\{X_{nj}^* = k\} = \binom{r+k-1}{k}(1-p)^k p^r, \quad k = 0, 1, 2, \dots \qquad (4.3.37)$$

where

$$\binom{r+k-1}{k} = \frac{\Gamma(r+k)}{\Gamma(r)\Gamma(k+1)}, \qquad (4.3.38)$$

and $\Gamma(r)$ is the gamma function given by

$$\Gamma(r) = \int_0^{\infty} t^{r-1} e^{-t}\, dt, \quad r > 0. \tag{4.3.39}$$

Theorem 4.3.13 (Mladenović and Vukmirović [165]) *Let us denote* $M_n^* = \max\{X_{n1}^*, \ldots, X_{nn}^*\}$, *where* $X_{n1}^*, \ldots, X_{nn}^*$ *are independent random variables with a common negative binomial distribution with parameters* $r \in (0, \infty)$ *and* $p = 1/n$, *and assume that* $u_n(x) = n(x + \ln n + (r-1)\ln\ln n - \ln\Gamma(r))$. *Then for every* $x \in \mathbb{R}$,

$$\lim_{n\to\infty} P\{M_n^* \leqslant u_n(x)\} = \exp(-e^{-x}). \tag{4.3.40}$$

Let us denote $\Delta_n(r, x) = P\{M_n^* \leqslant u_n(x)\} - \exp(-e^{-x})$. *The rate of convergence of maximum* M_n^*, *as* $n \to \infty$, *is given by:*

$$\Delta_n(r, x) \sim -(r-1)^2 e^{-x} \exp(-e^{-x}) \frac{\ln\ln n}{\ln n}, \quad \textit{if } r \neq 1. \tag{4.3.41}$$

Remark 4.3.14 The techniques employed in the proofs of Theorems 4.3.5 and 4.3.7 significantly use the fact that r is a positive integer and a combinatorial interpretation of the corresponding negative binomial distribution cannot be applied if r is not an integer. The proof of Theorem 4.3.13 can be found in Mladenović and Vukmirović [165].

4.3.3 The Problem of Collecting Pairs

The problem of waiting time until a given portion of pairs jj, $j \in \mathbb{N}_n$ is collected was studied in Mladenović [162]. The limit distributions were determined by the method of characteristic functions or using the extreme value theory for stationary sequences. We shall present the main results here.

Let $Z_1, Z_2, Z_3, \ldots$ be a sequence of independent random variables with a uniform distribution over set $\mathbb{N}_n = \{1, 2, \ldots, n\}$. We shall use the following notation:

$$X_{nj} = \min\{k : Z_{k-1} = Z_k = j\}, \quad j \in N_n \text{ is a fixed number,} \tag{4.3.42}$$

$$\widetilde{Y}_{nj} = \min\{k : Z_{k-1} = Z_k = a \text{ for some } a \in A \subset \mathbb{N}_n, |A| = j\}, \tag{4.3.43}$$

$$M_n = \max\{X_{n1}, \ldots, X_{nn}\}, \tag{4.3.44}$$

$$M_n^{(k)} = \text{the } k\text{-th maximum of random variables } X_{n1}, \ldots, X_{nn}, \tag{4.3.45}$$

where $|A|$ is the number of elements of a set A. Then X_{nj} is the waiting time until pair jj for some fixed $j \in N_n$ occurs as a run in process $Z_1, Z_2, \ldots$; $\widetilde{Y}_{nj}$ is the waiting time until some of pairs aa, where $a \in A$ and $|A| = j$, occur as a run in the same process; and M_n is the waiting time until all n pairs jj, $j \in \mathbb{N}_n$ occur.

Let Y_{nn} be the waiting time until the first pair $j_1 j_1$, where $j_1 \in \mathbb{N}_n$, occurs as a run in process $Z_1, Z_2, \ldots$ Let $Y_{n,n-1}$ be the waiting time for the second pair $j_2 j_2$, where $j_2 \in \mathbb{N}_1 \setminus \{j_1\}$, after the occurrence of the first pair, etc. Then $Y_{nj} \stackrel{d}{=} \widetilde{Y}_{nj}$ for every $j \in \mathbb{N}_n$, where $X \stackrel{d}{=} Y$ means that random variables X and Y have the same distribution. Let us denote by S_{n,a_n} the waiting time until the a_n of pairs jj, $j \in \mathbb{N}_n$ occur, i.e.,

$$S_{n,a_n} = Y_{nn} + Y_{n,n-1} + \cdots + Y_{n,n-a_n+1}, \quad a_n \in N_n. \tag{4.3.46}$$

It is obvious that the following relations hold:

$$Y_{n1} \stackrel{d}{=} X_{n1} \stackrel{d}{=} X_{n2} \stackrel{d}{=} \ldots \stackrel{d}{=} X_{nn}, \tag{4.3.47}$$

$$S_{nn} = Y_{nn} + Y_{n,n-1} + \cdots + Y_{n1} = \max\{X_{n1}, \ldots, X_{nn}\} = M_n, \tag{4.3.48}$$

$$S_{n,n-k+1} = M_n^{(k)}, \quad k \text{ is a fixed positive integer}, \tag{4.3.49}$$

$$P\{X_{n1} > m, X_{n2} > m, \ldots, X_{nj} > m\} = P\{Y_{nj} > m\}. \tag{4.3.50}$$

It is also clear that random variables $Y_{nn}, Y_{n,n-1}, \ldots, Y_{n1}$ are independent, but random variables $X_{n1}, X_{n2}, \ldots, X_{nn}$ are dependent. Let $F_n(x)$ be the common distribution function of random variables $X_{n1}, \ldots, X_{nn}$. As usual, $\Phi(x)$ is the standard normal distribution function. First, we shall give the exact distributions and related characteristics of random variables X_{nj} and Y_{nj} and results concerning the asymptotic behavior of the mean and variance of random variable S_{n,a_n}.

Theorem 4.3.15 (Mladenović [162])

(a) *The distribution of random variable X_{nj} is given by*

$$P\{X_{nj} = k\} = \sum_{s=0}^{[k/2]-1} \binom{k-s-2}{s} \left(1 - \frac{1}{n}\right)^{k-s-2} \frac{1}{n^{s+2}}, \quad k \geq 2. \tag{4.3.51}$$

(b) *If $u_n = n^2(x + \ln n)$, then the following equality holds:*

$$\lim_{n\to\infty} n(1 - F_n(u_n)) = \lim_{n\to\infty} nP\{X_{nj} > u_n\} = e^{-x}. \tag{4.3.52}$$

Proof

(a) It is easy to check that equality (4.3.51) holds for $k = 2$ and $k = 3$. The event $\{X_{nj} = k\}$, where $k > 3$, means that no two adjacent random variables among $Z_1, Z_2, \ldots, Z_{k-3}$ take the value j and $Z_{k-2} \neq j$, $Z_{k-1} = Z_k = j$. Denote by A_s the event that exactly s of random variables $Z_1, Z_2, \ldots, Z_{k-3}$ take the value

j, and no two of them that are adjacent take the value j. Then,

$$P(A_s) = \binom{k-s-2}{s}\left(1-\frac{1}{n}\right)^{k-3-s}\frac{1}{n^s}, \quad s \in \{0, 1, \ldots, [k/2]-1\}. \tag{4.3.53}$$

The following two equalities hold:

$$\{X_{nj} = k\} = \cup_{s=0}^{[k/2]-1}\{A_s, Z_{k-2} \neq j, Z_{k-1} = Z_k = j\}, \tag{4.3.54}$$

$$P\{Z_{k-2} \neq j, Z_{k-1} = Z_k = j\} = \left(1-\frac{1}{n}\right)\frac{1}{n^2}. \tag{4.3.55}$$

If $k > 3$, then equality (4.3.51) follows from (4.3.53)–(4.3.55).

(b) Using (4.3.51) we obtain that

$$1 - F_n(m) = \sum_{k=m+1}^{\infty}\frac{1}{n^2}\left(1-\frac{1}{n}\right)^{k-2}\sum_{s=0}^{[(k-2)/2)]}\binom{k-2-s}{s}\left(\frac{1}{n-1}\right)^s.$$

From problem 7(d), p. 76 of Riordan [193], the following identity holds:

$$\sum_{s=0}^{[r/2]}\binom{r-s}{s}x^s = \frac{1}{\alpha}\left\{\left(\frac{1+\alpha}{2}\right)^{r+1} - \left(\frac{1-\alpha}{2}\right)^{r+1}\right\}, \tag{4.3.56}$$

where $\alpha = (1+4x)^{1/2}$. The sum on the left side of identity (4.3.56) is related to the Chebyshev polynomials. For $x = 1/(n-1)$, we get

$$\alpha = \left(1+\frac{4}{n-1}\right)^{1/2} = \left(1+\frac{3}{n}\right)^{1/2}\left(1-\frac{1}{n}\right)^{-1/2},$$

and tail $1 - F_n(m)$ can be represented in the form

$$1 - F_n(m) = \frac{1}{n^2}\left(1-\frac{1}{n}\right)^{-1/2}\left(1+\frac{3}{n}\right)^{-1/2}\left\{\frac{q_1^m}{1-q_1} - \frac{q_2^m}{1-q_2}\right\}, \tag{4.3.57}$$

where

$$q_1 = \left(1-\frac{1}{n}\right)\frac{1+\alpha}{2} = 1 - \frac{1}{n^2} + \frac{5}{32n^3} + o\left(\frac{1}{n^3}\right), \tag{4.3.58}$$

$$q_2 = \left(1-\frac{1}{n}\right)\frac{1-\alpha}{2} = -\frac{1}{n} + \frac{1}{n^2} - \frac{5}{32n^3} + o\left(\frac{1}{n^3}\right). \tag{4.3.59}$$

Let us determine m from the condition $n(1 - F_n(m)) \to e^{-x}$ as $n \to \infty$. Using (4.3.57)–(4.3.59) this condition can be transformed as follows:

$$-\ln n - \frac{m}{n^2} + \frac{5m}{32n^3} + 2\ln n = -x + o(1), \quad \text{as } n \to \infty,$$

$$m = n^2(x + \ln n + o(1)), \quad \text{as } n \to \infty.$$

Consequently, (4.3.52) holds for $u_n = n^2(x + \ln n)$.

□

Theorem 4.3.16 (Mladenović [162])

(a) *The distribution of random variable Y_{nj} is given by*

$$P\{Y_{nj} = k\} = \frac{j}{n\{(n+1)^2 - 4j\}^{1/2}} \cdot \left\{ \left(\frac{t_1}{n}\right)^{k-1} - \left(\frac{t_2}{n}\right)^{k-1} \right\}, \quad k \geqslant 2, \tag{4.3.60}$$

where

$$t_1 = t_1(j) = \frac{n - 1 + \{(n+1)^2 - 4j\}^{1/2}}{2}, \tag{4.3.61}$$

$$t_2 = t_2(j) = \frac{n - 1 - \{(n+1)^2 - 4j\}^{1/2}}{2}. \tag{4.3.62}$$

(b) *Exact values of the mean and variance of random variable Y_{nj} are given by*

$$EY_{nj} = \frac{n^2 + n}{j}, \quad var\, Y_{nj} = \frac{n^4}{j^2}\left(1 + \frac{2}{n} - \frac{3j - 1}{n^2} - \frac{j}{n^3}\right). \tag{4.3.63}$$

Proof

(a) Let A be a subset of N_n, $|A| = j$, and let

$$a_0 = 1, \quad b_0 = 0. \tag{4.3.64}$$

For every positive integer l, let us consider the set S of all sequences of the form $c_1 c_2 \ldots c_l$, where $c_1, c_2, \ldots, c_l \in \mathbb{N}_n$, such that no sequence from S contains a subsequence of the form aa, $a \in A$. Let a_l be the number of sequences from S for which $c_l \in N_n \setminus A$, and b_l the number of sequences from S such that $c_l \in A$. Then, the following equalities hold:

$$a_1 = n - j = (n - j)(a_0 + b_0), \tag{4.3.65}$$

$$b_1 = j = ja_0 + (j - 1)b_0, \tag{4.3.66}$$

$$a_{k-1} = (a_{k-2} + b_{k-2}), \quad \text{for any } k \geq 2, \tag{4.3.67}$$

$$b_{k-1} = ja_{k-2} + (j - 1)b_{k-2}, \quad \text{for any } k \geq 2. \tag{4.3.68}$$

Denote $s_l = a_l + b_l$, for $l \geq 0$. It follows from (4.3.64)–4.3.68 that

$$s_0 = a_0 + b_0 = 1, \quad s_1 = a_1 + b_1 = n, \tag{4.3.69}$$

$$s_{k-1} = a_{k-1} + b_{k-1} = (n-1)s_{k-2} + (n-j)s_{k-3}, \quad \text{for } k \geq 3 \tag{4.3.70}$$

Hence, sequence (s_l) satisfies the linear difference equation (4.3.70) with initial conditions (4.3.69). It follows that

$$s_{k-1} = C_1 t_1^{k-1} + C_2 t_2^{k-1}, \quad \text{for } k \geq 1, \tag{4.3.71}$$

where $t_1 = t_1(j)$ and $t_2 = t_2(j)$ are given by (4.3.61) and (4.3.62). Using initial conditions (4.3.69) we obtain constants C_1 and C_2:

$$C_1 = \frac{1}{2}\left\{1 + \left(1 - \frac{4j}{(n+1)^2}\right)^{-1/2}\right\}, \tag{4.3.72}$$

$$C_2 = \frac{1}{2}\left\{1 - \left(1 - \frac{4j}{(n+1)^2}\right)^{-1/2}\right\}. \tag{4.3.73}$$

Note that

$$P\{Y_{nj} = k\} = b_{k-1}n^{-k}, \quad k \in \{2, 3, \dots\}, \tag{4.3.74}$$

$$b_{k-1} = s_{k-1} - a_{k-1} = s_{k-1} - (n-j)s_{k-2}. \tag{4.3.75}$$

Equalities (4.3.60) follow from (4.3.71)–(4.3.75).

(b) Formulas (4.3.63) can be obtained by calculations.

□

Theorem 4.3.17 (Mladenović [162]) *The asymptotic behavior of mean μ_n and variance σ_n^2 of random variable S_{n,a_n} is determined as follows:*

(a) *If $a_n \to \infty$ and $a_n/n \to 0$ as $n \to \infty$, then*

$$\mu_n = -n^2 \ln\left(1 - \frac{a_n}{n}\right) + o(na_n^{1/2}), \quad \sigma_n^2 \sim n^2 a_n \quad \text{as } n \to \infty. \tag{4.3.76}$$

(b) *If $a_n/n \to \lambda \in (0, 1)$ as $n \to \infty$, and $\lambda_0 = \lambda/(1-\lambda)$, then*

$$\mu_n = -n^2 \ln\left(1 - \frac{a_n}{n}\right) + o(n^{3/2}), \quad \sigma_n^2 \sim \lambda_0 n^3 \quad \text{as } n \to \infty. \tag{4.3.77}$$

(c) *If $a_n/n \to 1$ and $b_n = n - a_n \to \infty$ as $n \to \infty$, then*

$$\mu_n = -n^2 \ln\left(1 - \frac{a_n}{n}\right) + o(n^2 b_n^{-1/2}), \quad \sigma_n^2 \sim n^4/b_n \quad \text{as } n \to \infty. \tag{4.3.78}$$

(d) *As $n \to \infty$, the mean and variance of maximum $M_n = S_{nn}$ are given by*

$$EM_n = (n^2+n)(\ln n+\gamma)+\frac{n}{2}+\frac{5}{12}-\frac{1}{12n}+\frac{1}{120n^2}+o\left(\frac{1}{n^2}\right), \tag{4.3.79}$$

$$var\ M_n = \frac{\pi^2 n^4}{6}+\left(\frac{\pi^2}{3}-1\right)n^3-3n^2\ln n+O(n^2), \tag{4.3.80}$$

where $\gamma = 0.5772156649\ldots$ is the Euler constant.

Proof For every positive integer n let us denote

$$H_n = 1+\frac{1}{2}+\cdots+\frac{1}{n}, \quad H_n^{(2)} = 1+\frac{1}{2^2}+\cdots+\frac{1}{n^2}. \tag{4.3.81}$$

The following equalities hold:

$$H_n = \ln n+\gamma+\frac{1}{2n}-\frac{1}{12n^2}+\frac{1}{120n^4}-\frac{\varepsilon_n}{252n^6}, \tag{4.3.82}$$

$$H_n^{(2)} = \frac{\pi^2}{6}-\frac{1}{n}+\frac{\vartheta_n}{n(n+1)}, \tag{4.3.83}$$

where $0 < \varepsilon_n < 1$ and $0 < \vartheta_n < 1$. Equality (4.3.82) can be found in Graham, Knuth, and Patashnik [83], p. 480, and (4.3.83) can easily be proved. Since

$$ES_{n,a_n} = (n^2+n)(H_n - H_{n-a_n}), \tag{4.3.84}$$

relations concerning the asymptotic behavior of ES_{n,a_n} follow from (4.3.82) and (4.3.84). Similarly, using (4.3.82), (4.3.83) and the second of equalities (4.3.63), we get relations concerning the asymptotic behavior of $var\ S_{n,a_n}$. □

The next three theorems give the limiting distribution of random variable S_{n,a_n} for different asymptotic behaviors of sequence (a_n).

Theorem 4.3.18 (Mladenović [162]) *If $a_n = k$ for every $n \in N$, where k is a fixed positive integer, then random variable $n^{-1}S_{n,a_n}$ converges in distribution to a random variable whose characteristic function is*

$$f(t) = (1+t^2)^{-k/2}e^{ik\cdot\arctan t}. \tag{4.3.85}$$

If $a_n/n \to \lambda \in [0,1]$, $a_n \to \infty$ and $b_n = n - a_n \to \infty$ as $n \to \infty$, then $(S_{n,a_n} - \mu_n)/\sigma_n$ has asymptotically normal (0,1) distribution, where $\sigma_n^2 = var\ S_{n,a_n}$ and $\mu_n = ES_{n,a_n}$. Denote by ν_n and τ_n^2 the main terms of μ_n and σ_n^2 that are determined by (4.3.76)–(4.3.78). Using the Khintchine Lemma we can conclude that constants μ_n and σ_n^2 can be replaced in limit theorems by ν_n and τ_n^2, because

$\sigma_n/\tau_n \to 1$ and $(\mu_n - \nu_n)/\sigma_n \to 0$ as $n \to \infty$. Results are given by the following theorem in more detail.

Theorem 4.3.19 (Mladenović [162])

(a) *If $a_n \to \infty$ and $a_n/n \to 0$ as $n \to \infty$, then*

$$\lim_{n\to\infty} P\left\{\frac{S_{n,a_n} + n^2 \ln\left(1 - \frac{a_n}{n}\right)}{n a_n^{1/2}} \leqslant x\right\} = \Phi(x). \tag{4.3.86}$$

(b) *If $a_n/n \to \lambda \in (0, 1)$ as $n \to \infty$ and $\lambda_0 = \lambda/(1-\lambda)$, then*

$$\lim_{n\to\infty} P\left\{\frac{S_{n,a_n} + n^2 \ln\left(1 - \frac{a_n}{n}\right)}{\lambda_0^{1/2} n^{3/2}} \leqslant x\right\} = \Phi(x). \tag{4.3.87}$$

(c) *If $a_n/n \to 1$ and $b_n = n - a_n \to \infty$ as $n \to \infty$, then*

$$\lim_{n\to\infty} P\left\{\frac{S_{n,a_n} + n^2 \ln\left(1 - \frac{a_n}{n}\right)}{n^2 b_n^{-1/2}} \leqslant x\right\} = \Phi(x). \tag{4.3.88}$$

Theorem 4.3.20 (Mladenović [162]) *If $n - a_n + 1 = k$ for a fixed positive integer k and all positive integers n, then the limiting distribution of random variable $S_{n,a_n} = M_n^{(k)}$ is given by the following equality*

$$\lim_{n\to\infty} P\{M_n^{(k)} \leqslant n^2(x + \ln n)\} = e^{-e^{-x}} \sum_{s=0}^{k-1} \frac{e^{-sx}}{s!}. \tag{4.3.89}$$

Especially, maximum $M_n = S_{nn}$ has the Gumbel extreme value distribution asymptotically.

Let us denote $D = D(j) = \{(n+1)^2 - 4j\}^{1/2}$ and let $t_1 = t_1(j)$ and $t_2 = t_2(j)$ be given by (4.3.61) and (4.3.62). In the sequel we shall use the following series of expansions and the approximations that follow from them:

$$D = n\left(1 + \frac{1}{n} - \frac{2j}{n^2} + \frac{2j}{n^3} - \frac{2j + 2j^2}{n^4} + \frac{2j + 6j^2}{n^5} - \cdots\right), \tag{4.3.90}$$

$$D^{-1} = \frac{1}{n}\left(1 - \frac{1}{n} + \frac{1+2j}{n^2} - \frac{1+6j}{n^3} + \cdots\right) \tag{4.3.91}$$

$$t_1 = n\left(1 - \frac{j}{n^2} + \frac{j}{n^3} - \frac{j + j^2}{n^4} + \frac{j + 3j^2}{n^5} - \cdots\right), \tag{4.3.92}$$

$$t_2 = -1 + \frac{j}{n} - \frac{j}{n^2} + \frac{j + j^2}{n^3} - \frac{j + 3j^2}{n^4} - \cdots \tag{4.3.93}$$

$$n - t_1 = \frac{j}{n}\left(1 - \frac{1}{n} + \frac{1+j}{n^2} - \frac{1+3j}{n^3} - \cdots\right). \tag{4.3.94}$$

Proof of Theorem 4.3.18 Let us determine the characteristic function of random variable Y_{nj}. Using (4.3.60) and asymptotic relations (4.3.90)–(4.3.94) we obtain

$$\begin{aligned} f_{nj}(t) &= \frac{j}{nD}\left\{\sum_{k=2}^{\infty} e^{itk}\left(\frac{t_1}{n}\right)^{k-1} - \sum_{k=2}^{\infty} e^{itk}\left(\frac{t_2}{n}\right)^{k-1}\right\} \\ &= \frac{j}{t_1}\left(\frac{t_1}{n}\right)^2\left(1 + \frac{2}{n} + \frac{1-4j}{n^2}\right)^{-1/2}\left\{1 - \frac{t_2}{t_1}\cdot\frac{1-e^{it\frac{t_1}{n}}}{1-e^{it\frac{t_2}{n}}}\right\}\frac{e^{2it}}{n - t_1 e^{it}}. \end{aligned} \tag{4.3.95}$$

If $j/n \to 1$ as $n \to \infty$, then

$$\lim_{n\to\infty} \frac{j}{t_1}\left(\frac{t_1}{n}\right)^2\left(1 + \frac{2}{n} + \frac{1-4j}{n^2}\right)^{-1/2}\left\{1 - \frac{t_2}{t_1}\cdot\frac{1-e^{it/n\frac{t_1}{n}}}{1-e^{it/n\frac{t_2}{n}}}\right\} = 1,$$

and hence, the asymptotic behavior of $f_{Y_{nj}/n}(t)$ is given by

$$f_{Y_{nj}/n}(t) \sim \frac{e^{2it/n}}{n - t_1 e^{it/n}}, \quad n \to \infty.$$

The following relations hold:

$$\begin{aligned} |n - t_1 e^{it/n}|^2 &= \left|n - t_1 \cos\frac{t}{n} - it_1 \sin\frac{t}{n}\right|^2 = n^2 + t_1^2 - 2nt_1 \cos\frac{t}{n} \\ &= (n - t_1)^2 + 2nt_1\left(1 - \cos\frac{t}{n}\right) \to 1 + t^2, \quad n \to \infty,\ \frac{j}{n} \to 1; \end{aligned}$$

$$\begin{aligned} \arg f_{Y_{nj}/n}(t) \sim \arg\frac{e^{2it}}{n - t_1 e^{it/n}} &= \frac{2t}{n} - \arctan\frac{-t_1 \sin\frac{t}{n}}{n - t_1 \cos\frac{t}{n}} \\ &= \frac{2t}{n} - \arctan\frac{-t + o(1)}{1 + o(1)} \to \arctan t, \quad \text{as } n \to \infty, \end{aligned}$$

where we again use the assumption that $j/n \to 1$ as $n \to \infty$. Hence, $f_{Y_{nj}/n}(t) \to (1+t^2)^{-1/2}e^{i\cdot\arctan t}$ as $n \to \infty$. Consequently, if $a_n = k$, where k is a fixed positive integer, we get that $(Y_{nn} + \cdots + Y_{n,n-k+1})/n$ converges in distribution to a random variable whose characteristic function is given by (4.3.85). □

Proof of Theorem 4.3.19 Using (4.3.95) we obtain that the characteristic function of sum S_{n,a_n} can be represented in the following way:

$$f_{S_{n,a_n}}(t) = \prod_{j=n-a_n+1}^{n} f_{nj}(t) = P_{n1} \cdot P_{n2} \cdot P_{n3}(t) \cdot P_{n4}(t), \tag{4.3.96}$$

where P_{n1}, P_{n2}, $P_{n3}(t)$ and $P_{n4}(t)$ are given by

$$P_{n1} = \prod_{j=n-a_n+1}^{n} \frac{t_1(j)}{n}, \tag{4.3.97}$$

$$P_{n2} = \prod_{j=n-a_n+1}^{n} \left(1 + \frac{2}{n} + \frac{1-4j}{n^2}\right)^{-1/2}, \tag{4.3.98}$$

$$P_{n3}(t) = \prod_{j=n-a_n+1}^{n} \left\{1 - \frac{t_2(j)}{t_1(j)} \cdot \frac{1-e^{it} \cdot t_1(j)/n}{1-e^{it} \cdot t_1(j)/n}\right\}, \tag{4.3.99}$$

$$P_{n4}(t) = \prod_{j=n-a_n+1}^{n} \frac{e^{2it}}{\frac{n^2}{j}\left(1 - \frac{t_1}{n}e^{it}\right)}. \tag{4.3.100}$$

□

The remaining part of the proof is based on the following lemmas (for details *see* Mladenović [162]):

Lemma 4.3.21 *If $a_n/n \to \lambda \in [0, 1]$ as $n \to \infty$, then the following relation holds:*

$$\lim_{n\to\infty} P_{n1} = e^{-\lambda+\lambda^2/2}. \tag{4.3.101}$$

Lemma 4.3.22 *If $a_n/n \to \lambda \in [0, 1]$ as $n \to \infty$, then the following relation holds:*

$$\lim_{n\to\infty} P_{n2} = e^{\lambda-\lambda^2}. \tag{4.3.102}$$

Lemma 4.3.23 *If τ_n^2 is the main term of σ_n^2 determined by* (4.3.76)–(4.3.78), $\tau_n > 0$ *and $a_n \to \infty$ as $n \to \infty$, then for every real t the following equality holds:*

$$\lim_{n\to\infty} P_{n3}\left(\frac{t}{\tau_n}\right) = 1. \tag{4.3.103}$$

Lemma 4.3.24 *Let τ_n^2 be the main term of σ_n^2, $\tau_n > 0$, and suppose that $a_n \to \infty$ and $b_n = n - a_n \to \infty$ as $n \to \infty$. Then for every real t the following asymptotic relations hold as $n \to \infty$:*

$$\ln\left|P_{n4}\left(\frac{t}{\tau_n}\right)\right| \sim \frac{a_n^2}{2n^2} - \frac{n^4 t^2}{2\tau_n^2}(H_n^{(2)} - H_{n-a_n}^{(2)}), \tag{4.3.104}$$

$$\left|P_{n4}\left(\frac{t}{\tau_n}\right)\right| \to \exp\left(\frac{\lambda^2}{2} - \frac{t^2}{2}\right), \quad if \quad \frac{a_n}{n} \to \lambda \in [0, 1]. \tag{4.3.105}$$

Lemma 4.3.25 *If ν_n and τ_n^2 are main terms of the mean $\mu_n = ES_{n,a_n}$ and the variance $\sigma_n^2 = var\, S_{n,a_n}$, $\tau_n > 0$, and $S^*_{n,a_n} = \tau_n^{-1}(S_{n,a_n} - \nu_n)$, then*

$$\arg f_{S^*_{n,a_n}}(t) = o(1), \quad as\ n \to \infty. \tag{4.3.106}$$

Finally, using relations (4.3.96)–(4.3.105) we obtain that in all cases considered $f_{S^*_{n,a_n}}(t) \to e^{-t^2/2}$ as $n \to \infty$, and therefore the proof of Theorem 4.3.19 is completed. □

Proof of Theorem 4.3.20 is based on the following lemmas (for details *see* Mladenović [162]).

Lemma 4.3.26 *Let X^*_{nj}, $j \in \{1, 2, \ldots, n\}$, be independent random variables with the same probability distribution:*

$$P\{X^*_{nj} = k\} = \sum_{s=0}^{[k/2]-1} \binom{k-2-s}{s}\left(1 - \frac{1}{n}\right)^{k-s-2} \frac{1}{n^{s+2}}, \quad k \geq 2, \tag{4.3.107}$$

and let us denote $M^*_n = \max\{X^*_{n1}, \ldots, X^*_{nn}\}$. Then for every real x the following equality holds

$$\lim_{n\to\infty} P\{M^*_n \leq n^2(x + \ln n)\} = e^{-e^{-x}}. \tag{4.3.108}$$

Proof The asymptotic relations $P\{M^*_n \leqslant u_n\} \to e^{-\tau}$ as $n \to \infty$, and $n(1 - F_n(u_n)) \to \tau$ as $n \to \infty$, are equivalent. Hence, (4.3.108) follows from (4.3.52). □

Lemma 4.3.27 *If j is a fixed positive integer and $n \to \infty$, then the following asymptotic relation holds:*

$$P\{Y_{nj} > n^2(x + \ln n)\} = \frac{e^{-jx}}{n^j}\left\{1 + \frac{j(x + \ln n)}{n} + O\left(\left(\frac{\ln n}{n}\right)^2\right)\right\}. \tag{4.3.109}$$

Lemma 4.3.28 *Let $p_{nj} = P\{Y_{nj} > n^2(x + \ln n)\}$. For every positive integer $j \geqslant 2$ the following relation holds as $n \to \infty$:*

$$p_{nj} - p_{n,j-1}p_{n1} = \frac{e^{-jx}}{n^{j+1}} \cdot o(1). \tag{4.3.110}$$

Lemma 4.3.29 *Let x be a real number and k and l positive integers, such that $k+l \leq n$. Then there exists a positive constant $C(x)$ such that for $u_n = n^2(x + \ln n)$ the following inequality holds:*

$$\left|P\left(\bigcap_{i=1}^{k+l}\{X_{ni} \leqslant u_n\}\right) - P\left(\bigcap_{i=1}^{k}\{X_{ni} \leqslant u_n\}\right)P\left(\bigcap_{i=k+1}^{k+l}\{X_{ni} \leqslant u_n\}\right)\right|$$
$$\leqslant C(x)\min\{k,l\}\frac{1}{n^2} \leqslant \frac{C(x)}{n},$$

i.e., condition $D(u_n)$ is satisfied.

Lemma 4.3.30 *For $u_n = n^2(x + \ln n)$ condition $D'(u_n)$ is satisfied, i.e.,*

$$\lim_{k\to\infty}\limsup_{n\to\infty} n \cdot \sum_{j=2}^{[n/k]} P\{X_{n1} > u_n, X_{nj} > u_n\} = 0.$$

Now Theorem 4.3.20 follows from Lemmas 4.3.26, 4.3.29, 4.3.30, and Theorem 5.3.1 from Leadbetter, Lindgren, and Rootzén [143]. □

4.3.4 On the Rate of Convergence in Limit Theorems

Let X_n and Z be two random variables, F_n and F be the corresponding distribution functions, and μ_n and μ be the corresponding probability measures on the Borel σ-algebra $\mathcal{B}$. If $X_n \xrightarrow{d} Z$ as $n \to \infty$, i.e., $F_n(x) \to F(x)$ for every continuity point of F (weak convergence), we are interested in the closeness of F_n and F, or equivalently μ_n and μ.

Theorem 4.3.5 is related to the limiting distribution of waiting time M_n for all elements to appear r times in the coupon collector's problem. Items (b) and (c) of this theorem provided the rate of convergence of the distribution function of normalized maximum at fixed point x. The fact that very different rates of convergence of maxima can appear is confirmed once again.

Different methods have been suggested to measure the distance between two probability measures on the real line and corresponding distribution functions. The reader may find more about probability metrics in Gibbs and Su [74] and Rachev [189].

The Kolmogorov distance between μ_n and μ is defined by

$$d_K(\mu_n, \mu) = \sup_{x\in\mathbb{R}} |\mu_n(-\infty, x] - \mu(-\infty, x]| = \sup_{x\in\mathbb{R}} |F_n(x) - F(x)|.$$

Note that the convergence in *Kolmogorov metric* $d_K(\mu_n, \mu) \to 0$ as $n \to \infty$, implies weak convergence, i.e., if $d_K(\mu_n, \mu) \to 0$ as $n \to \infty$, then $F_n(x) \to F(x)$ for every continuity point of F. The converse statement is not true in general.

The Levy distance between μ_n and μ is defined by

$$d_L(\mu_n, \mu) = \inf\{\varepsilon > 0 : (\forall x \in \mathbb{R})\, F(x - \varepsilon) - \varepsilon \leqslant F_n(x) \leqslant F(x + \varepsilon) + \varepsilon\}.$$

Convergence in the Levy metric is equivalent to the weak convergence of a sequence of probability measures on $(\mathbb{R}, \mathcal{B})$.

The Total Variation Distance between μ_n and μ is defined by

$$d_{TV}(\mu_n, \mu) = \sup_{B \in \mathcal{B}} |\mu_n(B) - \mu(B)|.$$

Convergence in the total variation metric implies weak convergence, and the converse statement is not true in general. If μ_n and μ are determined by discrete random variables X_n and Z, defined on the same probability space $(\Omega, \mathcal{A}, P)$ with support $\mathbb{Z} = \{\dots, -1, 0, 1, \dots\}$, then the total variation distance $d_{TV}(\mu_n, \mu)$ can be represented as follows:

$$d_{TV}(\mu_n, \mu) = \frac{1}{2} \sum_{k \in \mathbb{Z}} |P\{X_n = k\} - P\{Z = k\}|.$$

The Stein-Chen Method In the 1970s Stein (*see* [38, 40, 215, 216]) introduced a new method to approximate the distribution of random variables. His approach was originally formulated in the context of the central limit theorem for dependent random variables when the limit distribution is normal. Chen [39] showed that Stein's approach can also be used for Poisson approximation. In fact, the Stein method enables an approximation of the total variation distance $d_{TV}(\mu_n, \mu)$ for every limit probability measure μ. Stein's method is a sophisticated approach to proving distributional limit theorems and at the same time bounds the approximation error. In order to show that a random variable X has the same asymptotic distribution as some other random variable Z using Stein's method, we should generally proceed as follows:

(a) Construct a so-called *characterizing operator* T for Z, with defining property $E((Tf)(Z)) = 0$, for any function f belonging to a fixed large class of suitable functions. Here are several characterizing operators for the given distributions:
  Standard normal distribution: $(Tf)(x) := f'(x) - xf(x)$.
  Poisson distribution $\mathcal{P}(\lambda)$, $\lambda > 0$: $(Tf)(k) := \lambda f(k+1) - kf(k)$.
  Gumbel distribution: $(Tf)(x) := f'(x) - (1 - e^{-x})f(x)$, etc.
(b) Construct a random variable X^* such that ordered pair (X, X^*) is an exchangeable pair, i.e., random vectors (X, X^*) and (X^*, X) are identically distributed (the joint distribution of (X, X^*) is a permutation invariant).
(c) Choose an operator α such that for any suitable real-valued function g, αg is a measurable antisymmetric function; let operator S be defined as $(Sg)(x) := E(\alpha g(X, X^*)|X = x)$, $x \in \mathbb{R}$, so that $|(Sg)(x) - Tg(x)| \leqslant \varepsilon_g$, where ε_g is a small error depending only on g (note that $E((Sg)(X)) = 0$).
(d) Take a function g from some class of functions and find the solution h of the so-called *Stein equation*: $(Th)(x) = g(x) - E(g(Z))$. Therefore,

$|E((Th)(X))|$ measures the distance between X and Z and we finally get $|E(g(X)) - E(g(Z))| \leqslant \varepsilon_h$.

Hence, the main idea of Stein's method is to prove the closeness of the distributions of X and Z by showing that their characterizing operators are close to each other. Now we shall provide some results on the rate of convergence in limit theorems related to the coupon collector's problem.

Collecting Pairs and Rates of Convergence In order to determine the upper bound of the Kolmogorov distance we use the very well-known uniform Berry–Esseen inequality for unbounded variables (*see*, for example, Chen, Goldstein and Chao [40], p. 53). In this way we specify the rate at which this convergence takes place.

Let $\xi_1, \xi_2, \ldots, \xi_n$ be mean zero, independent random variables defined on probability space $(\Omega, \mathcal{A}, P)$ with variances summing to one and finite third absolute moments. If $W = \sum_{i=1}^{n} \xi_i$, then the following estimate for the rate of convergence applies:

$$\sup_{z\in\mathbb{R}} |P\{W \leqslant z\} - \Phi(z)| \leqslant C_0\gamma, \tag{4.3.111}$$

where $\gamma = \sum_{i=1}^{n} E|\xi_i|^3$, and C_0 is an absolute constant. Shevtsova [211] showed that C_0 does not exceed value 0.5600. We shall consider here random variables

$$\xi_i = \frac{Y_{ni} - EY_{ni}}{(DS_{n,a_n})^{1/2}}, \qquad W = \sum_{i=n-a_n+1}^{n} \xi_i, \tag{4.3.112}$$

where random variables Y_{ni} are defined in Sect. 4.3.3, i.e., Y_{ni} is the waiting time until the $(n-i+1)$-th pair of the form jj occurs. Then, $S_{n,a_n} = Y_{nn} + Y_{n,n-1} + \cdots + Y_{n,n-a_n+1}$ is the waiting time until a_n of pairs jj, $j \in \mathbb{N}_n$, occur. Theorem 4.3.19 states that $(S_{n,a_n} - ES_{n,a_n})/\sigma_n$, where $\sigma_n^2 = \operatorname{Var} S_{n,a_n}$, has $\mathcal{N}(0, 1)$ asymptotic distribution if $a_n/n \to \lambda \in [0, 1]$, $a_n \to \infty$ and $b_n = n - a_n \to \infty$, as $n \to \infty$.

In the subsequent theorem we state that $\gamma = \sum_{i=n-a_n+1}^{n} E|\xi_i|^3$ is less than or equal to a particular expression, depending on the asymptotic behavior of sequence (a_n).

Theorem 4.3.31 (Glavaš and Mladenović [77]) *For the upper bound of γ the following asymptotics, as $n \to \infty$, hold:*

(a) *If $a_n \to \infty$ and $a_n/n \to 0$, as $n \to \infty$, then the bound behaves like*

$$\frac{2(6e^{-1} - 1)}{\sqrt{a_n}}. \tag{4.3.113}$$

(b) *If $a_n/n \to \lambda \in (0, 1)$, as $n \to \infty$, and $\lambda_0 = \lambda/(1-\lambda)$, then the bound behaves like*

$$\frac{2(6e^{-1}-1)}{\sqrt{\lambda_0 n}(1-\lambda)^2}. \tag{4.3.114}$$

(c) *If* $a_n/n \to 1$ *and* $b_n = n - a_n \to \infty$, *as* $n \to \infty$, *then the bound behaves like*

$$\frac{6e^{-1}-1}{\sqrt{b_n}}. \tag{4.3.115}$$

Let us now pay some attention to random variable M_n that represents the waiting time until all n pairs jj, $j \in \mathbb{N}_n$, appear for the first time. Using previous notation it is easy to see that M_n is, in fact, S_{nn}. Here we denote by X_{nj} the waiting time for pair jj, $j \in \mathbb{N}_n$, to appear for the first time as a run in process $Z_1, Z_2, \ldots$. Then: $M_n = \max\{X_{n1}, \ldots, X_{nn}\}$, where $X_{n1}, \ldots, X_{nn}$ are dependent random variables. The probability distribution of random variable X_{nj} is given by

$$P\{X_{nj} = k\} = \sum_{s=0}^{[k/2]-1} \binom{k-s-2}{s} \left(1 - \frac{1}{n}\right)^{k-s-2} \frac{1}{n^{s+2}}, \quad k \geqslant 2.$$

It was proved in Theorem 4.3.20 that M_n has the Gumbel extreme value distribution asymptotically as $n \to \infty$. Using a similar approach as in the proof of Theorem 4.3.5 we obtain the following result.

Theorem 4.3.32 *For any* $x \in \mathbb{R}$ *the following asymptotic relation holds:*

$$|P\{M_n \leqslant n^2(x + \ln n)\} - \exp(-e^{-x})| = O\left(\frac{\ln n}{n}\right) \quad \text{as } n \to \infty. \tag{4.3.116}$$

In Chatterjee, Diaconis, and Meckes [38] it is also shown that the order of error of Poisson approximations for the coupon collector's problem (in its simplest setting) using exchangeable pairs is far larger than $(\ln n)/n$, where $(\ln n)/n$ is the order of error obtained by the coupling approach. Posfai [187, 188] obtained many refinements of limit theorems related to the coupon collector's problem, considering Poisson approximations, normal approximations, and Gumbel-like approximations.

4.3.5 Point Processes Associated with the Coupon Collector's Problem

In this subsection we consider point processes determined by the times of first appearances of elements of $\mathbb{N}_n$. We shall prove that these processes, properly centered and normalized, converge to a non-homogeneous Poisson point process as $n \to \infty$. For general theory related to Poisson point processes *see* Kingman [131] and Last and Penrose [140].

Let X_{nj} be the number of trials necessary to obtain element $j \in \mathbb{N}_n$, and let X_{n1}^*, $\ldots, X_{nn}^*$ be the associated sequence of i.i.d. random variables, such that $X_{nj} \stackrel{d}{=} X_{nj}^*$. For any $x \in \mathbb{R}$ we shall use the following notation

$$F_n(x) := P\{X_{nj} \leqslant x\} = P\{X_{nj}^* \leqslant x\}, \tag{4.3.117}$$

$$u_n(x) := n(x + \ln n). \tag{4.3.118}$$

Now it is easy to see that for any $x \in \mathbb{R}$ and sufficiently large n the following equality holds

$$F_n(u_n(x)) = P\{X_{nj}^* \leqslant u_n(x)\} = 1 - \left(1 - \frac{1}{n}\right)^{[u_n(x)]}. \tag{4.3.119}$$

Let A_k be the event that some element from set $\mathbb{N}_n$ appears for the first time at the k-th experiment, i.e.,

$$A_k = \{\exists j \in \mathbb{N}_n : Z_1 \neq j, \ldots, Z_{k-1} \neq j, Z_k = j\}. \tag{4.3.120}$$

Let us define point processes N_n and N_n^* on $\mathbb{R}$ as follows:

$$\begin{aligned} N_n(\cdot) &= \sum_{k=1}^{\infty} \varepsilon_{(k/n)-\ln n}(\cdot) I_{A_k}, \\ &= \sum_{k=1}^{\infty} \varepsilon_{(k/n)-\ln n}(\cdot) \sum_{j=1}^{n} I_{\{X_{nj}=k\}} \end{aligned} \tag{4.3.121}$$

$$N_n^*(\cdot) = \sum_{k=1}^{\infty} \varepsilon_{(k/n)-\ln n}(\cdot) \sum_{j=1}^{n} I_{\{X_{nj}^*=k\}}, \tag{4.3.122}$$

where for every $x \in \mathbb{R}$, the measure ε_x on Borel σ-algebra $\mathcal{B}$ is given by

$$\varepsilon_x(B) = \begin{cases} 1, & \text{if } x \in B, \\ 0, & \text{if } x \notin B, \end{cases} \quad \text{for each } B \in \mathcal{B}. \tag{4.3.123}$$

Example 4.3.33 Let us consider a small example for $n = 6$ and choose elements from the set $\mathbb{N}_6 = \{1, 2, 3, 4, 5, 6\}$. Choosing an element of $\mathbb{N}_6$ can be considered the result of throwing a die. Point processes N_6 and N_6^* are then given by

$$N_6(\cdot) = \sum_{k=1}^{\infty} \varepsilon_{k/6-\ln 6}(\cdot) \sum_{j=1}^{6} I_{\{X_{6j}=k\}},$$

$$N_6^*(\cdot) = \sum_{k=1}^{\infty} \varepsilon_{k/6-\ln 6}(\cdot) \sum_{j=1}^{6} I_{\{X_{6j}^*=k\}}.$$

Realizations of point processes N_6 and N_6^* are the so-called point measures μ_6 and μ_6^* defined on Borel sets on the real line. Suppose that $X_{6,6} = 1$, i.e., 6 appears in the first throw of the die, and $X_{6,1} = 7$, $X_{6,2} = 4$, $X_{6,3} = 2$, $X_{6,4} = 13$, $X_{6,5} = 15$. Then the set of k's significant for this particular realization is $S = \{1, 2, 4, 7, 13, 15\}$. For $k \in S$ the expression $k/6 - \ln 6$ takes the following values: $-1.625\ldots$, $-1.458\ldots$, $-1.125\ldots$, $-0.625\ldots$, $0.374\ldots$, $0.708\ldots$. Hence, $\mu_6((-2,-1]) = 3$, $\mu_6((-1,0]) = 1$, and $\mu_6((0,1]) = 2$. On the other hand, independent random variables $X_{6,j}^*$, where $j \in \mathbb{N}_6$, can all take, for example, the value 1 with positive probability, and for such a realization we have $\mu_6^*((-2,-1]) = 6$, $\mu_6^*((-1,0]) = 0$, and $\mu_6^*((0,1]) = 0$. △

Note that two different elements from $\mathbb{N}_n$ cannot appear for the first time at the same trial. As a consequence we get that N_n is a simple point process on $\mathbb{R}$, i.e., $P\{N_n(\{x\}) \leqslant 1 \text{ for all } x \in \mathbb{R}\} = 1$. On the contrary, N_n^* is not a simple point process. The properties and asymptotic behavior of point processes N_n and N_n^* are given by the following theorems, *see* Glavaš and Mladenović [77].

Theorem 4.3.34 *For any real numbers c and d, where $c < d$, we have*

$$\lim_{n\to\infty} EN_n((c,d\,]) = \lim_{n\to\infty} EN_n^*((c,d\,]) = e^{-c} - e^{-d}. \tag{4.3.124}$$

Lemma 4.3.35 *For events A_k, $k \geqslant 1$, and random variables X_{nj} and X_{nj}^*, $j \in \mathbb{N}_n$ the following statements hold:*

(a) $P(A_k) = nP\{X_{nj} = k\} = \left(1 - \dfrac{1}{n}\right)^{k-1}$, *for each $k \in \mathbb{N}$.*

(b) $I_{A_k} = \sum_{j=1}^{n} I_{\{X_{nj}=k\}}$ *is the Bernoulli random variable that takes values 0 and 1 with probabilities $1 - P(A_k)$ and $P(A_k)$ respectively.*

(c) $\sum_{j=1}^{n} I_{\{X_{nj}^*=k\}}$ *is binomial random variable $\mathcal{B}(n,p)$ with parameters n and $p = P\{X_{nj}^* = k\}$.*

(d) *The following equalities hold:*

$$E\left(\sum_{j=1}^{n} I_{\{X_{nj}^*=k\}}\right) = E\left(\sum_{j=1}^{n} I_{\{X_{nj}=k\}}\right) = E(I_{A_k}) = P(A_k). \tag{4.3.125}$$

Proof Straightforward exercise. □

Proof of Theorem 4.3.34 For function $F_n(x) = P\{X_{nj} \leqslant x\}$, $u_n(x) = n(x + \ln n)$, and every real number x (*see*, for example, Holst [109] and Mladenović [159]) we have

$$\lim_{n\to\infty} n\{1 - F_n(u_n(x))\} = e^{-x}. \tag{4.3.126}$$

Using Lemma 4.3.35 and equality (4.3.126) we get that for real numbers $c < d$ the following relations hold:

$$\begin{aligned}
EN_n^*((c,d]) &= EN_n((c,d]) = E\left\{\sum_{k=r}^{\infty} \varepsilon_{(k/n)-\ln n}((c,d])I_{A_k}\right\} \\
&= \sum_{c<(k/n)-\ln n\leqslant d} P(A_k) \;=\; \sum_{u_n(c)<k\leqslant u_n(d)} P(A_k) \\
&= nP\{u_n(c) < X_{nj} \leqslant u_n(d)\} \\
&= n\{1 - F_n(u_n(c))\} - n\{1 - F_n(u_n(d))\} \\
&\to e^{-c} - e^{-d}, \quad \text{as } n \to \infty.
\end{aligned}$$

□

Theorem 4.3.36 *Let $c < d$ be two real numbers.*

(a) *For every $s \in \{0, 1, \ldots, n\}$ the following equality holds:*

$$P\{N_n^*((c,d]) = s\} = \binom{n}{s} p_n^s (1 - p_n)^{n-s}, \tag{4.3.127}$$

where $p_n = F_n(u_n(d)) - F_n(u_n(c))$, i.e., $N_n^*((c,d])$ has a binomial distribution with parameters n and p_n.

(b) *Random variable $N_n^*((c,d])$ has asymptotically Poisson distribution with parameter $\mu((c,d]) = e^{-c} - e^{-d}$, i.e., for every $s \in \{0, 1, 2, \ldots\}$,*

$$\lim_{n\to\infty} P\{N_n^*(c,d] = s\} = e^{-\mu((c,d])} \frac{\mu((c,d])^s}{s!}. \tag{4.3.128}$$

Proof

(a) Every random variable $X_{n1}^*, \ldots, X_{nn}^*$ takes a value in the interval $(u_n(c), u_n(d)]$ with probability p_n. Note also that

$$p_n = P\left\{\frac{X_{nj}^*}{n} - \ln n \in (c,d]\right\}. \tag{4.3.129}$$

Since $X_{n1}^*, \ldots, X_{nn}^*$ are independent random variables, equality (4.3.127) follows immediately from (4.3.122).

(b) Let us denote

$$r_n = n(c + \ln n) - [n(c + \ln n)], \tag{4.3.130}$$

$$s_n = n(d + \ln n) - [n(d + \ln n)]. \tag{4.3.131}$$

Then, $0 \leqslant r_n < 1$ and $0 \leqslant s_n < 1$. Using (4.3.119) we obtain that the following relations hold for large n:

$$\begin{aligned} p_n &= \left(1 - \frac{1}{n}\right)^{n(c+\ln n)-r_n} - \left(1 - \frac{1}{n}\right)^{n(d+\ln n)-s_n} \\ &= e^{(nc+n\ln n-r_n)\ln(1-1/n)} - e^{(nd+n\ln n-s_n)\ln(1-1/n)} \\ &= \frac{e^{-c} - e^{-d}}{n} + o\left(\frac{1}{n}\right) \quad \text{as } n \to \infty. \end{aligned} \tag{4.3.132}$$

Relation (4.3.128) now follows from the classical Poisson approximation theorem.

□

Theorem 4.3.37 *Let $c < d$ be two real numbers and let n be a positive integer such that $n(c + \ln n) > n$. Then the following equality holds:*

$$P\{N_n((c, d]) = 0\} = \sum_{k=1}^{n} \binom{n}{k} \left(\frac{k}{n}\right)^{[u_n(d)]} \sum_{j=0}^{k} (-1)^j \binom{k}{j} \left(1 - \frac{j}{k}\right)^{[u_n(c)]}. \tag{4.3.133}$$

Proof Since n is a positive integer such that $u_n(c) = n(c + \ln n) > n$, we obtain that

$$\begin{aligned} P\{N_n(c, d] = 0\} &= P\Bigg(\bigcap_{u_n(c)<s\leqslant u_n(d)} \{I_{A_s} = 0\}\Bigg) \\ &= \sum_{k=1}^{n} P\{B_{[u_n(c)]} = k\} \left(\frac{k}{n}\right)^{[u_n(d)]-[u_n(c)]}, \end{aligned} \tag{4.3.134}$$

where $B_{[u_n(c)]}$ is the number of distinct elements from $\mathbb{N}_n$ that occur until the $[u_n(c)]$-th trial, $\left(\frac{k}{n}\right)^{[u_n(d)]-[u_n(c)]}$ is the probability that only fixed k elements can occur in the interval $([u_n(c)], [u_n(d)]]$, and $[x]$ is the integer part of x. Next we have

$$P\{B_{[u_n(c)]} = k\} = \binom{n}{k} \left\{ \left(\frac{k}{n}\right)^{[u_n(c)]} - P(D_1 \cup D_2 \cup \cdots \cup D_k) \right\}, \tag{4.3.135}$$

where $\left(\frac{k}{n}\right)^{[u_n(c)]}$ is the probability that only elements from set $\{1, 2, \dots, k\}$ (not necessarily all of them) will occur until the $[u_n(c)]$-th trial, and D_s is the event that only elements from set $\{1, 2, \dots, k\} \setminus \{s\}$, $1 \leqslant s \leqslant k$, will occur until the $[u_n(c)]$-th trial. Then we have

$$P(D_s) = \left(\frac{k-1}{n}\right)^{[u_n(c)]}, \quad 1 \leqslant s \leqslant k, \tag{4.3.136}$$

$$P(D_s D_t) = \left(\frac{k-2}{n}\right)^{[u_n(c)]} \quad \text{for } s, t \in \{1, \dots k\},\ s \neq t, \tag{4.3.137}$$

....

Using the inclusion-exclusion principle we obtain

$$P\{B_{[u_n(c)]} = k\} = \binom{n}{k} \sum_{j=0}^{k} (-1)^j \binom{k}{j} \left(\frac{k-j}{n}\right)^{[u_n(c)]}. \tag{4.3.138}$$

Using (4.3.134) and (4.3.138) we get

$$P\{N((c,d]) = 0\} = \sum_{k=1}^{n} \binom{n}{k} \left(\frac{k}{n}\right)^{[u_n(d)]-[u_n(c)]} \sum_{j=0}^{k} (-1)^j \binom{k}{j} \left(\frac{k-j}{n}\right)^{[u_n(c)]}, \tag{4.3.139}$$

and (4.3.133) follows immediately from (4.3.139). Note that the sum $\sum_{j=0}^{k}$ on the right side of (4.3.138), and consequently the second sum in (4.3.139) and (4.3.133), can be replaced by $\sum_{j=0}^{k-1}$, because the addend obtained for $j = k$ is equal to 0. □

Theorem 4.3.38 (Glavaš and Mladenović [77]) *Let $(N_n)_{n \geqslant 1}$ be the sequence of point processes on $\mathbb{R}$ defined by* (4.3.121). *Let N be the Poisson random measure on $\mathbb{R}$ with mean measure μ determined by $\mu((c,d]) = e^{-c} - e^{-d}$, for all $-\infty < c < d < +\infty$. Then $N_n \Rightarrow N$, i.e., sequence $(N_n)_{n \geqslant 1}$ converges weakly to N as $n \to \infty$.*

Remark 4.3.39 If $c < d$ and n is sufficiently large then using (4.3.127) and (4.3.119) we obtain that

$$P\{N_n^*((c,d]) = 0\} = \left\{1 - \left(1 - \frac{1}{n}\right)^{[u_n(c)]} + \left(1 - \frac{1}{n}\right)^{[u_n(d)]}\right\}^n$$
$$= \sum_{k=0}^{n} \binom{n}{k} \left\{1 - \left(1 - \frac{1}{n}\right)^{[u_n(c)]}\right\}^k \left\{\left(1 - \frac{1}{n}\right)^{[u_n(d)]}\right\}^{n-k} \tag{4.3.140}$$

Equality (4.3.140) can be rewritten in the following form

$$P\{N_n^*((c,d]) = 0\} = \sum_{k=1}^{n} \binom{n}{k} \left(1 - \frac{1}{n}\right)^{[u_n(d)](n-k)} \times$$
$$\times \sum_{j=0}^{k} (-1)^j \binom{k}{j} \left(1 - \frac{1}{n}\right)^{[u_n(c)]j} + \left(1 - \frac{1}{n}\right)^{[u_n(d)]n}. \tag{4.3.141}$$

Bearing in mind (4.3.133), equality (4.3.141) will be used for comparison probabilities $P\{N_n((c,d]) = 0\}$ and $P\{N_n^*((c,d]) = 0\}$, as indicated in the following lemma.

Lemma 4.3.40 *Let $c < d$ and choose n such that $n(c + \ln n) > n$. Then the following equality holds:*

$$P\{N_n(c,d] = 0\} - P\{N_n^*(c,d] = 0\} = \sum_{k=1}^{n} \sum_{j=0}^{k} \delta_{kj}^{(n)} - \left(1 - \frac{1}{n}\right)^{[u_n(d)]n}, \tag{4.3.142}$$

where $\delta_{kj}^{(n)}$ is given by

$$\delta_{kj}^{(n)} = (-1)^j \binom{n}{k} \binom{k}{j} \times$$
$$\times \left\{ \left(\frac{k}{n}\right)^{[u_n(d)]} \left(\frac{k-j}{k}\right)^{[u_n(c)]} - \left(1 - \frac{1}{n}\right)^{[u_n(d)](n-k)+[u_n(c)]j} \right\}. \tag{4.3.143}$$

Proof Equality (4.3.142) with $\delta_{kj}^{(n)}$ given by (4.3.143) follows immediately from (4.3.133) and (4.3.141). □

Lemma 4.3.41 *For all c and d, where $c < d$, the following equality holds:*

$$\lim_{n \to \infty} \left(P\{N_n(c,d] = 0\} - P\{N_n^*(c,d] = 0\} \right) = 0. \tag{4.3.144}$$

Proof First note that $\delta_{n-i,j}^{(n)}$ from formula (4.3.143) can be represented in the form

$$\delta_{n-i,j}^{(n)} =$$
$$= (-1)^j \binom{n}{n-i} \binom{n-i}{j} \left\{ \left(1 - \frac{i}{n}\right)^{n(d+\ln n)-s_n} \left(1 - \frac{j}{n-i}\right)^{n(c+\ln n)-r_n} \right.$$
$$\left. - \left(1 - \frac{1}{n}\right)^{i(n(d+\ln n)-s_n)+j(n(c+\ln n)-r_n)} \right\}$$
$$= \frac{n!}{i!\,j!\,(n-i-j)!} (-1)^j (AB - C), \tag{4.3.145}$$

where

$$A = e^{(nd+n\ln n-s_n)\ln(1-i/n)}, \tag{4.3.146}$$

$$B = e^{(nc+n\ln n-r_n)\ln(1-j/(n-i))}, \tag{4.3.147}$$

$$C = e^{ind+in\ln n-is_n+jnc+jn\ln n-jr_n)\ln(1-1/n)}. \tag{4.3.148}$$

Let us consider the case $\frac{i}{n} \to 0$ and $\frac{j}{n} \to 0$ as $n \to \infty$, and additionally i and j are not too large in the sense that will be detailed later on. We shall use the following inequalities

$$-x - x^2 < \ln(1-x) < -x \quad \text{for } 0 < x < \frac{1}{2}, \tag{4.3.149}$$

$$\ln\left(1 - \frac{j}{n-i}\right) < -\frac{j}{n-i} = -\frac{j}{n}\left(1-\frac{i}{n}\right)^{-1} < -\frac{j}{n}\left(1+\frac{i}{n}\right), \tag{4.3.150}$$

$$-\frac{j}{n-i} - \left(\frac{j}{n-i}\right)^2 < \ln\left(1-\frac{j}{n-i}\right). \tag{4.3.151}$$

Note also that for large n, and i and j such that

$$0 < n^2 i - n(3i^2 + 2ij) + 2i^3 + i^2 j, \tag{4.3.152}$$

the following inequality also holds

$$\begin{aligned}
\frac{j}{n-i} + \left(\frac{j}{n-i}\right)^2 &= \frac{j}{n}\left(1-\frac{i}{n}\right)^{-1} + \frac{j^2}{n^2}\left(1-\frac{i}{n}\right)^{-2} \\
&< \frac{j}{n}\left(1+\frac{2i}{n}\right) + \frac{j^2}{n^2}\left(1+\frac{3i}{n}\right) \\
&= \frac{i}{n} + \frac{2ij+j^2}{n^2} + \frac{3ij^2}{n^3},
\end{aligned} \tag{4.3.153}$$

and consequently

$$-\frac{i}{n} - \frac{2ij+j^2}{n^2} - \frac{3ij^2}{n^3} < \ln\left(1-\frac{j}{n-i}\right). \tag{4.3.154}$$

If $i/n \to 0$ and $j/n \to 0$ as $n \to \infty$, then using (4.3.146) we obtain that for large n the following inequalities hold:

$$e^{(nd+n\ln n-s_n)(-i/n-i^2/n^2)} < A < e^{(nd+n\ln n-s_n)(-i/n)} \tag{4.3.155}$$

and consequently

$$\frac{e^{-id}}{n^i}e^{-i^2(d+\ln n)/n} < A < \frac{e^{-id}}{n^i}e^{i/n}. \tag{4.3.156}$$

Similarly, using (4.3.148) we obtain that C satisfies the following inequalities

$$\frac{e^{-id-jc}}{n^{i+j}}e^{-(id+i\ln n+jc+j\ln n)/n} < C < \frac{e^{-id-jc}}{n^{i+j}}e^{(i+j)/n}. \tag{4.3.157}$$

(a) If $i/n \to 0$ and $j/n \to 0$ as $n \to \infty$, and (4.3.152) holds, then using (4.3.147) and (4.3.154) we obtain that B satisfies the following inequalities

$$\frac{e^{-jc}}{n^j}e^{-(2ijc+j^2c+2ij\ln n+j^2\ln n)/n} < B < \frac{e^{-jc}}{n^j}e^{-(ijc+ij\ln n-j-ij)/n}. \tag{4.3.158}$$

Using (4.3.156)–(4.3.158) we obtain that

$$\frac{e^{-id-jc}}{n^{i+j}}D_1 < AB - C < \frac{e^{-id-jc}}{n^{i+j}}D_2, \tag{4.3.159}$$

where

$$D_1 = e^{-(i^2d+i^2\ln n+2ijc+j^2c+2ij\ln n+j^2\ln n)/n} - e^{(i+j)/n}, \tag{4.3.160}$$

$$D_2 = e^{-(ijc+ij\ln n-i-ij-1)/n} - e^{-(id+i\ln n+jc+j\ln n)/n}. \tag{4.3.161}$$

Note also that upper and lower bounds for D_1 and D_2 can be obtained using the following inequalities that are valid for $|x| \leqslant \frac{1}{2}$ and $|y| \leqslant \frac{1}{2}$:

$$x - y - (x^2 + y^2) < e^x - e^y < x - y + (x^2 + y^2). \tag{4.3.162}$$

(a1) If i and j are fixed positive integers and $n \to \infty$, then using (4.3.146)–(4.3.148) we obtain that

$$AB - C = \frac{e^{-id-jc}}{n^{i+j}}\left\{\frac{(i-i^2)(d+\ln n) + j(1-2i+j)(c+\ln n)}{2n} + o\left(\frac{\ln n}{n}\right)\right\} \tag{4.3.163}$$

and consequently, from (4.3.145) and (4.3.163), it follows that

$$\delta_{n-i,j}^{(n)} \sim \frac{(-1)^j e^{-id-jc}}{i!\,j!}\cdot\frac{(i-i^2)(d+\ln n) + j(1-2i+j)(c+\ln n)}{2n}. \tag{4.3.164}$$

The exceptions are cases $(i, j) \in \{(0, 0), (0, 1), (1, 0), (1, 1)\}$. If $(i, j) = (0, 0)$ or $(i, j) = (0, 1)$, then $\delta_{n-i,j}^{(n)} = 0$. If $(i, j) = (1, 0)$ or $(i, j) = (1, 1)$, then the second fraction on the right side of (4.3.164) is equal to 0 and should be replaced by a smaller value so that the asymptotic relation remains valid. Upper and lower bounds for $\delta^{(n)}_{n-i,j}$, similar to the expression on the right side of (4.3.164), can easily be obtained.

(a2) Suppose now that $i \sim \ln n$ and $j \sim \ln n$ as $n \to \infty$. Then, $AB - C$ satisfies (4.3.163). Using the Stirling formula $n! \sim n^n e^{-n}\sqrt{2\pi n}$ as $n \to \infty$ (*see*, for example, Sándor and Debnath [200]), we obtain the following asymptotic relations

$$\begin{aligned}
&\frac{n!}{i!\,j!\,(n-i-j)!}\cdot\frac{e^{-id+jc}}{n^{i+j}}\\
&\sim \frac{n^n\sqrt{n}}{2\pi\, i^i\, j^j\,(n-i-j)^{n-i-j}\sqrt{ij(n-i-j)}}\cdot\frac{e^{-id+jc}}{n^{i+j}}\\
&= \frac{1}{2\pi\, i^i\, j^j}\cdot\sqrt{\frac{n}{ij(n-i-j)}}\left(1-\frac{i+j}{n}\right)^{-(n-i-j)}\cdot\frac{e^{-id+jc}}{n^{i+j}}.
\end{aligned} \tag{4.3.165}$$

Similarly we can deduce bounds for $\delta^{(n)}_{n-i,j}$ in cases when one of the indices i and j is less than $\ln n$, while the other one is greater than $\ln n$.

(b) It is easy to prove that $(\ln n)^{\ln n}$ grows faster than any power n^α, where $\alpha > 0$ is a fixed real number. This fact will be used in order to determine how the powers i^i and j^j in the denominator of the first fraction on the right side of (4.3.165) influence the value of $\delta^{(n)}_{n-i,j}$.

(c) From the facts mentioned above we can deduce the following conclusions:

$$\sum_{0\leqslant i<\ln n}\ \sum_{0\leqslant j<\ln n} \delta^{(n)}_{n-i,j} \to 0 \quad \text{as} \quad n\to\infty, \tag{4.3.166}$$

$$\sum_{(i,j):\, i>\ln n \text{ or } j>\ln n} \delta^{(n)}_{n-i,j} \to 0 \quad \text{as} \quad n\to\infty. \tag{4.3.167}$$

and consequently we obtain that equality (4.3.144) holds.

□

Proof of Theorem 4.3.38 Let N be a Poisson random measure on $\mathbb{R}$ with mean measure μ determined by $\mu((c,d]) = e^{-c} - e^{-d}$, for all $-\infty < c < d < +\infty$. It follows from Theorem 4.3.34 that

$$\lim_{n\to\infty} EN_n((c,d]) = e^{-\mu((c,d])} = EN((c,d]). \tag{4.3.168}$$

Using Theorem 4.3.36 (b) we obtain that

$$\lim_{n\to\infty} P\{N_n^*((c,d]) = 0\} = e^{-\mu((c,d])}. \tag{4.3.169}$$

Equalities (4.3.169) and (4.3.144) imply that

$$\lim_{n\to\infty} P\{N_n((c,d])=0\} = e^{-\mu((c,d])} = P\{N((c,d])=0\}. \tag{4.3.170}$$

Now, it is easy to obtain that equality

$$\lim_{n\to\infty} P\{N_n(B)=0\} = P\{N(B)=0\} \tag{4.3.171}$$

holds for all $B=\bigcup_{i=1}^{k}(c_i,d_i]$, and $k=1,2,\ldots$ and Theorem 4.3.38 follows from Theorem 2.7.37. □

As a consequence of Theorem 4.3.38 we obtain the following statement:

Corollary 4.3.42 *For every Borel set $B\subset\mathbb{R}$ whose boundary has μ measure zero, i.e., $\mu(\partial B)=0$, the following equalities hold:*

$$\lim_{n\to\infty} P\{N_n(B)=s\} = e^{-\mu(B)}\cdot\frac{(\mu(B))^s}{s!}, \quad s\in\mathbb{N}_0. \tag{4.3.172}$$

If $B_1,\ldots,B_k\subset\mathbb{R}$ are disjoint Borel sets with $\mu(\partial B_i)=0$ for each i, then

$$(N_n(B_1),\ldots,N_n(B_k)) \xrightarrow{d} (N(B_1),\ldots,N(B_k)) \quad as \quad n\to\infty. \tag{4.3.173}$$

4.3.6 Coupon Collector's Problem and Generalized Pareto Distributions

In previous subsections we considered the limiting distribution of maximum M_n that represented the waiting time for some patterns to appear in a sequence of sampling elements of $\mathbb{N}_n$. In this subsection we shall give results concerning the asymptotic behavior of the excess of M_n over high thresholds. If the threshold is higher than the normalizing constant in the corresponding limit theorem for M_n, then the limit distribution of the excess of M_n over this threshold does not depend on the threshold.

Theorem 4.3.43 (Jocković and Mladenović [118]) *Let $(c_n)_{n\geqslant 1}$ be a sequence of real numbers such that $c_n\to+\infty$ as $n\to\infty$.*

(a) *Let M_n be the waiting time until all elements of $\mathbb{N}_n$ are sampled at least r times, where r is a positive integer, and let us denote*

$$a_n = \ln n + (r-1)\ln\ln n - \ln(r-1)!, \quad n\geqslant 1. \tag{4.3.174}$$

Then for every $x>0$ the following equality holds:

$$\lim_{n\to\infty} P\{M_n\leqslant n(x+c_n+a_n)\mid M_n>n(c_n+a_n)\} = 1-e^{-x}. \tag{4.3.175}$$

(b) *Let M_n be the waiting time until all pairs jj, $j \in \mathbb{N}_n$, are sampled. Then for every $x > 0$ the following equality holds:*

$$\lim_{n\to\infty} P\{M_n \leqslant n^2(x+c_n+\ln n) \mid M_n > n^2(c_n+\ln n)\} = 1-e^{-x}. \quad (4.3.176)$$

Remark 4.3.44 In both cases considered in Theorem 4.3.43 the limiting distribution of the excess is the standard exponential distribution and does not depend on sequence $(c_n)_{n\geqslant 1}$ if $c_n \to \infty$ as $n \to \infty$. If $c_n = c$ for all $n \in \mathbb{N}$ and some constant $c \in \mathbb{R}$, then the limits on the left-hand sides of (4.3.175) and (4.3.176) are again equal to each other. This common limit follows immediately from (4.3.6)–(4.3.7) and (4.3.89) for $k = 1$ but is not equal to the exponential distribution function.

Proof of Theorem 4.3.43

(a) First note that for every positive integer $k \geqslant nr$ the following equality holds (*see* Mladenović [159]):

$$P\{M_n \leqslant k\} = \sum_{m=0}^{n}(-1)^m\binom{n}{m}\sum_{s_m=0}^{r-1}\cdots\sum_{s_2=0}^{r-1}\sum_{s_1=0}^{r-1} p_{nm}(k, s_1, s_2, \ldots, s_m), \quad (4.3.177)$$

where

$$p_{nm}(k, s_1, s_2, \ldots, s_m) = \binom{k}{s_1}\binom{k-s_1}{s_2}\cdots\binom{k-s_s-s_2-\cdots-s_{m-1}}{s_m} \times$$
$$\times \left(\frac{n-m}{n}\right)^{k-s_1-s_2-\cdots-s_m}\left(\frac{1}{n}\right)^{s_1+s_2+\cdots+s_m}. \quad (4.3.178)$$

For $m \in \{0, 1, \ldots, n\}$ let us denote

$$a_{nk}(m) = \binom{n}{m}\sum_{s_m=0}^{r-1}\cdots\sum_{s_2=0}^{r-1}\sum_{s_1=0}^{r-1} p_{nm}(k, s_1, s_2, \ldots, s_m). \quad (4.3.179)$$

Let us determine the asymptotic behavior of $a_{n,[u_n]}(1)$ and $a_{n,[u_n]}(2)$ as $n \to \infty$, where $u_n = n(x + c_n + a_n)$ and $[\alpha]$ is the largest integer not greater than α. Using (4.3.177)–(4.3.179) for $m = 1$ we obtain that:

$$a_{n,[u_n]}(1) = n\sum_{s_1=0}^{r-1}\binom{[u_n]}{s_1}\left(1-\frac{1}{n}\right)^{u_n-r_n-s_1}\left(\frac{1}{n}\right)^{s_1}$$
$$= n\frac{(n\ln n)^{r-1}}{(r-1)!}\left(\frac{1}{n}\right)^{r-1}(1+o(1))\exp\left\{(u_n-r_n-s_1)\ln\left(1-\frac{1}{n}\right)\right\}$$

$$= \frac{n(\ln n)^{r-1}}{(r-1)!} \exp\Big\{\big(n(x + \ln n + (r-1)\ln\ln n - \ln(r-1)! + c_n)$$

$$-r_n - s_1\big)\Big(-\frac{1}{n} - \frac{1}{2n^2} + o\Big(\frac{1}{n^2}\Big)\Big)\Big\}(1+o(1))$$

$$= \frac{n(\ln n)^{r-1}}{(r-1)!} \cdot \frac{e^{-x}(r-1)!}{n(\ln n)^{r-1} e^{c_n}} \times$$

$$\times \exp\Big\{-\frac{\ln n + c_n}{2n} + o\Big(\frac{\ln n + c_n}{n}\Big)\Big\}(1+o(1))$$

$$= \frac{e^{-x}}{e^{c_n}} \exp\Big\{-\frac{\ln n + c_n}{2n} + o\Big(\frac{\ln n + c_n}{n}\Big)\Big\}(1+o(1)). \tag{4.3.180}$$

Similarly for $m = 2$ we get the following equalities:

$$a_{n,[u_n]}(2) = \binom{n}{2} \sum_{s_1=0}^{r-1} \sum_{s_2=0}^{r-1} \binom{[u_n]}{s_1} \binom{[u_n]-s_1}{s_2} \Big(1 - \frac{2}{n}\Big)^{u_n - r_n - s_1 - s_2} \frac{1}{n^{s_1+s_2}}$$

$$= \frac{e^{-2x}}{2\,e^{2c_n}} \exp\Big\{-\frac{2(\ln n + c_n)}{n} + o\Big(\frac{\ln n + c_n}{n}\Big)\Big\}(1+o(1)). \tag{4.3.181}$$

If $c_n = o(n)$ as $n \to \infty$, then using (4.3.180) and (4.3.181) we obtain the following asymptotic equalities for $a_{n,[u_n]}(1)$ and $a_{n,[u_n]}(2)$:

$$a_{n,[u_n]}(1) = \frac{e^{-x}}{e^{c_n}}(1+o(1)), \quad \text{as } n \to \infty; \tag{4.3.182}$$

$$a_{n,[u_n]}(2) = \frac{e^{-2x}}{2\,e^{2c_n}}(1+o(1)), \quad \text{as } n \to \infty. \tag{4.3.183}$$

In cases $c_n = O(n)$ or $c_n/n \to \infty$ as $n \to \infty$, we get

$$a_{n,[u_n]}(1) = \frac{e^{-x}}{e^{c_n} e^{c_n/(2n)}}(1+o(1)), \quad \text{as } n \to \infty; \tag{4.3.184}$$

$$a_{n,[u_n]}(2) = \frac{e^{-2x}}{2\,e^{2c_n} e^{2c_n/n}}(1+o(1)), \quad \text{as } n \to \infty. \tag{4.3.185}$$

The Bonferroni inequalities imply that

$$a_{n,[u_n]}(1) - a_{n,[u_n]}(2) \leqslant P\{M_n > u_n\} \leqslant a_{n,[u_n]}(1). \tag{4.3.186}$$

Now, (4.3.175) follows easily from relations (4.3.182)–(4.3.186).

(b) In this case we have $M_n = \max\{X_{n1}, \ldots, X_{nn}\}$, where random variables X_{nj}, $j \in \mathbb{N}_n$, are defined by (4.3.42). Recall that random variables $Y_{nn}, Y_{n,n-1}, \ldots, Y_{n1}$ are defined as follows: Y_{nn} is the waiting time until the first pair $j_1 j_1$, where $j_1 \in \mathbb{N}_n$, occurs; $Y_{n,n-1}$ is the waiting time for the second pair $j_2 j_2$, $j_2 \in \mathbb{N}_n \setminus \{j_1\}$, after the occurrence of the first pair $j_1 j_1$; etc. For every $j \in \mathbb{N}_n$ and every positive integer k we have

$$P\{X_{n1} > k, X_{n2} > k, \ldots, X_{nj} > k\} = P\{Y_{nj} > k\}. \tag{4.3.187}$$

The probability distribution of random variables Y_{nj} is given by (4.3.60)–(4.3.62). Using the inclusion-exclusion principle we obtain that

$$P\{M_n \leqslant k\} = \sum_{m=0}^{n} (-1)^m \binom{n}{m} P\{Y_{nm} > k\}, \tag{4.3.188}$$

where

$$P\{Y_{nm} > k\} = \frac{m}{n\sqrt{(n+1)^2 - 4m}} \times \\ \times \left\{ \left(\frac{t_1(m)}{n}\right)^k \frac{n}{n - t_1(m)} - \left(\frac{t_2(m)}{n}\right)^k \frac{n}{n - t_2(m)} \right\}, \tag{4.3.189}$$

and $t_1(m)$ and $t_2(m)$ are given by (4.3.61)–(4.3.62). The Bonferroni inequalities imply that

$$1 - nP\{Y_{n1} > k\} \leqslant P\{M_n \leqslant k\} \leqslant 1 - nP\{Y_{n1} > k\} + \binom{n}{2} P\{Y_{n2} > k\},$$

and consequently

$$nP\{Y_{n1} > k\} - \binom{n}{2} P\{Y_{n2} > k\} \leqslant P\{M_n > k\} \leqslant nP\{Y_{n1} > k\}. \tag{4.3.190}$$

For any fixed m the following equalities hold as $n \to \infty$:

$$t_1(m) = n - \frac{m}{n+1} + o\left(\frac{1}{n}\right), \quad t_2(m) = -1 + \frac{m}{n+1} + o\left(\frac{1}{n}\right). \tag{4.3.191}$$

It follows from (4.3.189) and (4.3.191) that $\left(\frac{t_2(m)}{n}\right)^k \frac{n}{n-t_2(m)}$ can be neglected in the asymptotic behavior of probability $P\{Y_{nm} > k\}$ as $n \to \infty$. Let us denote $u_n = n^2(x + c_n + \ln n)$, $r_n = u_n - [u_n]$ and determine the asymptotic behavior of $\binom{n}{m} P\{Y_{nm} > u_n\}$ for any fixed m as $n \to \infty$. Straightforward calculations show that

$$\frac{m}{n\sqrt{(n+1)^2-4m}}\cdot\frac{n}{n-t_1(m)}=1+o(1),\quad \text{as } n\to\infty; \tag{4.3.192}$$

and

$$\left\{\frac{t_1(m)}{n}\right\}^{[u_n]}=\left\{1-\frac{m}{n^2}+\frac{m}{n^3}-\frac{m+m^2}{n^4}+o\left(\frac{1}{n^4}\right)\right\}^{n^2(x+c_n+\ln n)-r_n}$$
$$=\frac{e^{-mx}e^{mc_n/n}}{n^m e^{mc_n}}(1+o(1)),\quad \text{as } n\to\infty. \tag{4.3.193}$$

Now using (4.3.189), (4.3.192), (4.3.193) and equality $P\{M_n > u_n\} = P\{M_n > [u_n]\}$, we obtain that

$$\binom{n}{m}P\{M_n > u_n\}=\binom{n}{m}\frac{e^{-mx}e^{mc_n/n}}{n^m e^{mc_n}}(1+o(1)),\quad \text{as } n\to\infty. \tag{4.3.194}$$

For $m = 1$ and $m = 2$ it follows from (4.3.194) that

$$nP\{Y_{n1} > u_n\}=\frac{e^{-x}e^{c_n/n}}{e^{c_n}}(1+o(1)), \tag{4.3.195}$$

$$\binom{n}{2}P\{Y_{n2} > u_n\}=\frac{e^{-2x}e^{2c_n/n}}{2e^{2c_n}}(1+o(1)). \tag{4.3.196}$$

Finally, equality (4.3.176) follows from (4.3.190) and (4.3.195)–(4.3.196). □

4.3.7 *Other Extensions*

This subsection reviews some extensions of the coupon collector's problem that are not presented in the previous sections. The general setting of the problem is given at the beginning of Sect. 4.3.

Baum and Billingsley [22] investigated the waiting time until a *given portion* of the elements of set $\mathbb{N}_n$ are sampled and determined all possible limiting distributions for such waiting times. For this purpose they represented the waiting time until b_n elements of $\mathbb{N}_n$ are sampled in the form $1+\Gamma_{(n-1)/n}+\Gamma_{(n-2)/n}+\cdots+\Gamma_{b_n/n}$, where $\Gamma_{k/n}$ is a random variable that has a geometric distribution with parameter k/n.

For further results related to the waiting time for all elements to appear r times *see* Foata, Han, and Lass [68], Foata and Zeilberger [69], and the references therein. They call it the *collector's brotherhood problem.* Foata, Han, and Lass [68] obtained the probability distribution generating function for random vector $(T, X_T^{(1)}, X_T^{(2)}, \ldots, X_T^{(r)})$, where $T := M_n$ is the waiting time for all elements of $\mathbb{N}_n$

to appear once, and $X_T^{(i)}$ is the number of the elements of set $\mathbb{N}_n$ that have occurred i times up to and including time T, $i \in \{1, \ldots, r\}$. When all elements of set $\mathbb{N}_n$ have occurred $p < r$ times, then some elements are still waiting to appear several times in order to obtain r full collections. Foata and Zeilberger [69] obtained several results related to generating functions of the probability distributions of the waiting times for these numbers of additional appearances.

Myers and Wilf [170] considered the problem when two collectors are simultaneously and independently seeking collections of all n elements of set $\mathbb{N}_n$. They determined the probability that the two collectors finish the game at the same trial. They also determined the probability that the two collectors are tied with each other for some initial number of trials, and after that the player who first gains the lead remains ahead throughout the game.

Xu and Tang [225] considered the following generalization of the coupon collector's problem. At each step a collector receives a d-subset of set $\mathbb{N}_n$ and chooses the least collected element (or one of the least collected elements) from this d-set. Let $D_{m,n}^{(d)}$ be the number of steps until every element of $\mathbb{N}_n$ is collected at least m times. For $d = 1$ we have the classical coupon collector's problem, and for $d = n$ there is no randomness, i.e., after mn steps every element is collected exactly m times, i.e., $D_{m,n}^{(n)} = mn$. Xu and Tang [225] obtained the lower and upper bounds for $E(D_{m,n}^{(d)})$:

$$\frac{\binom{n}{d}}{n^d} \cdot \frac{E(D_{m,n}^{(1)})}{d} \leqslant E(D_{m,n}^{(d)}) \leqslant \frac{E(D_{m,n}^{(1)})}{d} + mn\left(1 - \frac{1}{d}\right). \tag{4.3.197}$$

For fixed m and every $d > 1$, they also proved that the asymptotic behavior of the expected value of $D_{m,n}^{(d)}$ is given by the following equality:

$$\lim_{n\to\infty} \frac{E(D_{m,n}^{(d)}) - (n \ln n)/d}{n(m-1)(\ln \ln n)/d} = 1. \tag{4.3.198}$$

Note that the expression for $E(D_{m,n}^{(1)})$ is determined in Newman and Shepp [172], *see* also Theorem 4.3.4.

The following problem has also attracted considerable attention. At each step a collector receives an s-subset of set $\mathbb{N}_n$ and collects all s elements of this set. The assumption is that all $\binom{n}{s}$ s-subsets of $\mathbb{N}_n$ have the same probability to appear at each step. Let $W_{n,s,r}$ be the number of steps until all elements of $\mathbb{N}_n$ are collected at least r times. Among other results Stadje [214] obtained the probability distribution of random variable $W_{n,s,1}$. For each $k \geqslant 1$ and $j \in \mathbb{N}_n$, let A_{nj} be the event that after k steps element $j \in \mathbb{N}_n$ appears at most $r - 1$ times, and let

$$Z_{n,s,r} := \sum_{j=1}^{n} \mathbf{1}_{A_{nj}} \tag{4.3.199}$$

denote the number of elements of $\mathbb{N}_n$ that appear at most $r-1$ times after s steps, where $\mathbf{1}_{A_{nj}}$ is the indicator of event A_{nj}. For $x \in \mathbb{R}$ and $n > 1$ let us also denote

$$k_{n,s,r} := \left[\frac{n}{s}(x + \ln n + (r-1)\ln\ln n - \ln(r-1)!)\right], \tag{4.3.200}$$

where $[a]$ is notation for the greatest integer that is less than or equal to a. In the special case $r = 1$, it follows from Theorem 2 of Mikhailov [151] that for every $x \in \mathbb{R}$, the following asymptotic relation holds

$$Z_{n,s,1}(k_{n,s,1}(x)) \xrightarrow{d} Z \sim \mathcal{P}(e^{-x}), \quad \text{as } n \to \infty, \tag{4.3.201}$$

where $\mathcal{P}(\lambda)$, $\lambda > 0$, is notation for the Poisson distribution with parameter $\lambda > 0$, and $\xrightarrow{d}$ is the notation for convergence in distribution. Schilling and Henze [201] extended this result to the case $r > 1$. Namely they proved that

$$Z_{n,s,r}(k_{n,s,r}(x)) \xrightarrow{d} Z \sim \mathcal{P}(e^{-x}), \quad \text{as } n \to \infty. \tag{4.3.202}$$

Relation 4.3.202 implies that for each $x \in \mathbb{R}$,

$$\lim_{n\to\infty} P\left\{W_{n,s,r} \leqslant \frac{n}{s}\left(x+\ln n+(r-1)\ln\ln n-\ln(r-1)!\right)\right\} = e^{-e^{-x}}. \tag{4.3.203}$$

Relation 4.3.203 can also be considered an extension of Theorem 4.3.5 from the case $s = 1$ to the case $s > 1$.

As stated in Theorem 4.3.38, Glavaš and Mladenović [77] proved that point processes determined by the times of the first appearances of elements of $\mathbb{N}_n$ (properly centered and normalized) converge to a non-homogeneous Poisson point process as $n \to \infty$. Ilienko [114] extended this result to the case of the waiting time for all elements of $\mathbb{N}_n$ to appear r times, where $r \geqslant 2$, by using a poissonization idea from Holst [109].

The coupon collector's problem with unequal probabilities of sampling elements from set $\mathbb{N}_n$ was considered by several authors, *see* for example Anceaume, Busnel, and Sericola [9], Doumas and Papanicolaou [55], Neal [171], Schilling and Henze [201], and the references therein.

Mathematical models based on the coupon collector's problem have been used in very different areas. Kershenbaum, Freeberg, and Gammon [129] used the coupon collector model to estimate the vocal repertoire size in songbirds and in the communication of mammals with complex vocal communication systems. For an application of the coupon collector urn model with unequal probabilities in ecology and evolution, *see* Zoroa et al. [227].

4.4 The Polynomial Scheme

Let us consider a sequence of n independent trials with N possible outcomes E_1, ..., E_N. We suppose that every outcome has probability $1/N$. Let η_i be the number of occurrences of outcome E_i, and $\eta_{(i)}$ be the i-th order statistic for the sample $\eta_1, \ldots, \eta_N$, i.e., $\eta_{(1)} \leqslant \cdots \leqslant \eta_{(N)}$. In this section we formulate results related to the asymptotic behavior of extreme order statistics $\eta_{(N)}$ and $\eta_{(1)}$ that are obtained in Viktorova and Sevastyanov [230], Viktorova [229], and Kolchin [235]. *See* also Kolchin, Sevastyanov, and Chistyakov [237].

Let us denote $\alpha = n/N$, and

$$p_k = \frac{\alpha^k e^{-\alpha}}{k!}, \quad k = 0, 1, \ldots$$

As $n, N \to \infty$, the asymptotic distribution of $\eta_{(N)}$ depends on the asymptotic behavior of $\alpha/\ln N$ and Np_r. If $\alpha/\ln N \to 0$, then the support of the asymptotic distribution of $\eta_{(N)}$ consists of two nonnegative integers $r-1$ and r that are determined by the asymptotic relation $Np_r \to \lambda > 0$. If $\alpha/\ln N \to x > 0$, then the asymptotic distribution of $\eta_{(N)}$ is a discrete distribution with a countable support. If $\alpha/\ln N \to \infty$, then the asymptotic distribution of $\eta_{(N)}$ coincides with the Gumbel extreme value distribution. More precisely, the following theorems hold.

Theorem 4.4.1 (Viktorova and Sevastyanov [230]) *If $n, N \to \infty$, $\alpha/\ln N \to 0$, and $r = r(n, N)$ is chosen such that $r > \alpha$, and $Np_r \to \lambda$, where λ is a positive constant, then*

$$P\{\eta_{(N)} = r-1\} \to e^{-\lambda}, \quad P\{\eta_{(N)} = r\} \to 1 - e^{-\lambda}. \tag{4.4.1}$$

Theorem 4.4.2 (Viktorova and Sevastyanov [230]) *If $n, N \to \infty$, $\alpha/\ln N \to x$, and $r = r(n, N)$ is chosen such that $Np_r \to \lambda$, where x and λ are positive constants, then*

$$P\{\eta_{(N)} \leqslant r + k\} \to e^{-\lambda\gamma^{k+1}/(1-\gamma)}, \tag{4.4.2}$$

where $\gamma \in (0, 1)$ is a solution of the equation

$$\gamma + x(\ln\gamma - \gamma + 1) = 0. \tag{4.4.3}$$

Theorem 4.4.3 (Viktorova [229]) *If $n, N \to \infty$, $\alpha/\ln N \to \infty$, then*

$$P\left\{\frac{\eta_{(N)} - \alpha - \alpha u\left(\left(\ln N - \frac{1}{2}\ln\ln N\right)/\alpha\right)}{\sqrt{\alpha/(2\ln N)}} + \frac{1}{2}\ln 4\pi \leqslant z\right\} \to e^{-e^{-z}}, \tag{4.4.4}$$

where $u(w)$ is a positive function defined on the interval $0 < w < \infty$ by the following equation

$$-u + (1+u)\ln(1+u) = w. \tag{4.4.5}$$

The results from the papers by Viktorova and Sevastyanov [230] and Viktorova [229] were obtained using classical results related to the distributions of particles into cells. The method used in Kolchin [235] allows the reduction of the problem of the asymptotic behavior of extreme order statistics in polynomial schemes to corresponding problems for independent random variables. Results related to the asymptotic behavior of the minimal order statistic $\eta_{(1)}$ were proved in Kolchin [235].

Theorem 4.4.4 (Kolchin [235])
If $n, N \to \infty$ and $\alpha - \ln N \to -\infty$, then $P\{\eta_{(1)} = 0\} \to 1$.
Let us suppose that $n, N \to \infty$, and $\alpha = \ln N + r \ln\ln N + \ln r! + \gamma_N$, where r is a nonnegative integer. Then the following statements hold:
If $\gamma_N \to \gamma \in \mathbb{R}$, then

$$P\{\eta_{(1)} = r\} \to 1 - e^{-e^{-\gamma}}, \quad P\{\eta_{(1)} = r+1\} \to e^{-e^{-\gamma}}. \tag{4.4.6}$$

If $\gamma_N \to \infty$, and $\gamma_N - \ln\ln N \to -\infty$, then $P\{\eta_{(1)} = r+1\} \to 1$.
If $n, N \to \infty$, $\alpha/\ln N \to 1$, $(\alpha - \ln N)/\ln\ln N \to \infty$, and $r = r(n, N)$ is chosen such that $r < \alpha$, and $Np_r \to \lambda$, where λ is a positive constant, then

$$P\{\eta_{(1)} = r\} \to 1 - e^{-\lambda}, \quad P\{\eta_{(1)} = r+1\} \to e^{-\lambda}. \tag{4.4.7}$$

Theorem 4.4.5 (Kolchin [235]) *If $n, N \to \infty$, $\alpha/\ln N \to x > 1$, and $r = r(n, N)$ is chosen such that $r < \alpha$, and $Np_r \to \lambda$, where λ is a positive constant, then*

$$P\{\eta_{(1)} \leqslant r + k\} \to 1 - e^{-\lambda\beta^{k+1}/(\beta-1)}, \tag{4.4.8}$$

where β is the solution of the equation $\beta + x(\ln\beta - \beta + 1) = 0$, such that $1 < \beta < \infty$.

Theorem 4.4.6 (Kolchin [235]) *If $n, N \to \infty$, $\alpha/\ln N \to \infty$, then*

$$P\left\{\frac{\eta_{(1)} - \alpha + \alpha u\left(\left(\ln N - \frac{1}{2}\ln\ln N\right)/\alpha\right)}{\sqrt{\alpha/(2\ln N)}} - \frac{1}{2}\ln 4\pi \leqslant -z\right\} \to 1 - e^{-e^{-z}}, \tag{4.4.9}$$

where $u(w)$ is a function defined on the interval $0 \leqslant w < \infty$ by (4.4.5).

The asymptotic distributions of order statistics $\eta_{(N-m+1)}$ and $\eta_{(m)}$ were also obtained for every fixed m under conditions $n, N \to \infty$, and $\alpha/\ln N \to \infty$. We state here these results as they are given in Kolchin [235]. The normalizing constants for $\eta_{(N-m+1)}$ and $\eta_{(m)}$ are the same as in Theorems 4.4.3 and 4.4.6 that give the asymptotic distributions of extreme order statistics $\eta_{(N)}$ and $\eta_{(1)}$, respectively.

If $n, N \to \infty$, and $\alpha/\ln N \to \infty$, then we have

$$P\left\{\frac{\eta_{(N-m+1)} - \alpha - \alpha u\left(\left(\ln N - \frac{1}{2}\ln\ln N\right)/\alpha\right)}{\sqrt{\alpha/(2\ln N)}} + \frac{1}{2}\ln 4\pi \leqslant z\right\}$$

$$\to \sum_{k=0}^{m-1} \frac{e^{-kz}}{k!} e^{-e^{-z}}, \tag{4.4.10}$$

$$P\left\{\frac{\eta_{(m)} - \alpha + \alpha u\left(\left(\ln N - \frac{1}{2}\ln\ln N\right)/\alpha\right)}{\sqrt{\alpha/(2\ln N)}} - \frac{1}{2}\ln 4\pi \leqslant -z\right\}$$

$$\to 1 - \sum_{k=0}^{m-1} \frac{e^{-kz}}{k!} e^{-e^{-z}}, \tag{4.4.11}$$

where $u(w)$ is a function defined for $w \in [0, \infty)$ by equation (4.4.5).

4.5 Random Trees and Random Forests

4.5.1 The Height of a Random Tree

In this subsection we consider the set of trees with $m + 1$ different vertices, denoted by $0, 1, \ldots, m$, where 0 is considered to be the root. According to Cayley's theorem, the total number of such trees is equal to $(m + 1)^{m-1}$. Let us also assume that each tree has probability $(m+1)^{-(m-1)}$. In every tree each vertex v is connected with the root 0 by a unique path. The number of edges in such a path is called the *height of vertex v*. By convention, the height of the root is equal to 0. Let K_j be the set of all vertices with height equal to j in a randomly chosen tree. Let $\zeta_{m,j}$ be the number of vertices in set K_j, and Y_m be the height of the tree, i.e.,

$$Y_m = \max\{j \mid \zeta_{m,j} \neq 0\}. \tag{4.5.1}$$

We shall formulate here the results related to the asymptotic behavior of random variables $\zeta_{m,j}$ and Y_m

Theorem 4.5.1 (Stepanov [242]) *Let us assume that $m \to \infty$ and $j/\sqrt{m} \to \alpha$, where $0 < \alpha_1 \leqslant \alpha \leqslant \alpha_2 < \infty$. Then, random variable $\zeta_{m,j}/\sqrt{m}$ converges in distribution to a random variable whose distribution is uniquely determined by its moments a_r, $r \geqslant 1$.*

The expressions for moments a_r, $r \geqslant 1$, and the density of the limiting distribution are also provided in Stepanov [242].

Theorem 4.5.2 (Stepanov [242]) *The limiting distribution of random variable Y_n is given by the following equality*

$$\lim_{m\to\infty} P\left\{\frac{Y_m}{\sqrt{m}} \leqslant \alpha\right\} = \sum_{k=-\infty}^{\infty} (1-\alpha^2 k^2) e^{-\alpha^2 k^2/2}. \tag{4.5.2}$$

4.5.2 Random Forests

In this and the next two subsections we consider random forests and some of their extremal characteristics. Random forests may appear in connection with random mappings on finite sets, in connection with birth-death processes, in the mathematical modeling of telecommunications and multiservice networks, etc.

Let $\mathfrak{M}(n, N)$ be the set of all random forests containing $N+n$ vertices, of which N vertices are roots labeled $1, 2, \ldots, N$. The remaining n vertices that are not roots are also labeled $1, 2, \ldots, n$. Two forests are considered to be different if they contain at least one pair of different trees with roots labeled the same positive integer. According to Cayley's theorem the number of different trees with $k+1$ vertices of which one is the root is equal to $(k+1)^{k-1}$, *see* Moon [168]. Consequently, see Pavlov [238], the number of forests in $\mathfrak{M}(n, N)$ is given by

$$n! \sum_{k_1+\cdots+k_N=n,\, k_i \geqslant 0} \prod_{i=1}^{N} \frac{(k_i+1)^{k_i}}{(k_i+1)!} = N(N+n)^{n-1}. \tag{4.5.3}$$

Let us define the uniform probability distribution on set $\mathfrak{M}(n, N)$, i.e., we assume that every forest from this set has probability $N^{-1}(N+n)^{-n+1}$. Then different numerical characteristics of the forests from set $\mathfrak{M}(n, N)$ become random variables.

4.5.3 Maximum Tree Size in a Random Forest

Let us choose a random forest from set $\mathfrak{M}(n, N)$. Let ν_i be the size of a tree with root i, i.e., the number of vertices in this tree that are different from root i, $1 \leqslant i \leqslant N$, and let $\nu_{(1)} \leqslant \nu_{(2)} \leqslant \cdots \leqslant \nu_{(N)}$ be the corresponding order statistics. Results related to the asymptotic behavior of random variable $\nu_{(N)}$ as $n, N \to \infty$ were obtained in Pavlov [238]. *See* also Pavlov [239]. These results will be formulated under additional conditions related to the relation between n and N.

Let us first introduce the following notions. Let $\xi_1, \xi_2, \ldots, \xi_N$ be independent random variables with the same distribution given by

$$p_k = P\{\xi_1 = k\} = \frac{(k+1)^k}{(k+1)!} x^k e^{-\theta(x)}, \quad k \in \{0, 1, 2, \ldots\}, \tag{4.5.4}$$

where

$$x = \frac{n}{N+n} e^{-n/(N+n)}, \quad \theta(z) = \sum_{s=1}^{\infty} \frac{s^{s-1}}{s!} z^s. \tag{4.5.5}$$

Function $\theta(z)$ is defined for $|z| \leqslant 1/e$ and is moreover the solution of the equation $\theta(z)e^{-\theta(z)} = z$. Consequently, $\theta(x) = n/(N+n)$.

The joint distribution of sizes $\nu_1, \ldots, \nu_N$, *see* Pavlov [238], is given by

$$P\{\nu_1 = k_1, \ldots, \nu_N = k_N\} = P\{\xi_1 = k_1, \ldots, \xi_N = k_N \mid \xi_1 + \cdots + \xi_N = n\}.$$

Theorem 4.5.3 (Pavlov [238]) *Let us suppose that $n, N \to \infty$, and $n/N \to 0$. Let $r = r(n, N)$ be given such that $Np_r \to \infty$, $Np_{r+1} \to \lambda$, where λ is a nonnegative constant. Then we have*

$$P\{\nu_{(N)} = r\} \to e^{-\lambda}, \quad P\{\nu_{(N)} = r+1\} \to 1 - e^{-\lambda}. \tag{4.5.6}$$

Theorem 4.5.4 (Pavlov [238]) *Let us suppose that $n, N \to \infty$, and $n/N \to \gamma$, where γ is a positive constant. Let $r = r(n, N)$ be given such that $Np_r \to \lambda$, where λ is a positive constant. Then we have*

$$P\{\nu_{(N)} \leqslant r + k\} \to e^{-\lambda b^{k+1}(1-b)^{-1}}, \quad k \in \mathbb{Z}, \tag{4.5.7}$$

where $\mathbb{Z}$ is the set of integers and $b = \dfrac{\gamma}{1+\gamma} e^{1/(1+\gamma)}$.

Theorem 4.5.5 (Pavlov [238]) *Let us suppose that $n, N \to \infty$, $n/N \to \infty$, and $n/N^2 \to 0$. Then we have*

$$P\{\beta \cdot \nu_{(N)} - u \leqslant z\} \to e^{-e^{-z}}, \tag{4.5.8}$$

where

$$\beta = \beta(n, N) = -\ln\left(\frac{n}{N+n} e^{N/(N+n)}\right), \tag{4.5.9}$$

and $u = u(n, N)$ is chosen such that

$$\frac{N\beta^{1/2}}{u^{3/2}e^{u}} = \sqrt{2\pi}. \tag{4.5.10}$$

Theorem 4.5.6 (Pavlov [238]) *Let us suppose that $n, N \to \infty$, and $n/N^2 \to 1/a$, where a is a positive constant. Then*

$$P\left\{\frac{\nu(N)}{n} \leqslant z\right\} \to \tag{4.5.11}$$

$$a^{3/2}e^{a/2}\sum_{s=0}^{\infty}\frac{(-1)^s}{s!}\int_X \frac{a^s \exp\{-a^2/2(a-x_1-\cdots-x_s)\}\,dx_1\ldots dx_s}{(2\pi)^{s/2}\big(x_1\ldots x_s(a-x_1-\cdots-x_s)\big)^{3/2}},$$

where $X = X(s,z,a) = \{x_i \geqslant az,\ i = 1,\ldots,s,\ \sum_i x_i \leqslant a\}$, and for $s = 0$ the integral in formula of (4.5.11) *is considered to be equal to $a^{-3/2}e^{-a/2}$.*

4.5.4 The Height of a Random Forest

Again, let $\mathfrak{M}(n, N)$ be a set of random forests and let us define the uniform probability distribution on it. Every vertex $v \in \{1, 2, \ldots, n\}$ that is not a root is connected with exactly one root by a uniquely determined path. The number of edges of this path will be referred to as the height of vertex v. The height of a tree is the maximal height of the vertices that belong to this tree. The height of a forest, denoted by Y, is the maximal height of the trees in this forest.

Theorem 4.5.7 (Pavlov [239]) *Let us suppose that $n \to \infty$, $n/N^2 \to \infty$. Then for every fixed $x > 0$ the following relation holds:*

$$P\left\{\frac{Y}{\sqrt{n}} \leqslant x\right\} \to \sum_{k=-\infty}^{\infty}(1-k^2x^2)e^{-k^2x^2/2}. \tag{4.5.12}$$

Theorem 4.5.8 (Pavlov [239]) *Let us suppose that $n, N \to \infty$ such that $N/\sqrt{n} \to z$. Then uniformly related to z in the area $0 < z_0 \leqslant z \leqslant z_1 < +\infty$, and for every fixed $x > 0$, the following relation holds*

$$P\left\{\frac{Y}{\sqrt{n}} \leqslant x\right\} \to \frac{e^{z^2/2}}{z\sqrt{2\pi}}\int_{-\infty}^{\infty}\exp\left\{-iu - \frac{zf(x,u)}{x}\right\}du, \tag{4.5.13}$$

where

$$f(x,u) = \frac{x\sqrt{-2iu}\,\big(1+\exp(-x\sqrt{-2iu})\big)}{1-\exp(-x\sqrt{-2iu})}. \tag{4.5.14}$$

Theorem 4.5.9 (Pavlov [239]) *Let us suppose that n, $N \to \infty$ such that $n/N \to \infty$, and $n/N^2 \to 0$. Then for every fixed $x \in \mathbb{R}$ the following relation holds:*

$$P\left\{Y \cdot \ln\left(1 + \frac{N}{n}\right) - \ln\left(\frac{2N^2}{n}\right) \leqslant x\right\} \to e^{-e^{-x}}. \tag{4.5.15}$$

Let $F(s) = e^{\lambda(s-1)}$. For every positive integer r let us define the rth iteration $F^{*r}(s)$ of function $F(s)$, i.e., $F^{*1}(s) = F(s)$, $F^{*2}(s) = F(F(s))$, $F^{*3}(s) = F(F(F(s)))$, etc. The following theorem holds:

Theorem 4.5.10 (Pavlov [239]) *Let us suppose that $n, N \to \infty$ such that $n/N \to b$, where b is a positive constant, and let $t = t(n, N)$ be given such that $N\lambda^t \to \beta$, where β is a positive constant, and $\lambda = n/(N+n)$. Then, we have*

$$P\{Y < t + m\} \to e^{-K\beta(b/(1+b))^m}, \quad m \in \mathbb{Z}, \tag{4.5.16}$$

where constant K is given by

$$K = \lim_{r\to\infty} \frac{1 - F^{*r}(0)}{\lambda^r}, \tag{4.5.17}$$

Theorem 4.5.11 (Pavlov [239]) *Let us suppose that $n, N \to \infty$ such that $n/N \to 0$. Let $t = t(n, N)$ be given such that $N\lambda^t \to \infty$, $N\lambda^{t+1} \to \gamma$, where γ is a nonnegative constant, and $\lambda = n/(N+n)$. Then*

$$P\{Y = t\} \to e^{-\gamma}, \quad P\{Y = t+1\} \to 1 - e^{-\gamma}. \tag{4.5.18}$$

4.6 Random Partitions of Finite Sets

4.6.1 Partitions of Finite Sets

In this section we shall formulate some results related to the partitions and random partitions of finite sets.

Definition 4.6.1 A *partition* of a set S into k *blocks* is a collection of sets $\{S_1, S_2, \ldots, S_k\}$ such that:

(a) $S_j \neq \varnothing$, for every $j \in \{1, 2, \ldots, k\}$.
(b) $S_i \cap S_j = \varnothing$, for all $i \neq j$, $i, j \in \{1, 2, \ldots, k\}$.
(c) $S = S_1 \cup S_2 \cup \cdots \cup S_k$.

The number of elements of any block S_j is called the *size* of this block. Every partition $\pi = \{S_1, S_2, \ldots, S_k\}$ of set S into k blocks generates $k!$ ordered partitions of set S into k blocks. Any ordered partition is a permutation of the blocks of partition π.

Let $\alpha_1, \alpha_2, \ldots, \alpha_k$ be positive integers such that $\alpha_1 + \alpha_2 + \cdots + \alpha_k = n$, and let us introduce the following notation:

- $B_n(\alpha_1, \alpha_2, \ldots, \alpha_k)$—the number of ordered partitions $(S_1, S_2, \ldots, S_k)$ of an n-set S into k blocks, such that for each $j \in \{1, 2, \ldots, k\}$, $|S_j| = \alpha_j$.
- B_{nk}—the number of partitions of an n-set into k blocks.
- B_n—the number of all partitions of an n-set.

The numbers B_{nk} and B_n are defined for all $n, k \in \mathbb{N}$. It is obvious that $B_{nk} = 0$ for $k > n$. By definition $B_{00} = B_0 = 1$. The numbers B_{nk}, $n, k \in \mathbb{N}_0$ are called *Stirling numbers of the second kind.* The number B_n is called the nth *Bell number.*

The following formulas hold for Bell numbers:

$$B_n = \frac{1}{e} \sum_{k=0}^{\infty} \frac{k^n}{k!}, \tag{4.6.1}$$

$$\sum_{n=0}^{\infty} \frac{B_n x^n}{n!} = \exp(e^x - 1). \tag{4.6.2}$$

The asymptotic behavior of B_n as $n \to \infty$ is given by

$$B_n = \frac{1}{\sqrt{r+1}} \exp\left\{n\left(r + \frac{1}{r} - 1\right) - 1\right\}(1 + o(1)), \tag{4.6.3}$$

where r is the unique solution as $n \to \infty$ of the equation $re^r = n$. The asymptotic behavior of r as $n \to \infty$ is given by the following formula:

$$r = \ln n - \ln \ln n + \sum_{k=0}^{\infty} \sum_{l=0}^{\infty} c_{kl} \frac{(\ln \ln n)^{l+1}}{(\ln n)^{k+l+1}}. \tag{4.6.4}$$

More about partitions of finite sets and the properties of Stirling numbers of the second kind and Bell numbers can be found in Andrews [11], Moser and Wyman [169], Harper [103], Graham, Knuth and Patashnik [83], Sachkov [241], Mladenović [164], and Spivey [213].

4.6.2 Random Partitions

Let S be an n-set, and let us suppose that every partition of set S has probability $1/B_n$. Let X_n be the number of blocks of a random partition of set S. The probability distribution of random variable X_n is given by

$$P\{X_n = k\} = \frac{B_{nk}}{B_n}, \quad k \in \{0, 1, \ldots, n\}. \tag{4.6.5}$$

The mathematical expectation and variance of random variable X_n are given by the following equalities:

$$E(X_n) = \frac{B_{n+1}}{B_n} - 1, \tag{4.6.6}$$

$$\mathrm{Var}(X_n) = \frac{B_{n+2}}{B_n} - \left(\frac{B_{n+1}}{B_n}\right)^2 - 1, \tag{4.6.7}$$

and their asymptotic behavior as $n \to \infty$, *see* Sachkov [198], is given by

$$E(X_n) \sim \frac{n}{r} \sim \frac{n}{\ln n}, \quad \mathrm{Var}(X_n) \sim \frac{n}{r^2} \sim \frac{n}{(\ln n)^2}, \tag{4.6.8}$$

where $re^r = n$ and random variable $(\mathrm{Var}\, X_n)^{-1/2} \cdot (X_n - EX_n)$ as $n \to \infty$ has asymptotically normal $\mathcal{N}(0, 1)$ distribution.

Let M_n and m_n be respectively the maximal and minimal block sizes of a random partition of an n set. Let us also denote by $\nu_n(l)$ the number of blocks of size l in the random partition of an n-set. The following asymptotic results hold:

Theorem 4.6.2 $\lim_{n\to\infty} P\{\nu_n(l) > 0\} = 1$ *for every fixed* $l \in \mathbb{N}$.

Theorem 4.6.3 $\lim_{n\to\infty} P\{m_n = 1\} = 1$.

Theorem 4.6.4 *As* $n \to \infty$ *uniformly for all* x *in every bounded interval, the following relation holds:*

$$P\{M_n - [er - \ln\sqrt{2\pi er} - \ln(e-1)] \leqslant x\} - e^{-e^{-[x]+\delta}} \to 0, \tag{4.6.9}$$

where $\delta = \{er - \ln\sqrt{2\pi er} - \ln(e-1)\}$, $re^r = n$, $[a]$ *is notation for the integer part of* a, *and* $\{a\} = a - [a]$ *is the fractional part of* a.

Theorem 4.6.3 is an immediate consequence of Theorem 4.6.2. The proofs of Theorems 4.6.2 and 4.6.4 can be found in Sachkov [198]. The distribution of random variable M_n is concentrated in the neighborhood of point $[er - \log\sqrt{2\pi er} - \log(e - 1)]$, and in this domain it is close to the Gumbel extreme value distribution.

4.7 Geometric Properties of a Sample of Random Vectors

Sequences of random vectors have also been investigated extensively in probabilistic and statistical literature. A sample from such a sequence can be considered the finite set of points in the corresponding space, for example, $\mathbb{R}^d$ for some $d \geqslant 2$, the unit ball in $\mathbb{R}^d$ or the unit sphere surface. Some geometric characteristics of such a finite set may be of particular interest, for example, diameter, convex hull, etc. In this section we present some asymptotic results related to these characteristics.

4.7.1 Spherically Symmetric Distributions

Let $\boldsymbol{x} \in \mathbb{R}^d$ and Q be an $d \times d$ orthogonal matrix. Then, $||\boldsymbol{x}|| = ||Q\boldsymbol{x}||$, and hence the points $\boldsymbol{x}$ and $Q\boldsymbol{x}$ lie on the same sphere centered at the origin. A nonempty set $A \subset \mathbb{R}^d$ is *spherically symmetric* if $\boldsymbol{x} \in A$ implies that $Q\boldsymbol{x} \in A$ for each orthogonal $d \times d$ matrix Q. An n-dimensional random vector $\boldsymbol{X}$ has a *spherically symmetric distribution* if for every orthogonal $d \times d$ matrix Q vectors $\boldsymbol{X}$ and $Q\boldsymbol{X}$ have the same distribution.

Examples of spherically symmetric distributions include the multivariate standard normal distribution, the multivariate extension of t-distribution, uniform distribution inside the unit ball, uniform distribution on the unit sphere surface, and others. More about spherically symmetric distributions and testing for spherical symmetry can be found in Chmielewski [45], Koltchinskii and Li [134], and Henze, Hlávka, and Meintanis [105].

4.7.2 Diameter of a Random Sample in the Unit Ball

The distribution of the distance between random points in a hypersphere was investigated in Hammersley [99], Lord [144], and Alagar [3].

Mayer and Molchanov [149] investigated the asymptotic behavior of the maximum interpoint distance for a random sample of n i.i.d. points in the unit d-dimensional ball, where $d \geqslant 2$, and when the points have spherically symmetric distribution. Note that the maximum interpoint distance for a random sample can be considered the diameter of the set of points determined by this random sample.

Theorem 4.7.1 (Mayer and Molchanov [149]) *Let $diam P_n$ be the diameter of the convex hull P_n of n independent points that are distributed uniformly in the d-dimensional unit ball B, where $d \geqslant 2$. The limit distribution of $diam P_n$ is given by*

$$\lim_{n\to\infty} P\left\{n^{4/(d+3)}(2 - diam P_n) \leqslant t\right\}$$
$$= 1 - \exp\left\{-\frac{2^d d\Gamma\left(\frac{d}{2}+1\right)}{\sqrt{\pi}(d+1)(d+3)\Gamma\left(\frac{d+1}{2}\right)} t^{(d+3)/2}\right\}, \tag{4.7.1}$$

where $t > 0$ and $\Gamma(x) = \int_0^\infty s^{x-1}e^{-s}\,ds$ is the gamma function.

4.7.3 Diameter in a Sample of Vectors with Spherically Symmetric Distribution in $\mathbb{R}^n$

Jammalamadaka and Janson [116] investigated the maximum interpoint distance in a sample of random vectors with spherically symmetric distribution in $\mathbb{R}^d$, where $d \geqslant 2$. Special cases of this problem have previously been studied by several authors, *see*, for example, Matthews and Rukhin [148] (they considered symmetric normal distribution), Henze and Klein [106] (Kotz distribution), Appel, Najim and Russo [13] (uniform distribution in a ball), Appel and Russo [12] (uniform distribution on a sphere), Mayer and Molchanov [149] (uniform distribution in a ball or on a sphere), and the references therein. Note also that Jammalamadaka and Janson [115] investigated the minimum interpoint distance.

Let $(X_n)_{n\geqslant 1}$ be a sequence of i.i.d. random vectors in $\mathbb{R}^d$ with a spherically symmetric distribution, where $d \geqslant 2$. Let us denote

$$M_n = \max_{1\leqslant i\leqslant n} |X_i|, \tag{4.7.2}$$

$$M_n^{(2)} = \max_{1\leqslant i<j\leqslant n} |X_i - X_j|, \tag{4.7.3}$$

where $|\cdot|$ is the Euclidean distance.

Theorem 4.7.2 (Jammalamadaka and Janson [116]) *Let us suppose that $d \geqslant 2$ and that $X, X_1, X_2, \ldots$ are i.i.d. random vectors in $\mathbb{R}^d$ with a spherically symmetric distribution such that for some sequences $(a_n)_{n\geqslant 1}$ and $(b_n)_{n\geqslant 1}$ of positive numbers with $b_n = o(a_n)$,*

$$P\{|X| > a_n + tb_n\} = \frac{1+o(1)}{n} e^{-t}, \tag{4.7.4}$$

as $n \to \infty$, for all $t = t_n$ with $|t| \leqslant \frac{d-1}{2} \ln \frac{a_n}{b_n}$. Let us denote

$$c_d := \frac{(d-1)2^{d-4}}{\sqrt{\pi}} \Gamma\left(\frac{d}{2}\right). \tag{4.7.5}$$

Then, the limiting distribution of $M_n^{(2)}$ is given by

$$\lim_{n\to\infty} P\left\{\frac{M_n^{(2)} - 2a_n}{b_n} + \frac{d-1}{2} \ln \frac{a_n}{b_n} - \ln\ln \frac{a_n}{b_n} - \ln c_d \leqslant x\right\} = e^{-e^{-x}}. \tag{4.7.6}$$

Hence, under the conditions stated in Theorem 4.7.2 the properly centered and normalized maximum $M_n^{(2)}$ has the Gumbel extreme value distribution asymptotically as $n \to \infty$.

Remark 4.7.3 The assumption $b_n = o(a_n)$, i.e., $\ln(a_n/b_n) \to \infty$, as $n \to \infty$, implies that (4.7.4) holds for every fixed t, and consequently

$$\lim_{n\to\infty} P\left\{\frac{M_n - a_n}{b_n} \leqslant x\right\} = e^{-e^{-x}}. \tag{4.7.7}$$

Moreover, relations (4.7.6) and (4.7.7) imply that

$$\frac{2M_n - M_n^{(2)}}{b_n \ln(a_n/b_n)} \xrightarrow{P} \frac{d-1}{2}. \tag{4.7.8}$$

Jammalamadaka and Janson [116] also provided two special versions of the main results given by Theorem 4.7.2, and examples that cover some special spherically symmetric distributions. The first of these examples is related to the standard multivariate normal distribution that was previously considered by Matthews and Rukhin [148], and Henze and Klein [106].

Example 4.7.4 Let us suppose that $(X_i)_{i\geqslant 1}$ is a sequence of i.i.d. random vectors with a standard multivariate normal distribution in $\mathbb{R}^d$, where $d \geqslant 2$. The density function is given by $f(\boldsymbol{x}) = (2\pi)^{-d/2} e^{-|\boldsymbol{x}|^2/2}$. Then the random variable

$$\sqrt{2\ln n}M_n^{(2)} - \left(4\ln n + \frac{d-3}{2}\ln\ln n + \ln\ln\ln n + \ln\frac{(d-1)2^{(d-7)/2}}{\sqrt{\pi}\Gamma(d/2)}\right)$$

has the standard Gumbel extreme value distribution asymptotically. △

Example 4.7.5 (Henze and Klein [106]) Let us suppose that $(X_i)_{i\geqslant 1}$ is a sequence of i.i.d. random vectors that have a symmetric Kotz-type distribution in $\mathbb{R}^d$, where $d \geqslant 2$, with density

$$f(\boldsymbol{x}) = \frac{\kappa^{d/2+b-1}\Gamma(d/2)}{\pi^{d/2}\Gamma(d/2+b-1)} |\boldsymbol{x}|^{2(b-1)} e^{-\kappa|\boldsymbol{x}|^2},$$

where $b \in \mathbb{R}$ and $\kappa > 0$. Then the random variable

$$\sqrt{4\kappa\ln n}M_n^{(2)} - \left(4\ln n + \frac{4b+d-7}{2}\ln\ln n + \ln\ln\ln n + \ln\frac{(d-1)2^{(d-7)/2}\Gamma(d/2)}{\sqrt{\pi}\,\Gamma^2(d/2+b-1)}\right)$$

has the Gumbel extreme value distribution asymptotically as $n \to \infty$. △

4.7.4 *The Distance Between a Convex Set and an Interior Random Convex Hull*

The convex hull of a set of random points in $\mathbb{R}^d$, $d \geqslant 2$, has been investigated by many authors, *see*, for example, Bräker and Hsing [32], Bräker, Hsing, and Bingham [33], Hsing [111], Dümbgen and Walter [56], and the references therein. Bräker and Hsing [32] considered the area and perimeter of a random convex hull in a bounded convex set. Bräker, Hsing, and Bingham [33] obtained the limiting distribution of the Hausdorff distance $\Delta_n = \Delta(K, K_n)$ between a convex set K and the convex hull of a sample of random points from K. Here we will provide one of the results from Bräker, Hsing, and Bingham [33], where the Gumbel extreme value distribution appears.

There is a biological motivation for considering the convex hull of points, for example in $\mathbb{R}^2$. The set $K \subset \mathbb{R}^2$ we are interested in is the territorial range of an animal and the points from K that form the sample are the positions $X_1, X_2, \dots, X_n \in K$ of the animal (tagged with a radio transmitter) that are recorded during some period of time. Then the convex hull K_n of points $X_1, X_2, \dots, X_n$ is an *approximation* of set K. We are interested in the asymptotic distribution of the Hausdorff distance $\Delta_n = \Delta(K, K_n)$, as $n \to \infty$. Recall that the *Hausdorff metric* is given by

$$\Delta(A, B) := \inf\{\varepsilon > 0 \mid A \subset B_\varepsilon, B \subset A_\varepsilon\},$$

where A_ε is notation for the ε-neighborhood of set A.

More formally, let $K \subset \mathbb{R}^2$ be a bounded convex set and assume that the boundary ∂K has length L and parametrization

$$t \mapsto \boldsymbol{c}(t), \quad \text{where } t = \text{length from } \boldsymbol{c}(0) \text{ to } \boldsymbol{c}(t).$$

We assume that the parametrization is positively oriented, i.e., counterclockwise. We also assume that the curvature $\kappa(t) := |\ddot{\mathbf{c}}(t)|$ is well-defined and bounded away from 0 and ∞, and that κ has a bounded derivative. Let us define the tangent and normal vectors at $\boldsymbol{c}(t)$ as follows:

$$\boldsymbol{e}_1(t) = \dot{\mathbf{c}}(t), \quad \boldsymbol{e}_2(t) = (\kappa(t))^{-1}\ddot{\mathbf{c}}(t).$$

Let $\boldsymbol{X}_1, \boldsymbol{X}_2, \dots, \boldsymbol{X}_n$ be points that are randomly and independently sampled from K. In fact, we assume that $\boldsymbol{X}_1, \boldsymbol{X}_2, \dots, \boldsymbol{X}_n$ are i.i.d. random vectors with density $f(\boldsymbol{x})$, $\boldsymbol{x} \in K$. Additionally, we assume that $f(\boldsymbol{x})$ decays at a polynomial rate as $\boldsymbol{x}$ tends to ∂K, i.e.,

$$\frac{f(\boldsymbol{c}(t) + h\boldsymbol{e}_2(t))}{h^\alpha} \to g(t), \quad h \downarrow 0,$$

uniformly in t, for some $\alpha > -1$, where $g(t)$ is a continuous function bounded away for 0 and ∞. The case when $\alpha = 0$ corresponds to the uniform distribution of X_i over K. Let us introduce the following notation

$$\gamma = 1(3+2\alpha),$$
$$\lambda(t) = \frac{\sqrt{\kappa(t)}}{g(t)}, \quad 0 \leqslant t < L, \quad \lambda_0 := \max \lambda(t), \quad \Lambda_0 := \operatorname{argmax} \lambda(t).$$

Theorem 4.7.6 (Bräker, Hsing, and Bingham [33]) *Let us suppose that*

$$\lim_{n\to\infty} (\ln n)^{\nu_n} \int_0^L \exp\left\{ - \gamma_n \left(\frac{\lambda_0}{\lambda(t)} - 1 \right) \ln n \right\} g(t)\, dt = \mu \in (0, \infty),$$

where $(\nu_n)_{n\geqslant 1}$ is a bounded sequence of nonnegative constants, and

$$\gamma_n = \left(\gamma + (1 - \gamma - \nu_n) \frac{\ln \ln n}{\ln n} \right).$$

Then for every $x \in \mathbb{R}$ the following asymptotic relation holds

$$\lim_{n\to\infty} P\{\Delta_n \leqslant \tau_n(x)\} = e^{-d_1 e^{-d_2 x}},$$

where $\tau_n(x) = n^{-2\gamma}((c_n \ln n)^{2\gamma} + x(\ln n)^{2\gamma-1})$,

$$c_n = \frac{c_0 \gamma_n}{\gamma}, \quad d_1 = \frac{c_0^{1-\gamma} \mu}{1+\alpha}, \quad d_2 = \frac{1}{2c_0^{2\gamma}}, \quad c_0 = \frac{\lambda_0}{2\sqrt{2}\, B(1/2, \alpha+1)},$$

and $B(a, b) = \int_0^1 x^{a-1}(1-x)^{b-1}\, dx$ for $a, b > 0$.

Remark 4.7.7 If we denote $x = (y + \ln d_1)/d_2$, then the statement of Theorem 4.7.6 is equivalent to

$$\lim_{n\to\infty} P\{\Delta_n \leqslant a_n y + b_n\} = e^{-e^{-y}}, \quad y \in \mathbb{R},$$

where constants a_n and b_n are given by

$$a_n = \frac{2c_0^{2\gamma}}{n^{2\gamma}(\ln n)^{1-2\gamma}},$$
$$b_n = b^{(1)} \left(\frac{\ln n}{n} \right)^{2\gamma} + b^{(2)} \frac{\ln \ln n}{n^{2\gamma}(\ln n)^{1-2\gamma}} + b^{(3)} \frac{1}{n^{2\gamma}(\ln n)^{1-2\gamma}},$$

with

$$b^{(1)} = c_0^{2\gamma}, \quad b^{(2)} = 2c_0^{2\gamma}\frac{1-\gamma}{\lambda_0}, \quad b^{(3)} = 2c_0^{2\gamma}\ln\left(\frac{1}{\alpha+1}c_0^{1-\gamma}\int_{\Lambda_0} g(t)\,dt\right).$$

Exercises and Supplements

Random Permutations
In the following problems we use notation $\mathbb{N}_n = \{1, 2, \ldots, n\}$, and Ω_n is the set of all permutations of $\mathbb{N}_n$.

4.1 Let $n = mk + 1$ be a positive integer. Prove that every permutation of $\mathbb{N}_n$ contains an increasing subsequence of length $m + 1$ or contains a decreasing subsequence of length $k + 1$. If $n = k^2 + 1$ every permutation of set $\mathbb{N}_n$ contains a monotone subsequence of length $k + 1$.

4.2

(a) Let $\omega \in \Omega_n$ be a permutation in the second canonical representation. Let $f(\omega)$ be the permutation in one line representation obtained from ω by deleting parentheses. Prove that $f : \Omega_n \to \Omega_n$ is a bijection.
(b) Let i and j be two distinct elements of $\mathbb{N}_n$, and let $\omega \in \Omega_n$ be a randomly chosen permutation of $\mathbb{N}_n$. Prove that the probability that i and j belong to the same cycle of ω is 1/2.

4.3 Let $i, k \in \mathbb{N}_n$. Prove that the probability that element i is included in a cycle of length k is $1/n$.
Hint Let us consider a permutation $\omega = a_1a_2\ldots a_n$ in the second canonical representation, choose i to be equal to n, and suppose that n is included in a cycle of length k. Then in the second canonical representation, $a_{n-k+1} = n$. The number of permutations with $a_{n-k+1} = n$ is $(n-1)!$.

4.4

(a) Let us suppose that k is a positive integer such that $n/2 < k \leqslant n$. What is the probability that a randomly chosen permutation of set $\mathbb{N}_n$ has a cycle of length k?
(b) What is the probability that a randomly chosen permutation of set $\mathbb{N}_n$ has a cycle of length greater that $n/2$?

4.5 Prove Theorem 4.2.11.
Hint Random variables $X_{n1}, \ldots, X_{nn}$ satisfy Lindeberg's condition.

4.6 Prove Theorems 4.2.12–4.2.15.

Random Functions
Let Ω_n be the set of all functions $f : \mathbb{N}_n \to \mathbb{N}_k$, where $\mathbb{N}_n = \{1, 2, \ldots, n\}$ and $k = k(n)$. Let us suppose that every function $f \in \Omega_n$ has probability k^{-n}. Let X_{nj}, $1 \leqslant j \leqslant n$, be random variables that are defined as follows: for every $f \in \Omega_n$ and

$j \in \mathbb{N}_n$,

$$X_{nj}(f) = |f(j) - f(j+1)|, \quad \text{where } f(n+1) = f(1).$$

Let us also define: $M_n = \max\{X_{n1}, \dots, X_{nn}\}$, $m_n = \min\{X_{n1}, \dots, X_{nn}\}$. In the next three exercises we formulate results related to random variables X_{nj}, $1 \leqslant j \leqslant n$, M_n and m_n, *see* Mladenović [160].

4.7 Prove that for every $j \in \mathbb{N}_n$ the distribution of random variable X_{nj} is given by:

$$P\{X_{nj} = 0\} = \frac{1}{k}, \quad P\{X_{nj} = l\} = \frac{2(k-l)}{k^2}, \quad l \in \{1, 2, \dots, k-1\}.$$

4.8

(a) If $n/k^2 \to 0$ as $n \to \infty$, then we have

$$\lim_{n\to\infty} P\left\{M_n \leqslant \frac{xk}{\sqrt{n}} + k\right\} = \begin{cases} e^{-x^2}, & \text{if } x < 0, \\ 1, & \text{if } x \geqslant 0. \end{cases}$$

(b) If $n/k^2 \to \alpha$ as $n \to \infty$, where $0 < \alpha < \infty$, then we have

$$\lim_{n\to\infty} P\{M_n \leqslant k - 1 + x\} = \begin{cases} e^{-[x]([x]-1)\alpha}, & \text{if } x < 0, \\ 1, & \text{if } x \geqslant 0, \end{cases}$$

where $[x]$ is the integer part of x.

(c) If $n/k^2 \to \infty$ as $n \to \infty$, then constants $a_n > 0$ and $b_n \in \mathbb{R}$ cannot be determined such that random variable $(M_n - b_n)/a_n$ has a nondegenerate limit distribution.

4.9

(a) If $n/k \to 0$ as $n \to \infty$, then we have

$$\lim_{n\to\infty} P\left\{m_n \leqslant x \cdot \frac{k}{2n}\right\} = \begin{cases} 0, & \text{if } x \leqslant 0, \\ 1 - e^{-x}, & \text{if } x > 0. \end{cases}$$

(b) If $n/k \to \beta$ as $n \to \infty$, where $0 < \beta < \infty$, then we have

$$\lim_{n\to\infty} P\{m_n \leqslant x\} = \begin{cases} 0, & \text{if } x < 0, \\ 1 - e^{-(2[x]+1)\beta}, & \text{if } x \geqslant 0, \end{cases}$$

where $[x]$ is the integer part of x.

(c) If $n/k \to \infty$ as $n \to \infty$, then constants $a_n > 0$ and $b_n \in \mathbb{R}$ cannot be determined such that random variable $(m_n - b_n)/a_n$ has a nondegenerate limit distribution.

The Coupon Collector's Problem

Let $Z_1, Z_2, Z_3, \ldots$ be a sequence of independent random variables with a common uniform distribution over the set $\mathbb{N}_n = \{1, 2, \ldots, n\}$, i.e., Z_i is the result of the i-th random sampling of an element of $\mathbb{N}_n$. For each $j \in \mathbb{N}_n$ let us denote $X_{nj} = \min\{k : Z_k = j\}$.

4.10 Let us denote $M_n = \max\{X_{n1}, \ldots, X_{nn}\}$. Prove that for every $k \geqslant n$ the following equalities hold:

$$P\{M_n = k\} = \sum_{i=0}^{n-2} (-1)^i \binom{n-1}{i} \left(1 - \frac{i+1}{n}\right)^{k-1},$$

$$P\{M_n \leqslant k\} = \sum_{m=0}^{n} (-1)^m \binom{n}{m} \left(1 - \frac{m}{n}\right)^{k}.$$

4.11 Let k be a fixed positive integer. Let $M_n^{(k)}$ be the waiting time until $n - k + 1$ elements of set $\mathbb{N}_n$ are sampled. Prove that the limiting distribution of $M_n^{(k)}$ is given by

$$\lim_{n\to\infty} P\{M_n^{(k)} \leqslant n(x + \ln n)\} = e^{-e^{-x}} \cdot \sum_{l=0}^{k-1} \frac{(e^{-x})^l}{l!}.$$

4.12 Baum and Billingsley [22]: Let $W_n = W_n(a_n)$ be the waiting time until a_n+1 elements are sampled, where $0 \leqslant a_n < n$. Obviously, for $a_n = n-1$, we have $W_n = M_n = \max\{X_{n1}, \ldots, X_{nn}\}$. Let $b_n = n - a_n$, and let μ_n and σ_n^2 be the mean and variance of W_n. Prove that the following statements give the asymptotic behavior of appropriately centered and scaled random variable W_n for different asymptotic behaviors of (a_n):

(a) If $a_n/\sqrt{n}$ converges to 0, then $W_n - a_n - 1$ converges in probability to 0.
(b) If $a_n/\sqrt{n}$ converges to a positive constant λ, then $W_n - a_n - 1$ converges in distribution to the Poisson random variable with mean $\lambda^2/2$.
(c) If $a_n/\sqrt{n}$ and b_n both tend to infinity, then $(W_n - \mu_n)/\sigma_n$ converges in distribution to the normal $\mathcal{N}(0, 1)$ random variable.
(d) If $b_n = b$, where b is a fixed positive integer, then $\exp\{-(n^{-1}W_n - \ln 2n)\}$ converges in distribution to the chi-square random variable with $2b$ degrees of freedom.

4.13 Prove that statement (d) of Exercise 4.12 is equivalent to the statement of Exercise 4.11.

4.14 Let $W_{n,m}$ be the waiting time until $n-m$ distinct elements of $\mathbb{N}_n$ are sampled, i.e., $W_{n,m} = Z_{n,n} + Z_{n,n-1} + \cdots + Z_{n,m+1}$, where $Z_{n,n} = 1$, and random variable $Z_{n,i}$ has a gamma distribution with parameter i/n, where $i \in \{m+1, \ldots, n-1\}$, and $m \in \{0, 1, \ldots, n-1\}$. Let us denote

$$F_{n,m}(x) = P\left\{\frac{1}{n} W_{n,m} - \sum_{k=m+1}^{n} \frac{1}{k} \leqslant x\right\}, \quad x \in \mathbb{R}.$$

Then the following statements hold:

(a) Erdös and Rényi [60]: (the statement is equivalent to (4.3.3)): If

$$\gamma = \lim_{n\to\infty} \left(\sum_{k=1}^{n} \frac{1}{k} - \ln n\right) = 0.577215\ldots,$$

then for every $x \in \mathbb{R}$,

$$\lim_{n\to\infty} F_{n,0}(x) = F_0(x) := e^{-e^{-(x+\gamma)}}.$$

(b) Baum and Billingsley [22]: If $C_m = \gamma - \sum_{k=1}^{m} \frac{1}{k}$, then for every $x \in \mathbb{R}$,

$$\lim_{n\to\infty} F_{n,m}(x) = F_m(x) := \frac{1}{m!} \int_{-\infty}^{x} e^{(m+1)(y+C_m)} e^{-e^{-(y+C_m)}}\, dy$$
$$= \frac{1}{m!} \int_{-\infty}^{x+C_m} e^{(m+1)y} e^{-e^{-y}}\, dy.$$

(c) Csörgö [48]: There exists a positive constant D_m depending only on m, such that

$$\sup_{x\in\mathbb{R}} |F_{n,m}(x) - F_m(x)| \leqslant D_m \frac{\ln n}{n}.$$

4.15 Mladenović [155]: Let M_n and S_n be the maximum and the second largest among $X_{n1}, \ldots, X_{nn}$. The limiting distribution of random vector (S_n, M_n) is given by $\lim_{n\to\infty} P\{S_n \leqslant n(x + \ln n), M_n \leqslant n(y + \ln n)\} = G(x, y)$, where

$$G(x, y) = \begin{cases} \exp(-e^{-x}) \cdot (1 - e^{-y} + e^{-x}), & \text{if } x < y, \\ \exp(-e^{-y}), & \text{if } x \geqslant y. \end{cases}$$

4.16 Mladenović [155]: Let $(\alpha_n)_{n\geqslant 1}$ and $(\beta_n)_{n\geqslant 1}$ be two sequences of positive integers such that $\alpha_n, \beta_n \in \mathbb{N}_n$ for all n. Let us introduce the following notation:

$A_n = \{1, 2, \ldots, \alpha_n\}$, $B_n = \{n-\beta_n+1, n-\beta_n+2, \ldots, n\}$, $M(A_n) = \max\{X_{nj} \mid j \in A_n\}$, $M(B_n) = \max\{X_{nj} \mid j \in B_n\}$, and $T_n = \min\{M(A_n), M(B_n)\}$. Then the following statements hold:

(a) If $\alpha_n \to \infty$ as $n \to \infty$, then the limiting distribution of random variable $M(A_n)$ is given by:

$$\lim_{n\to\infty} P\{M(A_n) \leqslant n(x + \ln \alpha_n)\} = \exp(-e^{-x}).$$

(b) If $\alpha_n/n \to \alpha > 0$ and $\beta_n/n \to \beta > 0$ as $n \to \infty$, then the limiting distribution of random variable T_n is given by:

$$\lim_{n\to\infty} P\{T_n \leqslant n(x + \ln n)\} = \exp(-\alpha e^{-x}) + \exp(-\beta e^{-x}) - \exp(-\mu e^{-x}),$$

where $\mu = \min\{\alpha + \beta, 1\}$.

(c) If $\alpha_n \to \infty$, $\alpha_n/n \to 0$, and $\beta_n/\alpha_n \to \gamma > 0$ as $n \to \infty$, then

$$\lim_{n\to\infty} P\{T_n \leqslant n(x + \ln \alpha_n)\} = \exp(-e^{-x}) + \exp(-\gamma e^{-x}) - \exp(-(1+\gamma)e^{-x}).$$

Random Partitions

4.17 Let $n, k \in \mathbb{N}$ and $n \geqslant k$. Prove that for Stirling numbers of the second kind B_{nk}, and Bell numbers B_n the following equalities hold:

$$B_{nk} = \frac{1}{k!} \sum_{\alpha_1+\alpha_2+\cdots+\alpha_k=n} \frac{n!}{\alpha_1!\alpha_2!\cdots\alpha_k!},$$

$$B_{nk} = \frac{1}{k!} \sum_{j=0}^{k} (-1)^{k-j} \binom{k}{j} j^n,$$

$$B_n = \sum_{k=1}^{n} B_{nk} = \sum_{k=1}^{n} \frac{1}{k!} \sum_{j=0}^{k} (-1)^{k-j} \binom{k}{j} j^n.$$

4.18 Prove that for Bell numbers the following recursive relation holds:

$$B_{n+1} = \sum_{k=0}^{n} \binom{n}{k} B_k, \quad n \geqslant 0.$$

Geometric Properties of a Sample of Random Vectors

4.19 Jammalamadaka and Janson [116]: Let us suppose that $X, X_1, X_2, \ldots$ are i.i.d. random vectors in $\mathbb{R}^d$, where $d \geqslant 2$, with a spherically symmetric distribution such that

$$P\{|X| > x\} = e^{-g(x)+o(1)} \quad \text{as } x \to \infty,$$

for some twice differentiable function $g(x)$ such that, as $x \to \infty$,

$$xg'(x) \to \infty, \qquad \frac{g''(x)}{(g'(x))^2} \ln^2(xg'(x)) \to 0,$$

and a_n and b_n are such that as $n \to \infty$, $a_n \to \infty$ and

$$\begin{aligned} g(a_n) &= \ln n + o(1), \\ b_n &= \frac{1 + o(1/\ln(a_n g'(a_n)))}{g'(a_n)}. \end{aligned}$$

Let $M_n^{(2)}$ be given by (4.7.3). Then relation (4.7.6) holds true.
Hint Prove that the statement is a special version of Theorem 4.7.2.

4.20 Jammalamadaka and Janson [116]: Let us suppose that $X_1, X_2, \ldots$ are i.i.d. random vectors in $\mathbb{R}^d$, where $d \geqslant 2$, with a spherically symmetric distribution such that, as $|\boldsymbol{x}| \to \infty$,

$$f(\boldsymbol{x}) \sim c|\boldsymbol{x}|^{\alpha} e^{-\beta|\boldsymbol{x}|^{\gamma}},$$

for some $c, \beta, \gamma > 0$ and $\alpha \in \mathbb{R}$. If

$$c_d' = \frac{(d-1)2^{d-2}\pi^{d-1/2}}{\Gamma(d/2)},$$

then

$$\begin{aligned} (\beta^{1/\gamma}\gamma \ln^{1-1/\gamma} n) \cdot M_n^{(2)} - \Big\{ & 2\gamma \ln n + \left(2\frac{\alpha+d}{\gamma} - \frac{d+3}{2}\right) \ln\ln n \\ & + \ln\ln\ln n + \ln\left(c_d' \beta^{-2(\alpha+d)/\gamma} \gamma^{-(d+3)/2} c^2\right) \Big\} \end{aligned}$$

converges in distribution to the standard Gumbel extreme value distribution.
Hint Prove that the statement is a special version of Theorem 4.7.2.

4.21 Prove that the result provided in Remark 4.7.7 follows from Theorem 4.7.6 after substitution of the variable given by $x = (y + \ln d_1)/d_2$.

References

1. Adamović, D.D. (1966a): Sur quelques propriétés des fonctions à croissance lente de Karamata I. *Matematički Vesnik*, **3** (18), 123–136.
2. Adamović, D.D. (1966b): Sur quelques propriétés des fonctions à croissance lente de Karamata II. *Matematički Vesnik*, **3** (18), 161–172.
3. Alagar, V.S. (1976): The distribution of the distance between random points. *Journal of Applied Probability*, **13**, 558–566.
4. Albin, J.M.P. and Samorodnitsky, G. (2004): On overload in a storage model, with a self-similar and infinitely divisible input. *The Annals of Applied Probability*, **14**, 820–844.
5. Aldous, D. and Diaconis, P. (1995): Hammersley's Interacting Particle Process and Longest Increasing Subsequences. *Probability Theory and Related Fields*, **103**, 199–213.
6. Aldous, D. and Diaconis, P. (1999): Longest Increasing Subsequences: From Patience Sorting and the Baik-Deift-Johansson Theorem. *Bulletin (New Series) of the American Mathematical Society*, **36**(4), 413–432.
7. Aljančić, S. and Arandjelović, D. (1977): O-regularly varying functions. *Publications de l'Institut Mathématique*, **22** (36), 5–22.
8. Aljančić, S. and Karamata, J. (1956): Pravilno promenljive funkcije i Frullani-ev integral. *Recueil des travaux de l'Académie Serbe des Sciences.* **L**, **5**, 239–248.
9. Anceaume, E., Busnel, Y. and Sericola, B. (2015): New results on a generalized coupon collector problem using Markov chains. *Journal of Applied Probability*, **52**(2), 405–418.
10. Anderson, C.W. (1970): Extreme value theory for a class of discrete distribution with applications to some stochastic processes. *Journal of Applied Probability*, **7**, 99–113.
11. Andrews, G.E. (1976): *The Theory of Partitions*, Addison-Wesley Publishing Company, London, Amsterdam, Ontario, Sydney, Tokyo.
12. Appel, M.J.B. and Russo, R.P. (2006): Limiting distributions for the maximum of a symmetric function on a random point set. *Journal of Theoretical Probability*, **19**, No. 2, 365–375.
13. Appel, M.J.B., Najim, C.A. and Russo, R.P. (2002): Limit laws for the diameter of a random point set. *Advances in Applied Probability*, **34**, 1–10.
14. Auld, G. and Papastathopoulos, I. (2021): Extremal clustering in non-stationary random sequences. *Extremes*, **24**, 725–752.
15. Avakumović, V.G. (1936): O jednom O-inverznom stavu. *Rad Jugoslovenske Akademije Znanosti i Umjetnosti* (*Razreda matematičko-prirodoslovnoga* 79) **254**, 169–186.
16. Baik, J., Deift, P. and Johansson, K. (1999): On the distribution of the length of the longest increasing subsequence of random permutations. *Journal of the American Mathematical Society.* **12**(4), 1119–1178.

P. Mladenović, *Extreme Values In Random Sequences*,
Springer Series in Operations Research and Financial Engineering,
https://doi.org/10.1007/978-3-031-57412-2

17. Balkema, A.A. and de Haan, L. (1972): On R. von Mises' condition for the domain of attraction of $\exp(-e^{-x})$. *The Annals of Mathematical Statistics*, **43**, 1352–1354.
18. Balkema, A.A. and de Haan, L. (1974): Residual life-time at great age. *The Annals of Probability.* **2**, 7928–804.
19. Barbour, A.D., Holst, L. and Janson, S.(1992): *Poisson Approximation*. Oxford University Press, Oxford (1992).
20. Basrak, B., Davis, R.A. and Mikosh, T. (2002): A characterization of multivariate regular variation. *The Annals of Applied Probability.* **12**, 908–920.
21. Basrak, B. and Segers, J. (2009): Regularly varying multivariate time series. *Stochastic Processes and Their Applications*, **119**, 1055–1080.
22. Baum, L.E. and Billingsley, P. (1965): Asymptotic distributions for the coupon collector's problem. *The Annals of Mathematical Statistics*, **36**, 1835–1839.
23. Beirlant, J., Teugels, J. L. and Vynckier, P. (1996): *Practical Analysis of Extreme Values*. Leuven University Press, Leuven.
24. Békéssy, A. (1964): On classical occupancy problems II, Sequential occupancy, *Magy. Tud. Akad. Mat. Kutató Int. Közl.* **9**, 133–141.
25. Berman, S.M.. (1964): Limit theorems for the maximum term in stationary sequences, *The Annals of Mathematical Statistics*, **35**, 502–516.
26. Billingsley, P. (1986): *Probability and Measure*. John Willey & Sons, New York, Chichester, Brisbane, Toronto, Singapore.
27. Billingsley, P. (1999): *Convergence of Probability Measures*, Second Edition, John Willey & Sons, Inc. New York, Chichester, Weinheim, Brisbane, Singapore, Toronto.
28. Bingham, N., Goldie, C. and Teugels, J. (1987): *Regular Variation*. Encyclopedia of Mathematics and Its Applications, Vol. 27. Cambridge University Press, Cambridge UK.
29. Bojanić, R. and Seneta, E. (1973): A unified theory of regularly varying sequences. *Mathematische Zeitschrift*, **134**, 302–315.
30. Bóna, M. (2004): *Combinatorics of Permutations*. Chapman & Hall/ CRR, Boca Raton, London, New York, Washington, D.C.
31. Bradley, R.C. (2005): Basic properties of strong mixing conditions. A survey and some open questions. *Probability Survey*, **2**, 107–144.
32. Bräker, H. and Hsing, T. (1998): On the area and perimeter of a random convex hull in a bounded convex set. *Probabability Theory and Related Fields*, **111**, 517–550.
33. Bräker, H., Hsing, T. and Bingham, N.H. (1998): On the Hausdorff distance between a convex set and an interior random convex hull. *Advances in Applied Probability*, **30**, 295–316.
34. Breiman, L. (1965): On some limit theorems similar to the arc-sin law. *Theory of Probability and Its Applications*, **10**, 323–331.
35. Buldygin, V.V., Klesov, O.I and Steinebach, J.G. (2002): Properties of a subclass of Avakumović functions and their generalized inverses. *Ukrainian Mathematical Journal*, **54**, 149–169.
36. Buriticá, G., Meyer, N., Mikosch, T. and Wintenberger, O. (2021): Some variations on the extremal index, *Zap. Nauchn. Sem. POMI*, 501, 52–77.
37. Canto e Castro, L., de Haan, L. and Temido, M.G. (2001): Rarely observed sample maxima. *Theory of Probability and Its Applications*, **42**(4), 779–782.
38. Chatterjee, P. Diaconis and E. Meckes (2005): Exchangeable pairs and Poisson approximation. *Probabability Survey*, **2**, 64–106.
39. Chen, L.H.Y. (1975): Poisson approximation for dependent trials. *The Annals of Probability*, **3**, 534–545.
40. Chen, L.H.Y., Goldstein, L. and Chao, Q.-M. (2011): *Normal approximation by Stein's method*. Springer-Verlag, Berlin Heidelberg.
41. Chernick, M. R. (1981): A limit theorem for the maximum of autoregressive processes with uniform marginal distributions. *The Annals of Probability*, **9**, 145–149.
42. Chernick, M. R. and Davis, R.A. (1982): Extremes in autoregressive processes with uniform marginal distributions. *Statistics and Probability Letters*, **1**, 85–88.

43. Chernick, M. R., Hsing, T. and McCormick, W. (1991): Calculating the extremal index for a class of stationary sequences. *Advances in Applied Probability*, **6**, 711–731.
44. Cheong, C.K. and Teugels, J.F. (1972): General solidarity theorems for semi-Markov processes. *Journal of Applied Probability*, **9**, 789–802.
45. Chmielewski, M.A. (1981): Elliptically symmetric distributions: A review and bibliography. *International Statistical Review*. **49**, 67–74.
46. Cline, D.B.H. (1986): Convolution tails, product tails and domains of attraction. *Probability Theory and Related Fields*, **72**, 529–557.
47. Coles, S. (2001): *An Introduction to Statistical Modeling of Extreme Values*, Springer-Verlag London Limited.
48. Csörgö, S. (1993): A rate of convergence for coupon collectors. *Acta Scientiarum Mathematicarum (Szeged)*, **57**, 337–351.
49. Davis, R. A. (1979): Maxima and minima of stationary sequences. *The Annals of Probability*, **7**, 453–460.
50. Davis, R.A. and Hsing, T. (1995): Point process and partial sum convergence for weakly dependent random variables with infinite variance, *The Annals of Probability*, **23**, 879–917.
51. Davis, R. and Resnick, S. (1985): Limit theory for moving averages of random variables with regularly varying tail probabilities. *The Annals of Probability*, **13**, 179–195.
52. Davis, R. and Resnick, S. (1988): Extremes of moving averages of random variables from the domain of attraction of the double exponential distribution. *Stochastic Processes and their Applications*, **30**, 41–68.
53. Davis, R. and Resnick, S. (1996): Limit theory for bilinear processes with heavy-tailed noise. *The Annals of Applied Probability*, **6**, 1191–1210.
54. Denisov, D. and Zwart, B. (2007): On a theorem of Breiman and a class of random difference equations. *Journal of Applied Probability*, **44**, 1031–1046.
55. Doumas, A.V. and Papanicolaou, V.G. (2013): Asymptotics of the rising moments for the coupon collector's problem. *Electronic Journal of Probability*, **8**:15.
56. Dümbgen, L. and Walter, G. (1996): Rates of convergence for random approximations of convex sets. *Advances in Applied Probability*, **28**, 384–393.
57. Eastoe, E.F. and Tawn, J.A. (1986): Modelling non-stationary extremes with application to surface level ozone. *Journal of the Royal Statistical Society, Series C, Applied Statistics*, **58**, Part 1, 25–45.
58. Embrechts, P., Klüppelberg C and Mikosch, T. (1997): *Modelling Extremal Events*. Springer, Berlin Heidelberg.
59. Embrechts, P. and Omey, E. (1984): A property of longtailed distributions. *Journal of Applied Probability*, **21**, 80–87.
60. Erdös, P. and Rényi, A. (1961): On a classical problem of probability theory. *Magy.Tud.Akad. Mat. Kutató Int. Közl.* **6**(1–2), 215–220.
61. Erdös, P. and Spencer, J. (1974): *Probabilistic Methods in Combinatorics*, Academic Press, New York and London.
62. Erdös, P. and Szekeres, G. (1935): A combinatorial theorem in geometry. *Compositio Mathematica*, **2**, 463–470.
63. Falk, M., Hüsler, J. and Reiss, R.-D. (2011): *Laws of Small Numbers: Extremes and Rare Events*, Birkhäuser, Berlin.
64. Feller, W. (1971): *An Introduction to Probability Theory and Its Applications*, Vol. 1, Wiley, New York.
65. Ferro, C.A.T. and Segers, J. (2003): Inference for clusters of extreme values, *Journal of the Royal Statistical Society. Series B (Statistical Methodology)*, **65**, Part 2, 545–556.
66. Fisher, R.A. and Tippett, L.H.C. (1928): Limiting forms of the frequency distribution of the largest or smallest member of a sample. *Mathematical Proceedings of the Cambridge Philosophical Society*, **24**, 180–190.
67. Flatto, L. (1982): Limit theorems for some random variables associated with urn models. *The Annals of Probability*, **10**, 927–934.

68. Foata, D., Han. G.-N. and Lass, B. (2001): Les nombres hyperharmoniques et la fratrie du collectionneur de vignettes. *Séminaire Lotharingien de Combinatoire*, **B47a**, 20pp.
69. Foata, D. and Zeilberger, D. (2003): The collector's brotherhood problem using the Newman-Shepp symbolic method. *Algebra Universalis*, **49**, 387–395.
70. Fréchet, M. (1927): Sur la loi de probabilité de l'écart maximum. *Annales de la Société Polonaise de Mathématique*, **6**, 93–116.
71. Galambos, J. (1978): The Asymptotic Theory of Extreme Order Statistics. John Wiley & Sons, New York-Chichester-Brisbane-Toronto.
72. Galambos, J. and Seneta, E. (1973): Regularly varying sequences. *Proceedings of the American Mathematical Society*, **41**, 110–116.
73. Geluk, J.L. and Haan, L. de (1987): *Regular Variation, Extensions and Tauberian Theorems.* Centre for Mathematics and Computer Science, Amsterdam.
74. Gibbs, A.L. and Su, F.E. (2002): On choosing and bounding probability metrics, *International Statistical Review*, **70**, No. 3, 419–435.
75. Glavaš, L. (2015): New examples of partial samples from the uniform AR(1) process and asymptotic distributions of extremes, *Filomat*, **129**(10), 2289–2299.
76. Glavaš, L., Jocković and Mladenović P. (2018): Maximum of the sum of consecutive terms in random permutations, *Journal of Statistical Planning and Inference*, **197**, 15–20.
77. Glavaš, L. and Mladenović P. (2018): New limit results related to the coupon collector's problem, *Studia Scientiarum Mathematicarum Hungarica*, **55**(1), 115–140.
78. Glavaš, L. and Mladenović P. (2020): Extreme values of linear processes with heavy-tailed innovations and missing observations, *Extremes*, **23**, 547–567.
79. Glavaš, L. and Mladenović P. (2024): Extremes in incomplete samples from moving averages of random variables from the domain of attraction of the Gumbel distribution, *Statistics and Probability Letters*, **207**, 110012.
80. Glavaš, L., Mladenović P. and Samorodnitsky, G. (2017): Extreme values of the uniform order 1 autoregressive processes and missing observations, *Extremes*, **20**, 671–690.
81. Gnedenko, B.V. (1943): Sur la distribution limite du terme maximum d'une série aléatoire. *The Annals of Mathematics*, **44**, 423–453.
82. Goldie, C.M. (1991): Implicit renewal theory and tails of solutions of random equations, *The Annals of Applied Probability*, **1**, 126–166.
83. Graham, R.L., Knuth, D.E. and Patashnik, O. (1994): *Concrete Mathematics. Second Edition.* Addison–Wesley Publishing Company, New York.
84. Grey, D.R. (1994): Regular variation in the tail behavior of solutions of random difference equations. *The Annals of Applied Probability*, **4**, 169–183.
85. Griffits, R.C. (1988): On the distribution of points in a Poisson Dirichlet process. *Journal of Applied Probability*, **25**, 336–345.
86. Gumbel, E.J. (1958): *Statistics of Extremes.* Columbia University Press, New York.
87. Haan, L. de (1970): *On Regular Variation and its Application to the Weak Convergence of Sample Extremes.* Mathematical Centre Tracts 32, Amsterdam.
88. Haan, L. de (1971): A form of regular variation and its application to the domain of attraction of the double exponential distribution. *Zeitschrift für Wahrscheinlichkeitstheorie und verwandte Gebiete*, **17**, 241–258.
89. Haan, L. de (1974): Equivalence classes of regularly varying functions. *Stochastic Processes and their Applications*, **2**, 243–259.
90. Haan, L. de (1976): Sample extremes: an elementary introduction. *Statistica Neerlandica*, **30**, 161–172.
91. Haan, L. de and Ferreira, A. (2006): *Extreme Value Theory: An Introduction*, Springer, New York.
92. Haan, L. de, Resnick, S.I., Rootzén, H. and de Vries, C.G. (1989): Extremal behavior of solutions to a stochastic difference equation with applications to ARCH processes. *Stochastic Processes and their Applications*, **32**, 213–224.
93. Hall, P. (1979): On the rate of convergence of normal extremes. *Journal of Applied Probability*, **16**, 433–439.

94. Hall, A. and Hüsler, J. (2006): Extremes of stationary sequences with failures. *Stochastic Models*, **1**, 537–557.
95. Hall, A. and Scotto, M.G. (2003): Extremes of sub-sampled integer-valued moving average models with heavy-tailed innovations. *Statistics and Probability Letters*, **63**, 97–105.
96. Hall, A. and Scotto, M.G. (2008): On the extremes of randomly sub-sampled time series. *Revstat – Statistical Journal*, **6**, 151–164.
97. Hall, A. and Scotto, M.G. and Ferreira, H. (2004): On the extremal behaviour of generalised periodic sub-sampled moving average models with regularly varying tails. *Extremes*, **7**, 149–160.
98. Hall, A. and Temido, M.G. (2009): On the max-semistable limit of maxima of stationary sequences with missing values. *Journal of Statistical Planning and Inference*, **139**, 875–890.
99. Hammersley, J.M. (1950): The distribution of the distance in a hypersphere. *The Annals of Mathematical Statistics*, **21**, 447–452.
100. Hammersley, J.M. (1972): A few seedlings of research. *Proceedings of the Sixth Berkeley Symposium on Mathematical Statistics and Probability*, Vol. 1, 345–394, University of California Press.
101. *Handbook of Mathematical Functions With Formulas, Graphs, and Mathematical Tables.* (Edited by Abramowitz, M. and Stegun, I.A.) National Bureau of Standards, Appl. Math. Series, **55**, Issued 1964.
102. Hardy, G.H. and Wright, E.M. (1945): *An Introduction to the Theory of Numbers.* Calderon Press, Oxford.
103. Harper, L.H. (1967): Stirling behavior is asymptotically normal. *The Annals of Mathematical Statistics*, **38**, No. 2, 410–414.
104. Hastings, S.P. and McLeod, J.B. (1980): A boundary value problem associated with the second Painlevé transcedent and the Korteweg de Vries equation. *Archive for Rational Mechanics and Analysis*, **73**, 31–51.
105. Henze, N., Hlávka, Z. and Meintanis, S.G. (2014): Testing for spherical symmetry via the empirical characteristic function. *Statistics*, **48**, No. 6, 1282–1296.
106. Henze, N. and Klein, T. (1996): The limit distribution of the largest interpoint distance from a symmetric Kotz sample. *Journal of Multivariate Analysis*. **57**, 228–239.
107. Holst, L. (1971): Limit theorems for some occupancy and sequential occupancy problems. *The Annals of Mathematical Statistics*, **42**, 1671–1680.
108. Holst, L. (1977): Some asymptotic results for occupancy problems. *The Annals of Probability*, **5**, 1028–1035.
109. Holst, L. (1986): On birthday, collectors, occupancy and other classical urn problems. *International Statistical Review*, **54**, 15–27.
110. Holst, L. (2001): The Poisson-Dirichlet distribution and its relatives revisited. *Can be found on the* researchgate.net.
111. Hsing, T. (1994): On the asymptotic distribution of the area outside a random convex hull in a disk. *The Annals of Applied Probability*, **4**, No. 2, 478–493.
112. Hüsler, J. (1986): Extreme values of non-stationary random sequences. *Journal of Applied Probability*, **23**, No. 4, 937–950.
113. Hüsler J. and Piterbarg, V. (2004): Limit theorem for maximum of the storage process with fractional Brownian motion as input. *Stochastic Processes and their Applications*, **114**, 231–250.
114. Ilienko, A. (2019): Convergence of point processes associated with coupon collector's and Dixie cup problems. *Electronic Communications in Probability*, **24**, No. 51, 1–9.
115. Jammalamadaka, S.R. and Janson, S. (1986): Limit theorems for a triangular scheme of U-statistics with applications to inter-point distances. *The Annals of Probability*. **14**, No. 4, 1347–1358.
116. Jammalamadaka, S.R. and Janson, S. (2015): Asymptotic distribution of the maximum interpoint distance in a sample of random vectors with spherically symmetric distribution. *The Annals of Applied Probability*. **25**, No. 6, 3571–3591.

117. Janssen, A. and Segers, J. (2014): Markov tail chains. *Journal of Applied Probability*, **51**, 1133–1153.
118. Jocković, J. and Mladenović, P. (2011): Coupon collector's problem and generalized Pareto distributions, *Journal of Statistical Planning and Inference*, 141, 2348–2352.
119. Jocković, J. and Mladenović, P. (2014): Coupon collector's problem and its extensions in extreme value framework, *Statistics and Its Interface*, **7**(3), 381–388.
120. Kallenberg, O. (1983): *Random Measures*, 3rd edition. Akademie-Verlag, Berlin.
121. Kallenberg, O. (1983): *Random Measures, Theory and Applications*, Springer International Publishing Switzerland.
122. Karamata, J. (1930a): Sur un mode de croissance régulière des fonctions. *Mathematica* (Cluj), **3**, 38–53.
123. Karamata, J. (1930b): Sur certain 'Tauberian theorems' de M. M. Hardy et Littlewood. *Mathematica*, (Cluj), **3** 33–48.
124. Karamata, J. (1933): Sur un mode de croissance régulière. Théorèmes fondamentaux. *Bulletin de la Société Mathématique de France*, **61**, 55–62.
125. Karamata, J. (1936): Primedba na prethodni rad g. V. Avakumovića, sa obradom jedne klase funkcija koje se javljaju kod inverznih stavova zbirljivosti. *Rad Jugoslovenske Akademije Znanosti i Umjetnosti* (*Razreda matematičko-prirodoslovnoga* 79), **254**, 188–200.
126. Karamata, J. (1962): Some theorems concerning slowly varying functions. *Mathematics Research Center*, Technical Reports, **369**, Madison, Wisconsin.
127. Kendall, M.G. and Stuart, A. (1977): *The Advanced Theory of Statistics, Vol. I: Distribution Theory*. Macmillan, Inc., New York.
128. Kesten, H. (1973): Random difference equations and renewal theory for products of random matrices, *Acta Mathematica*, **131**, 207–248.
129. Kershenbaum, A., Freeberg, T.M. and Gammon, D.E. (2015): Estimating vocal repertoire size is like collecting coupons: A theoretical framework with heterogeneity in signal abundance, *Journal of Theoretical Biology*, **373**, 1–11.
130. Kingman, J.F.C. (1975): Random discrete distributions. *Journal of the Royal Statistical Society. Series B*, **37**, 1–15.
131. Kingman, J.F.C. (1993): *Poisson Processes*. Calderon Press, Oxford.
132. Klüppelberg, C. (1988): Subexponential distributions and integrated tails. *Journal of Applied Probability*, **25**, 132–141.
133. Klüppelberg, C. (1989): Subexponential distributions and characterizations of related classes. *Probability Theory and Related Fields*, **82**, 259–269.
134. Koltchinskii, V.I. and Li, L (1998): Testing for spherical symmetry of a multivariate distribution. *Journal of Multivariate Analysis*. **65**, 228–244.
135. Kotz, S. and Nadarajah, S. (2000): *Extreme Value Distributions, Theory and Applications*. Imperial College Press, London.
136. Krajka, T. (2011): The asymptotic behavior of maxima of complete and incomplete samples from stationary sequences. *Stochastic Processes and Their Applications*, **121**, 1705–1719.
137. Krajka, T.K. and Rychlik, Z. (2014): The limiting behavior of sums and maximus of i.i.d. random variables from the viewpoint of different observers. *Probability and Mathematical Statistics*, **34**, 237–252.
138. Kudrov, A.V. and Piterbarg, V.I. (2007): On maxima of partial samples in Gaussian sequences with pseudo-stationary trends. *Lithuanian Mathematical Journal*, **47**, 48–56.
139. Lao, W. and Mayer, M. (2008): U-max-statistics. *Journal of Multivariate Analysis*. **99**, 2039–2052.
140. Last, G. and Penrose, M. (2018): *Lectures on the Poisson Process*. Cambridge University Press, Cambridge.
141. Leadbetter, M.R. (1974): On extremes values in stationary sequences. *Zeitschrift für Wahrscheinlichkeitstheorie und verwandte Gebiete*, **28**, 289–303.
142. Leadbetter, M.R. (1983): Extremes and local dependence of stationary sequences. *Zeitschrift für Wahrscheinlichkeitstheorie und verwandte Gebiete*, **65**, 291–306.

143. Leadbetter, M.R., Lindgren, G. and Rootzén, H. (1983): *Extremes and Related Properties of Random Sequences and Processes*. Springer-Verlag, New York–Heidelberg–Berlin.
144. Lord, R.D. (1954): The distribution of distance in a hypersphere. *The Annals of Mathematical Statistics*, **24**, 794–798.
145. Loynes, R.M. (1965): Extremes values in uniformly mixing stationary stochastic processes. *The Annals of Mathematical Statistics*, **36**, 993–999.
146. Marcus, M. and Pinsky, M. (1969): On the domain of attraction of $e^{-e^{-x}}$. *Journal of Mathematical Analysis and Applications*, **28**, 440–449.
147. Markovich, N.M. (2017): Clusters of extremes: modeling and examples, *Extremes*, **20**, 519–538.
148. Matthews, P.C. and Rukhin, A.L. (1993): Asymptotic distribution of the normal sample range. *The Annals of Applied Probability*. **3**, No. 2, 454–466.
149. Mayer, M. and Molchanov, I. (2007): Limit theorems for the diameter of a random sample in the unit ball. *Extremes*. **10**, 129–150.
150. Mejzler, D.G. (1949): On a theorem of B.V. Gnedenko. *Sb. Trudov Inst. Mat. Akad. Nauk. Ukrain. R.S.R.* **12**, 31–35.
151. Mikhailov, V.G. (1977): A Poisson limit theorem in the scheme of group disposal of particles. *Theory of Probability and Its Applications*, **22**, 152–156.
152. Mikosch, T. (2009): *Non-Life Insurance Mathematics. An Introduction with the Poisson Process*. Springer-Verlag, Berlin–Heidelberg.
153. Mises, R. von (1936): La distribution de la plus grande de n valeurs. In: *Selected Papers* II (American Mathematical Society), 271–294.
154. Mittal, Y. (1978): Maxima of partial samples in Gaussian sequences. *The Annals of Probability*, **6**, 421–432.
155. Mladenović, P. (1999a): Limit theorems for the maximum terms of a sequence of random variables with marginal geometric distributions. *Extremes*, 2:4 (1999), 405–419.
156. Mladenović, P. (1999b): Extreme values of the sequences of independent random variables with mixed distributions, *Matematički Vesnik*, **51**, 29–37.
157. Mladenović, P. (2002a): A note on random permutations and extreme value distributions, *Proceedings of the Japan Academy*, **78**, Ser. A, No. 8, 157–160.
158. Mladenović, P. (2002b): Double arrays of random variables and extreme value distributions, *Theory of Stochastic Processes* (Kiev), **8**(24), No. 3–4, 300–308.
159. Mladenović, P. (2005): A generalization of the Mejzler-de Haan theorem, *Теория вероятностей и ее применения*, **50**, 177–189. Reprinted in: *Probability Theory and its Applications*, 50, 141–153 (2006).
160. Mladenović, P. (2006): On extreme steps of a random function on finite sets, *Indian Journal for Pure and Applied Mathematics*, **37**(2), 89–98.
161. Mladenović, P. and Piterbarg, V. (2006): On asymptotic distribution of maxima of complete and incomplete samples from stationary sequences. *Stochastic Processes and Their Applications*, **116**, 1977–1991.
162. Mladenović, P. (2008): Limit distributions for the problem of collecting pairs, *Bernoulli*, 14(2), 419–439.
163. Mladenović, P. (2009): Maximum of a partial sample in the uniform AR(1) processes, *Statistics and Probability Letters*, **79**, 1414–1420.
164. Mladenović, P. (2019): *Combinatorics. A Problem-Based Approach*. Springer Nature Switzerland.
165. Mladenović, P. and Vukmirović, J. (2010): Rates of convergence in certain limit theorems for extreme values, *Journal of Mathematical Analysis and Applications*, **363**, 287–295.
166. Mladenović, P. and Živadinović (2015): Uniform AR(1) processes and maxima on partial samples, *Communications in Statistics – Theory and Methods*, **44**, 2546–2563.
167. Mohan, N.R. and Ravi, S. (1993): Max domains of attraction of univariate and multivariate p-max stable laws. *Theory of Probability and Its Applications*, **37**, 632–643.

168. Moon, J. (1967): Various proofs of Cayley's formula for counting trees, in *A Seminar on Graph Theory*, Holt, New York, pp. 70–78.
169. Moser, L. and Wyman, M. (1955): An asymptotic formula for the Bell numbers. *Transactions of the Royal Society of Canada*, **XLIX**, Series III, 49–54.
170. Myers, A.N. and Wilf, H.S. (2003): Some new aspects of the coupon collector's problem. *SIAM Journal on Discrete Mathematics*, **17**, 1–17.
171. Neal, P. (2008): The generalised coupon collector problem. *Journal of Applied Probability*, **45**(3), 621–629.
172. Newman, D.J. and Shepp, L. (1960): The double dixie problem. *The American Mathematical Monthly*, **67**, 58–61.
173. Norros, I. (1994): A storage model with self-similar input. *Queueing Systems*, **16**, 387–396.
174. Novak, S.Y. (2012): *Extreme Value Methods with Applications to Finance*, CRC Press Taylor & Francis Group, Boca Raton, London, New York.
175. Oldyzko, A.M. and Rains, E.N. (1993): On longest increasing subsequences in random permutations. https://www-users.cse.umn.edu/~odlyzko/doc/incr.subseq.pdf
176. Pancheva, E.I. (1985): Limit theorems for extreme order statistics under nonlinear normalization. In *Stability Problems for Stochastic Models* (V.V. Kalashnikov and V.M. Zolotarev eds.), Lecture Notes in Math. **1155**, 248–309, Springer, Berlin.
177. Pancheva, E. (1992): Multivariate max-stable distributions. *Theory of Probability and Its Applications*, **18**, 679–705.
178. Parameswaran, S. (1961): Partition functions whose logarithms are slowly oscillating. *Transactions of the American Mathematical Society* **100**, 217–240.
179. Peng, Z., Weng, Z, and Nadarajah, S. (2010): Almost sure limit theorems of extremes of complete and incomplete samples of stationary sequences. *Extremes*. **13**, 463–480.
180. Perfekt, R. (1994): Extremal behaviour of stationary Markov chains with applications. *The Annals of Applied Probability*, **4**, 529–548.
181. Piterbarg, V.I. (1996): *Asymptotic Methods in the Theory of Gaussian Processes and Fields*. AMS Translations of Mathematical Monographs, 148, Providence, Rhode Island.
182. Piterbarg, V.I. (2001): Large deviations of a storage process with fractional Brownian motion as input. *Extremes*, **4(2)**, 147–164.
183. Piterbarg, V.I. (2004): Discrete and continuous time extremes of Gaussian processes. *Extremes*, **7**, 161–177.
184. Pitman, E.J.G. (1980): Subexponential distribution functions. *Journal of the Australian Mathematical Society, Ser. A*, **29**, 337–347.
185. Pitman, J. (2002): Poisson–Dirichlet and GEM invariant distributions for split-and-merge transformations of an interval partition. *Combinatorics. Probability and Computing*. **11**, 501–514.
186. Pitman, J. and Yor, M. (1997): The two-parameter Poisson-Dirichlet distribution derived from a stable subordinator. *The Annals of Probability*, **25**, No. 2, 855–900.
187. Pósfai, A. (2009): Poisson approximation in a Poisson limit theorem inspired by coupon collecting. *Journal of Applied Probability*, **46**, 585–592.
188. Posfai, A. (2010): *Approximation Theorems Related to the Coupon Collector's Problem*. Ph.D. Thesis, University of Szeged, Bolyai Institute.
189. Rachev, S.T. (1991): *Probability Metrics and the Stability of Stochastic Models*. Wiley.
190. Reiss, R.-D. and Thomas, M. (2007): *Statistical Analysis of Extreme Values with Applications to Insurance, Finance, Hydrology and Other Fields*. 3rd Edition, Birkhäuser, Basel.
191. Resnick, S.I. (1987): *Extreme Values, Regular Variation and Point Processes*. Springer-Verlag, New York-Berlin-Heidelberg-London-Paris-Tokyo.
192. Resnick, S.I. (2008): *Extreme Values, Regular Variation and Point Processes*. Springer Science+Business Media, New York.
193. Riordan, R. (1968): *Combinatorial Identities*. John Wiley & Sons, New York–London–Sydney.
194. Robert, C.Y. (2010): On the asymptotic distribution of maxima of stationary sequences subject to random failures or censoring. *Statistics and Probability Letters*, **80**, 134–142.

195. Robinson, M.E. and Tawn, J.A. (2000): Extremal analysis of processes sampled at different frequencies. *Journal of the Royal Statistical Society. Series B*, **62**, Part 1, 117–135.
196. Rootzén, H. (1978): Extremes of moving averages of stable processes. *The Annals of Probability* **6**, 847–869.
197. Rosenblatt, M. (1956): A central limit theorem and a strong mixing condition. *Proceedings of the National Academy of Sciences of the United States of America* **42**, 43–47.
198. Sachkov, V.N. (1997): *Probabilistic Methods in Combinatorial Analysis*. Cambridge University Press, Cambridge.
199. Samuel–Cahn, E. (1974): Asymptotic distributions for occupancy and waiting time problems with positive probability of falling through the cells. *The Annals of Probability*, **2**, 515–521.
200. Sándor, J. and Debnath, L. (2000): On certain inequalities involving the constant e and their applications. *Journal of Mathematical Analysis and Applications*, **249**, 569–582.
201. Schilling, J. and Henze, N. (2021): Two Poisson limit theorems for the coupon collector's problem with group drawings. *Journal of Applied Probability*, **58**(4), 966–977.
202. Scotto, M.G. (2005): Extremes of a class of deterministic sub-sampled processes with applications to stochastic difference equations. *Stochastic Processes and Their Applications*, **115**, 417–434.
203. Scotto, M. and Ferreira, H. (2003): Extremes of deterministic sub-sampled moving averages with heavy-tailed innovations. *Applied Stochastic Models in Business and Industry*, **19**, 303–313.
204. Scotto, M.G. and Turkman, K.F. (2002): On the extremal behavior of sub-sampled solutions of stochastic difference equations. *Portugaliae Mathematica*, **24**, 579–590.
205. Scotto, M.G., Turkman, K.F. and Anderson, C.W. (2003): Extremes of some sub-sampled time series. *Journal of Time Series Analysis*, **24**, 579–590.
206. Sebastião, J.R., Martins, A.P., Pereira, L. and Ferreira, H. (2010): Clustering of upcrossings of high values. *Journal of Statistical Planning and Inference*, **140**, 1003–1012.
207. Segers, J. (2003): Functionals of clusters of extremes, *Advances in Applied Probability*, **35**, 1028–1045.
208. Segers, J. (2005): Approximate distributions of clusters of extremes, *Statistics and Probability Letters*, **74**, 330–336.
209. Segers, J. (2007): Multivariate regular variation of heavy-tailed Markov chains. Discussion Paper, **0703**, Institut de statistique, Université catholique de Louvain.
210. Seneta, E. (1976): *Regularly Varying Functions*, Lecture Notes in Mathematics 508, Springer-Verlag, Berlin-Heidelberg-New York. Springer Nature Switzerland.
211. Shevtsova, I.G. (2010): An improvement of convergence rate estimates in the Lyapunov theorem. *Doklady Mathematics*, **82**(3), 862–864.
212. Shneina, E. and Božin, V. (2013): Extreme values of an infinite mixture of normally distributed variables. *Filomat*, **27:5**, 909–916.
213. Spivey, M.Z. (2008): Extreme values of an infinite mixture of normally distributed variables. *Journal of Integer Sequences*, **11**, Article 08.2.5.
214. Stadje, W. (1990): The collector's problem with group drawings. *Advances in Applied Probability*, **22**, 866–882.
215. Stein, C. (1978): Asymptotic evaluation of the number of Latin rectangles, *Journal of Combinatorial Theory, Series A*, **25**, 38–49.
216. Stein, C. (1986): *Approximate Computation of Expectations*. Institute of Mathematical Statistics Lecture Notes – Monograph Series, Vol. 7, Hayward, California.
217. Tan, Z. and Wang. Y. (2012): Some asymptotic results on extremes of incomplete samples. *Extremes*, **15**, 319–332.
218. Tang, Q. (2008): From light tails to heavy tails through a multiplier. *Extremes*, **11**, 379–391.
219. Temido, M.G. and Canto e Castro (2003): Max-semistable laws in extremes of stationary random sequences. *Theory of Probability and Its Applications*, **47**, 365–374.
220. Ulam, S.M. (1961): *Monte Carlo calculations in problems of mathematical physics* in *Modern Mathematics for the Engineers*, Beckenbach, E.F. ed., McGraw-Hill, 261–281.

221. Vershik, A.M. and Kerov, S.V. (1985): Asymptotic behavior of the maximum and generic dimensions of irreducible representations of the symmetric group, *Functional Analysis and Its Applications*, **19**(1), 21–31.
222. Vervaat, W. (1979): On a stochastic difference equation and a representation of non-negative infinitely divisible random variables. *Advances in Applied Probability*, **11**, 750–783.
223. Waterson, G.A. (1976): The stationary distribution of the infinitely many neutral alleles diffusion model. *Journal of Applied Probability*, **13**, 639–651.
224. Weissman, I. and Cohen, U. (1995): The extremal index and clustering of high values for derived stationary sequences, *Journal of Applied Probability*, **32**, 972–981.
225. Xu, W. and Tang, K. (2011): A generalized coupon collector problem. *Journal of Applied Probability*, **48**, No. 4, 1081–1094.
226. de Zea Bermudez, P. and Kotz, S. (2010): Parameter estimation of the generalized Pareto distributions. *Journal of Statistical Planning and Inference*, Part I, **140**, 1353–1373; Part II, **140**, 1374–1388.
227. Zoroa, N., Lesigne, E., Frenández-Sáez, Zoroa, P. and Casas, J. (2017): The coupon collector urn model with unequal probabilities in ecology and evolution. *Journal of The Royal Society Interface*, **14**, 1353–1373; Part II, **140**, 20160643.
228. Бари, Н.К., Стечкин, С.Б. (1956): Наилучшие приближения и дифференциальные свойства двух сопряженных функций, *Тр. ММО*, том **5**, 483–522. [Bari, N.K., Stechkin, S.B. (1956): Best approximations and differential properties of two conjugate functions, *Tr. Mosk. Mat. Obs.*, Volume **5**, 483–522.]
229. Викторова, И. И. (1969): Об асимптотическом поведении максимума в равновероятной полиномиальной схеме, *Математические заметки*, **5**, вып. 3, 305–316. [Viktorova, I. I. (1969): On asymptotic behavior of extreme order statistics in an equiprobable polynomial scheme, *Mathematical Notes*, No. 3, 331–338.]
230. Викторова, И. И., Севастьянов, Б. А. (1967): О предельном поведении максимума в полиномиальной схеме, *Математические заметки*, **1**, вып. 3, 331–338. [Viktorova, I. I. and Sevastyanov, B. A. (1967): On the limiting behavior of extreme order statistics in polynomial scheme, *Mathematical Notes*, No. 3, 331–338.]
231. Гнеденко, Б.В., Колмогоров, А.Н. (1949): *Предельные распределения для сумм независимых случайных величин.* Гостехиздат, Москва. [Gnedenko, B.V., Kolmogorov, A.N. (1949): Limit Distributions For Sums Of Independent Random Variables.]
232. Гончаров, В. Л. (1944): Из области комбинаторики, *Известия АН СССР, Сер. матем*, **8**, No. 1, с. 3–48. [Goncharov, V.L. (1944): From the field of combinatorics, *Izvestiya AN SSSR, Series Math.*, **8**, No. 1, pp. 3–48.]
233. Ибрагимов, И.А., Линник, Ю.В. (1965): *Независимые и стационарно связанные величины.* Наука, Москва. [Ibragimov, I.A. and Linnik, Yu. V. (1971): Independent and Stationary Sequences of Random Variables, English Trans, edited J.F.C. Kingman, Independent and Stationary Sequences of Random Variables, Wolters-Noordhoff Publishing, Groningen, The Netherlands.]
234. Колмогоров, А.Н., Фомин, С.В. (1989): *Элементы теории функций и функционального анализа.* Наука, Москва. [Kolmogorov, A.N., Fomin, S.V. (1989): Elements of the Theory of Functions and Functional Analysis. Nauka, Moskva.]
235. Колчин, В. Ф. (1969): О предельном поведении крайных членов вариационного ряда в полиномиальной схеме, *Теория вероятностей и ее применения*, **14**, вып. 3, 476–487. [Kolchin, V.F. (1969): On the limiting behaviour of extreme order statistics in a polynomial scheme, *Probability Theory and Its Applications*, **14**, No. 3, 476–487.]
236. Колчин, В. Ф. (1978): Момент вырождения ветвящегося процесса и высота случайного дерева, *Математические заметки*, **24**, вып.

6, 859–870. [Kolchin, V. F. (1978): The moment of degeneration of a birth-and-death process and the height of a random tree, *Mathematical Notes*, **24**, No. 6, 859–870.]

237. Колчин, В.Ф., Севастьянов, Б.А., Чистяков, В.П. (1976): *Случайные размещения.* Наука, Москва. [Kolchin, V.F., Sevastyanov, B.A. and Chistyakov, V.P. (1978): *Random Allocations*, English Translation, Editor A.V. Balakrishnan, V.H. Winston & Sons, Washington, D.C.]
238. Павлов, Ю. Л. (1977): Асимптотическое распределение максимального объема дерева в случайном лесе, *Теория вероятностей и ее применения*, **22**, вып. 3, 523–533. [Pavlov, Yu. L. (1977): The asymptotic distribution of maximum tree size in a random forest, *Probability Theory and Its Applications*, **22**, No. 3, 523–533.]
239. Павлов, Ю. Л. (1983): Предельные распределения высоты случайного леса, *Теория вероятностей и ее применения*, **28**, вып. 3, 449–457. [Pavlov, Yu. L. (1983): Limit distributions of the height of a random forest, *Probability Theory and Its Applications*, **28**, No. 3, 449–457.]
240. Прудников, А.П., Брычков, Ю.А. Маричев, О.И. (1981): *Интегралы и ряды. Элементарные функции.* Наука, Москва. [Prudnikov, A.P., Brychkov, Yu.A. Marichev, O.I. (1981): *Integrals and Series. Elementary Functions.* Nauka, Moscow.]
241. Сачков, В.Н. (1978): *Введение в комбинаторные методы дискретной математики. Наука, Москва.* [Sachkov, V.N. (1982): *Introduction to Combinatorial Methods of Discrete Mathematics*, Nauka, Moscow.]
242. Степанов, В. Е. (1969): О распределении числа вершин в слоях случайного дерева, *Теория вероятностей и ее применения*, **14**, вып. 1, 64–77. [Stepanov, V.E. (1969): On the distribution of the number of vertices in strata of a random tree, *Probability Theory and Its Applications*, **14**, No. 1, 64–77.]

Index

P. Mladenović, *Extreme Values In Random Sequences*,
Springer Series in Operations Research and Financial Engineering,
https://doi.org/10.1007/978-3-031-57412-2

The manufacturer's authorised representative in the EU is Springer Nature Customer Service Centre GmbH, Europaplatz 3, 69115 Heidelberg, Germany. If you have any concerns regarding our products, please contact ProductSafety@springernature.com

Printed and bound by CPI Group (UK) Ltd, Croydon, CR0 4YY
30/07/2026
02182055-0002